Dynamical Systems and Random Processes

CONTEMPORARY MATHEMATICS

736

Dynamical Systems and Random Processes

16th Carolina Dynamics Symposium
April 13–15, 2018
Agnes Scott College, Decatur, Georgia

Jane Hawkins
Rachel L. Rossetti
Jim Wiseman
Editors

AMS
AMERICAN MATHEMATICAL SOCIETY

2010 *Mathematics Subject Classification.* Primary 37D40, 37F10, 37B10, 34A99, 35J47, 37B20, 37B25, 55-04, 37A50.

Library of Congress Cataloging-in-Publication Data

Names: Carolina Dynamics Symposium (16th : 2018 : Decatur, Ga.) | Hawkins, Jane, 1954– editor. | Rossetti, Rachel L., 1986– editor. | Wiseman, Jim, 1974– editor.

Title: Dynamical systems and random processes : Carolina Dynamics Symposium, April 13–15, 2018, Agnes Scott College, Decatur, Georgia / Jane Hawkins, Rachel L. Rossetti, Jim Wiseman, editors.

Description: Providence, Rhode Island : American Mathematical Society, [2019] | Series: Contemporary mathematics ; volume 736 | "The 16th Carolina Dynamics Symposium"–Preface. | Includes bibliographical references.

Identifiers: LCCN 2019011700 | ISBN 9781470448318 (alk. paper)

Subjects: LCSH: Geometry, Differential–Congresses. | Differentiable dynamical systems–Congresses. | Random dynamical systems–Congresses. | Stochastic processes–Congresses. | AMS: Dynamical systems and ergodic theory – Dynamical systems with hyperbolic behavior – Dynamical systems of geometric origin and hyperbolicity (geodesic and horocycle flows, etc.). msc | Dynamical systems and ergodic theory – Complex dynamical systems – Polynomials; rational maps; entire and meromorphic functions. msc | Dynamical systems and ergodic theory – Topological dynamics – Symbolic dynamics. msc | Ordinary differential equations – General theory – None of the above, but in this section. msc | Partial differential equations – Elliptic equations and systems – Second-order elliptic systems. msc | Dynamical systems and ergodic theory – Topological dynamics – Notions of recurrence. msc | Dynamical systems and ergodic theory – Topological dynamics – Lyapunov functions and stability; attractors, repellers. msc | Algebraic topology – Explicit machine computation and programs (not the theory of computation or programming). msc | Dynamical systems and ergodic theory – Ergodic theory – Relations with probability theory and stochastic processes. msc

Classification: LCC QA641 .C348 2018 | DDC 515/.39–dc23

LC record available at https://lccn.loc.gov/2019011700

Contemporary Mathematics ISSN: 0271-4132 (print); ISSN: 1098-3627 (online)

DOI: https://doi.org/10.1090/conm/736

∞ The paper used in this book is acid-free and falls within the guidelines
established to ensure permanence and durability.
Visit the AMS home page at `https://www.ams.org/`

10 9 8 7 6 5 4 3 2 1 24 23 22 21 20 19

Contents

Preface

The first Carolina Dynamics Symposium, originally called the Carolina Dynamics Seminar, was held at the University of North Carolina at Charlotte on April 12, 2003 and organized by Douglas Shafer of UNC Charlotte. What started as an informal gathering of mathematicians interested in dynamical systems continued at the College of Charleston in 2004, then moved to Chapel Hill in 2005. The meetings were so beneficial to the participants' research that the group has met every year since, each time including new Ph.D.'s and a broader range of fields relating to dynamics as well. We have added at least one talk aimed at undergraduates at each meeting as well and, with the addition of NSF support, have been able to bring in speakers from farther away.

In the intervening years the symposium has also been hosted by Clemson, Meredith, Davidson, UNC Asheville, Furman, and Agnes Scott, with many of the venues hosting repeat conferences. Georgia, Maryland, Virginia, and Tennessee send participants too, so despite its name, the participants reflect the richness of the field of dynamics throughout the southeastern U.S.

The articles contained in this volume are by the participants of the most recent meeting at Agnes Scott College, the 16th Carolina Dynamics Symposium and the second one held at Agnes Scott. Also included are a few papers from some collaborators who have been involved with Carolina Dynamics in some capacity for many years. The papers cover a wide swath of topics in dynamics and randomness and reflect a sampling of the breadth of the field. The editors are grateful for NSF Grant DMS-1600746, which supported many of the authors who participated in the event at Agnes Scott.

Jane Hawkins
Rachel L. Rossetti
Jim Wiseman

List of Participants

Julia Barnes
Western Carolina University

Barrett Brister
Georgia State University

Emily Burkhead
University of North Carolina
at Chapel Hill

Jim Campbell
University of Memphis

Kevin Daley
Georgia State University

Sarah Day
College of William and Mary

Albert Fathi
Georgia Tech

Sarah Frick
Furman University

Joanna Furno
University of Houston

Jane Hawkins
University of North Carolina
at Chapel Hill

Jonathan Hulgan
Oxford College, Emory

Sam Kaplan
University of North Carolina Asheville

Lorelei Koss
Dickinson College

Jeffrey Lawson
Western Carolina University

Max Lenk
Georgia Tech

Kevin McGoff
University of North Carolina Charlotte

Claire Merriman
University of Illinois
at Urbana-Champaign

Donna Molinek
Davidson College

Nicholas Ormes
University of Denver

Andy Parrish
Eastern Illinois University

Karl Petersen
University of North Carolina
at Chapel Hill

Predrag Punosevac
Carnegie Mellon University

Eric Roberts
University of California, Merced

Joseph Rosenblatt
IUPUI

Rachel Rossetti
Agnes Scott

Shrey Sanadhya
University of Iowa

Martin Schmoll
Clemson University

Douglas Shafer
University of North Carolina Charlotte

Howie Weiss
Georgia Tech

Jim Wiseman
Agnes Scott

Undergraduate participants from Oxford College, Emory Univ: Emily Rexer, Justin Burton

Undergraduate participants from Agnes Scott: Emily Smith, Yihan Jiang, Yutong Wang, Laura Stordy, Jessa Rhea, Angela Hong, Yuemin Zuo, Huiming Zang

Group photo: 16th Carolina Dynamics Symposium, Agnes Scott College, April 2018

Contemporary Mathematics
Volume **736**, 2019
https://doi.org/10.1090/conm/736/14831

Chain recurrence and strong chain recurrence on uniform spaces

Ethan Akin and Jim Wiseman

In Memory of John Mather

ABSTRACT. We extend the theory of the chain relation (based on Conley's notion of chain recurrence) and the strong chain relation (based on Easton's strong chain recurrence) from the setting of continuous maps on compact spaces to general relations on uniform spaces, following the barrier function approach introduced by Fathi and Pageault. We consider equivalent characterizations of these relations, Lyapunov functions, and attractors.

1. Introduction

Let f be a continuous map on a compact metric space (X,d). If $\epsilon \geq 0$ then a sequence $\{x_0,\dots,x_n\}$ with $n \geq 1$ is an ϵ chain for f if $\max_{i=1}^n \ d(f(x_{i-1}),x_i) \leq \epsilon$ and a strong ϵ chain for f if $\Sigma_{i=1}^n \ d(f(x_{i-1}),x_i) \leq \epsilon$. Thus, a 0 chain is just an initial piece of an orbit sequence.

The Conley chain relation $\mathcal{C}f$ consists of those pairs $(x,y) \in X \times X$ such that there is an ϵ chain with $x_0 = x$ and $x_n = y$ for every $\epsilon > 0$. The Easton, or Aubry-Mather, strong chain relation $\mathcal{A}_d f$ consists of those pairs $(x,y) \in X \times X$ such that there is a strong ϵ chain with $x_0 = x$ and $x_n = y$ for every $\epsilon > 0$. As the notation indicates, $\mathcal{C}f$ is independent of the choice of metric, while $\mathcal{A}_d f$ depends on the metric. See [**5**] and [**6**].

Fathi and Pageault have studied these matters using what they call *barrier functions*, [**11**], [**7**] and their work has been sharpened in [**12**], [**13**]. $M_d^f(x,y)$ is the infimum of the ϵ's such that there is an ϵ chain from x to y and $L_d^f(x,y)$ is the infimum of the ϵ's such that there is a strong ϵ chain from x to y. Thus, $(x,y) \in \mathcal{C}f$ iff $M_d^f(x,y) = 0$ and $(x,y) \in \mathcal{A}_d f$ iff $L_d^f(x,y) = 0$.

Our purpose here is to extend these results in two ways.

First, while our interest focuses upon homeomorphisms or continuous maps, it is convenient, and easy, to extend the results to relations, following [**1**].

A relation $f : X \to Y$ is just a subset of $X \times Y$ with $f(x) = \{y \in Y : (x,y) \in f\}$ for $x \in X$. Let $f(A) = \bigcup_{x \in A} \ f(x)$ for $A \subset X$. So f is a mapping when $f(x)$ is a singleton set for every $x \in X$, in which case we will use the notation $f(x)$ for both the singleton set and the point contained therein. For example, the identity map

The second author was supported by a grant from the Simons Foundation (282398, JW).

on a set X is $1_X = \{(x,x) : x \in X\}$. If X and Y are topological spaces then f is a closed relation when it is a closed subset of $X \times Y$ with the product topology.

The examples $\mathcal{C}f$ and $\mathcal{A}_d f$ illustrate how relations arise naturally in dynamics.

The other extension is to non-compact spaces. This has been looked at in the past; see [**9**] and [**11**]. However, the natural setting for the theory is that of uniform spaces as described in [**10**] and [**4**], and reviewed in Section 2.3 below and in more detail in [**2**, Appendix B].

A uniform structure $\mathcal{U}$ on a set X is a collection of relations on X which satisfy various axioms so as to generalize the notion of metric space. A uniformity $\mathcal{U}$ is equivalently given by its *gage* $\Gamma(\mathcal{U})$, the set of pseudo-metrics d on X with the metric uniformity $\mathcal{U}(d)$, generated by ϵ-neighborhoods of the diagonal, contained in $\mathcal{U}$. The use of covers in [**11**] and continuous real-valued functions in [**9**] is equivalent to certain choices of uniformity.

Once we leave the compact category, the chain relation $\mathcal{C}_d f$, as well as the Aubry-Mather relation $\mathcal{A}_d f$, depends on the metric.

Example 1.1. Consider the translation map $f : \mathbb{R} \to \mathbb{R}$ given by $f(x) = x+1$. If we give $\mathbb{R}$ the usual metric, then there is no recurrent behavior. But if we think of $\mathbb{R}$ as the unit circle in $\mathbb{R}^2$ minus the point $(0,1)$ and give it the usual metric inherited from $\mathbb{R}^2$, then both $\mathcal{C}f$ and $\mathcal{A}_d f$ are equal to $\mathbb{R} \times \mathbb{R}$.

Example 1.2. Let $g : \mathbb{R}^2 \to \mathbb{R}^2$ be the translation map $g(x,y) = (x+1,y)$. Again, with the usual metric, there is no recurrent behavior. Construct a metric d_1 as follows: Take the unit square $[0,1] \times [0,1]$ and identify the points $(0,1/2)$ and $(1,1/2)$; consider $\mathbb{R}^2$ as the subset $(0,1) \times (0,1)$ and let d_1 be the inherited metric. Then $\mathcal{C}g$ is the entire space $(0,1) \times (0,1)$, while a point $((x,y),(x,y))$ is in $\mathcal{A}_{d_1} g$ iff $y = 1/2$. (In fact, $((x_1,1/2),(x_2,1/2))$ is in $\mathcal{A}_{d_1} g$ for any x_1 and x_2.)

We can construct another metric d_2 in a similar way. Take the unit square $[0,1] \times [0,1]$ and now identify the points $(0,y)$ and $(1,y)$ for all y (giving a cylinder). Again consider $\mathbb{R}^2$ as the subset $(0,1) \times (0,1)$ and let d_2 be the inherited metric. We still have $\mathcal{C}g = (0,1) \times (0,1)$, but now the point $((x_1,y_1),(x_2,y_2))$ is in $\mathcal{A}_{d_2} g$ iff $y_1 = y_2$.

The paper is organized as follows. Section 2 contains definitions and background information on relations, pseudo-metrics, and uniform spaces.

In Section 3, we define the barrier functions m_d^f and ℓ_d^f of a relation f on a set X with respect to a pseudo-metric d and we describe their elementary properties. We use a symmetric definition which allows a jump at the beginning as well as the end of a sequence. (In [**2**], we show that the alternative definitions yield equivalent results in cases which include when f is a continuous map.)

In Section 4, we describe the properties of the Conley relation $\mathcal{C}_d f = \{(x,y) : m_d^f(x,y) = 0\}$ and the Aubrey-Mather relation $\mathcal{A}_d f = \{(x,y) : \ell_d^f(x,y) = 0\}$. Following [**1**] we regard $\mathcal{C}_d$ and $\mathcal{A}_d$ as operators on the set of relations on X. We observe that each of these operators is idempotent.

In Section 5, we consider Lyapunov functions. With the pseudo-metric d fixed, a Lyapunov function L for a relation f on X is a continuous map $L : X \to \mathbb{R}$ such that $(x,y) \in f$ implies $L(x) \leq L(y)$, or, equivalently, $f \subset \leq_L$ where $\leq_L = \{(x,y) : L(x) \leq L(y)\}$. Notice that we follow [**1**] in using Lyapunov functions which increase, rather than decrease, on orbits. Following [**11**] and [**7**] we show that the barrier functions can be used to define Lyapunov functions. If g is a relation on

X with $f \subset g$ and $z \in X$ then $x \mapsto m_d^g(x,z)$ is a Lyapunov function for $\mathcal{C}_d f$ and $x \mapsto \ell_d^g(x,z)$ is a Lyapunov function for $\mathcal{A}_d f$. Even when f is a map, it is convenient to use associated relations like $g = f \cup 1_X$ or $g = f \cup \{(y,y)\}$ for y a point of X.

In Section 6, we turn to uniform spaces. The Conley relation $\mathcal{C}_{\mathcal{U}} f$ is the intersection of $\{\mathcal{C}_d f : d \in \Gamma(\mathcal{U})\}$ and $\mathcal{A}_{\mathcal{U}} f$ is the intersection of $\{\mathcal{A}_d f : d \in \Gamma(\mathcal{U})\}$. Thus, $(x,y) \in \mathcal{C}_{\mathcal{U}} f$ iff $m_d^f(x,y) = 0$ for all $d \in \Gamma(\mathcal{U})$ and similarly $(x,y) \in \mathcal{A}_{\mathcal{U}} f$ iff $\ell_d^f(x,y) = 0$ for all $d \in \Gamma(\mathcal{U})$. While the gage definition is convenient to use, we show that each of these relations has an equivalent description which uses the uniformity directly. Each of these is a closed, transitive relation which contains f. We let $\mathcal{G} f$ denote the smallest closed, transitive relation which contains f, so that $f \subset \mathcal{G} f \subset \mathcal{A}_{\mathcal{U}} f \subset \mathcal{C}_{\mathcal{U}} f$.

If L is a uniformly continuous Lyapunov function for f then it is automatically a Lyapunov function for $\mathcal{A}_{\mathcal{U}} f$. If X is Hausdorff and we let L vary over all uniformly continuous Lyapunov functions for f then $1_X \cup \mathcal{A}_{\mathcal{U}} f = \bigcap_L \ \leq_L$. That is, if $(x,y) \notin 1_X \cup \mathcal{A}_{\mathcal{U}} f$, then there exists a uniformly continuous Lyapunov function L such that $L(x) > L(y)$. If, in addition, X is second countable, then there exists a uniformly continuous Lyapunov function L such that $1_X \cup \mathcal{A}_{\mathcal{U}} f = \leq_L$. If X is Hausdorff and we let L vary over all Lyapunov functions for $\mathcal{C}_{\mathcal{U}} f$ then $1_X \cup \mathcal{C}_{\mathcal{U}} f = \bigcap_L \ \leq_L$. If, in addition, X is second countable, then there exists a Lyapunov function L such that $1_X \cup \mathcal{C}_{\mathcal{U}} f = \leq_L$. These results use the barrier function Lyapunov functions developed in the preceding section.

For the Conley relation there are special results. A set A is called $\mathcal{U}$ *inward* for a relation f on $(X, \mathcal{U})$ if for some $U \in \mathcal{U}$ $(U \circ f)(A) \subset A$. A continuous function $L : X \to [0,1]$ is called an *elementary Lyapunov function* if $(x,y) \in f$ and $L(x) > 0$ imply $L(y) = 1$. For a $\mathcal{U}$ uniformly continuous elementary Lyapunov function L the sets $\{x : L(x) > \epsilon\}$ for $\epsilon \geq 0$ are open $\mathcal{U}$ inward sets. On the other hand, if A is a $\mathcal{U}$ inward set, then there exists a $\mathcal{U}$ uniformly continuous elementary Lyapunov function L such that $L = 0$ on $X \setminus A$ and $L = 1$ on $f(A)$. Each set $\mathcal{C}_{\mathcal{U}} f(x)$ is an intersection of inward sets. If A is an open $\mathcal{U}$ inward set then it is $\mathcal{C}_{\mathcal{U}} f$ $^+$invariant and the maximum $\mathcal{C}_{\mathcal{U}} f$ invariant subset A_∞ is called the associated *attractor*.

In [**2**] we obtain additional results on the effect of various continuity assumptions, compactness and compactifications, and recurrence and transitivity.

2. Definitions and background

2.1. Relations. For a relation $f : X \to Y$ the *inverse relation* $f^{-1} : Y \to X$ is $\{(y,x) : (x,y) \in f\}$. Thus, for $B \subset Y$, $f^{-1}(B) = \{x : f(x) \cap B \neq \emptyset\}$. We define $f^*(B) = \{x : f(x) \subset B\}$. These are equal when f is a map.

If $f : X \to Y$ and $g : Y \to Z$ are relations then the *composition* $g \circ f : X \to Z$ is $\{(x,z) :$ there exists $y \in Y$ such that $(x,y) \in f$ and $(y,z) \in g\}$. That is, $g \circ f$ is the image of $(f \times Z) \cap (X \times g)$ under the projection $\pi_{13} : X \times Y \times Z \to X \times Z$. As with maps, composition of relations is clearly associative.

The *domain* of a relation $f : X \to Y$ is

$$Dom(f) \ = \ \{x : f(x) \neq \emptyset\} \ = \ f^{-1}(Y). \tag{2.1}$$

We call a relation *surjective* if $Dom(f) = X$ and $Dom(f^{-1}) = Y$, i.e. $f(X) = Y$ and $f^{-1}(Y) = X$.

If $f_1 : X_1 \to Y_1$ and $f_2 : X_2 \to Y_2$ are relations, then the product relation $f_1 \times f_2 : X_1 \times X_2 \to Y_1 \times Y_2$ is $\{((x_1,x_2),(y_1,y_2)) : (x_1,y_1) \in f_1, (x_2,y_2) \in f_2\}$.

We call f a *relation on* X when $X = Y$. In that case, we define, for $n \geq 1$ $f^{n+1} = f \circ f^n = f^n \circ f$ with $f^1 = f$. By definition, $f^0 = 1_X$ and $f^{-n} = (f^{-1})^n$. If $A \subset X$, then A is called f ${}^+$*invariant* if $f(A) \subset A$ and f *invariant* if $f(A) = A$. In general, for $A \subset X$, the *restriction to* A is $f|A = f \cap (A \times A)$. If u is a real-valued function on X we will also write $u|A$ for the restriction of u to A, allowing context to determine which meaning is used.

The *cyclic set* $|f|$ of a relation f on X is $\{x \in X : (x,x) \in f\}$.

A relation f on X is *reflexive* if $1_X \subset f$, *symmetric* if $f^{-1} = f$ and *transitive* if $f \circ f \subset f$.

2.2. Pseudo-metrics. If d is a pseudo-metric on a set X and $\epsilon > 0$, then we define the relations $V^d_\epsilon = \{(x,y) : d(x,y) < \epsilon\}$ and $\bar{V}^d_\epsilon = \{(x,y) : d(x,y) \leq \epsilon\}$ on X. Thus, for $x \in X$, $V^d_\epsilon(x)$ (or $\bar{V}^d_\epsilon(x)$) is the open (resp. closed) ball centered at x with radius ϵ.

A pseudo-ultrametric d on X is a pseudo-metric with the triangle inequality strengthened to $d(x,y) \leq \max(d(x,z), d(z,y))$ for all $z \in X$. A pseudo-metric d is a pseudo-ultrametric iff the relations V^d_ϵ and $\bar{V}^d_\epsilon$ are equivalence relations for all $\epsilon > 0$.

If (X_1, d_1) and (X_2, d_2) are pseudo-metric spaces then the product $(X_1 \times X_2, d_1 \times d_2)$ is defined by

$$d_1 \times d_2((x_1,x_2),(y_1,y_2)) \quad = \quad \max(d_1(x_1,y_1), d_2(x_2,y_2)).$$

Thus, $V^{d_1 \times d_2}_\epsilon = V^{d_1}_\epsilon \times V^{d_2}_\epsilon$ and $\bar{V}^{d_1 \times d_2}_\epsilon = \bar{V}^{d_1}_\epsilon \times \bar{V}^{d_2}_\epsilon$.

Throughout this work, all pseudo-metrics are assumed bounded. For example, on $\mathbb{R}$ we use $d(a,b) = \min(|a-b|, 1)$. Thus, if A is a non-empty subset of X the *diameter*, *diam*$(A) = \sup\{d(x,y) : x, y \in A\}$, is finite.

For metric computations, the following will be useful.

Lemma 2.1. *Let* $a_1, a_2, b_1, b_2 \in \mathbb{R}$. *With* $a \vee b = \max(a,b)$ *and* $a \wedge b = \min(a,b)$*:*

$$\begin{aligned} |a_1 \vee b_1 - a_2 \vee b_2|, |a_1 \wedge b_1 - a_2 \wedge b_2| \quad &\leq \quad |a_1 - a_2| \vee |b_1 - b_2|. \\ (a_1 \vee b_1) \wedge (a_2 \vee b_1) \wedge (a_1 \vee b_2) \wedge (a_2 \vee b_2) \quad &= \quad (a_1 \wedge a_2) \vee (b_1 \wedge b_2). \end{aligned} \tag{2.2}$$

Proof: First, we may assume without loss of generality that $a_1 \vee b_1 \geq a_2 \vee b_2 = a_2$ and so that $a_2 \geq b_2$. If $a_1 \vee b_1 = a_1$ then $|a_1 \vee b_1 - a_2 \vee b_2| = a_1 - a_2$. If $a_1 \vee b_1 = b_1$ then $|a_1 \vee b_1 - a_2 \vee b_2| = b_1 - a_2 \leq b_1 - b_2$. For the $\wedge$ estimate, observe that $a \wedge b = -(-a) \vee (-b)$.

For the second, factor out b_1 and b_2 to get $(a_1 \vee b_1) \wedge (a_2 \vee b_1) = (a_1 \wedge a_2) \vee b_1$, and $(a_1 \vee b_2) \wedge (a_2 \vee b_2) = (a_1 \wedge a_2) \vee b_2$. Then factor out $a_1 \wedge a_2$. □

2.3. Uniformities. A *uniform structure*, or *uniformity*, $\mathfrak{U}$ on X is a collection of relations satisfying the following conditions:

- $1_X \subset U$ for all $U \in \mathfrak{U}$.
- $U_1, U_2 \in \mathfrak{U}$ implies $U_1 \cap U_2 \in \mathfrak{U}$.
- If $U \in \mathfrak{U}$ and $W \supset U$, then $W \in \mathfrak{U}$.
- $U \in \mathfrak{U}$ implies $U^{-1} \in \mathfrak{U}$.
- If $U \in \mathfrak{U}$, then there exists $W \in \mathfrak{U}$ such that $W \circ W \subset U$.

The first condition says that the relations are reflexive and the next two imply that they form a filter. The last two conditions are the uniform space analogues of symmetry and the triangle inequality for pseudo-metrics.

As mentioned earlier, a uniformity $\mathcal{U}$ is equivalently given by its gage $\Gamma(\mathcal{U})$, the set of pseudo-metrics d on X (bounded by stipulation) with the metric uniformity $\mathcal{U}(d)$, generated by $\{V_\epsilon^d : \epsilon > 0\}$, contained in $\mathcal{U}$. To a uniformity there is an associated topology and we say that $\mathcal{U}$ is compatible with a topology on X if the uniform topology agrees with the given topology on X. A topological space admits a compatible uniformity iff it is completely regular. A completely regular space X has a maximum uniformity $\mathcal{U}_M$ compatible with the topology. Any continuous function from a completely regular space X to a uniform space is uniformly continuous from $(X, \mathcal{U}_M)$.

A completely regular, Hausdorff space is called a *Tychonoff space*. A compact Hausdorff space X has a unique uniformity consisting of all neighborhoods of the diagonal 1_X.

We will need the following lemma.

Lemma 2.2. *Let $\{d_1, d_2, \dots\}$ be a sequence in $\Gamma(\mathcal{U})$ with d_i bounded by $K_i \geq 1$. If $\{a_1, a_2, \dots\}$ is a summable sequence of positive reals then $d = \Sigma_{i=1}^{\infty} (a_i/K_i)d_i$ is a pseudo-metric in $\Gamma(\mathcal{U})$.*

Proof: Dividing by $\Sigma_{i=1}^{\infty} (a_i/K_i)$ we can assume the sum is 1. Given $\epsilon > 0$ choose N so that $\Sigma_{i=N+1}^{\infty} a_i < \epsilon/2$. Then $\bigcap_{k=1}^{N} V_{\epsilon/2}^{d_k} \subset V_\epsilon^d$. □

3. Barrier functions

Let f be a relation on a pseudo-metric space (X, d). That is, f is a subset of $X \times X$ and d is a pseudo-metric on the non-empty set X.

Let $f^{\times n}$ be the $n-$fold product of copies of f, i.e. the space of sequences in f of length $n \geq 1$, so that an element of $f^{\times n}$ is a sequence $[a, b] = (a_1, b_1), (a_2, b_2), ..,$ (a_n, b_n) of pairs in f. If $[a, b] \in f^{\times n}, [c, d] \in f^{\times m}$, then the *concatenation* $[a, b] \cdot [c, d] \in f^{\times(n+m)}$ is the sequence of pairs $(x_i, y_i) = (a_i, b_i)$ for $i = 1, \dots, n$ and $(x_i, y_i) = (c_{i-n}, d_{i-n})$ for $i = n+1, \dots, n+m$.

Define for $(x, y) \in X \times X$ and $[a, b] \in f^{\times n}$ the *xy chain-length of* $[a, b]$ (with respect to d) to be the sum

$$d(x, a_1) + \Sigma_{i=1}^{n-1} d(b_i, a_{i+1}) + d(b_n, y) \tag{3.1}$$

and the *xy chain-bound of* $[a, b]$ (with respect to d) to be

$$\max(d(x, a_1), d(b_1, a_2), \dots, d(b_{n-1}, a_n), d(b_n, y)). \tag{3.2}$$

For $(x, y) \in X \times X$, define

$$\begin{aligned} \ell_d^f(x, y) &= \inf \{d(x, a_1) + \Sigma_{i=1}^{n-1} d(b_i, a_{i+1}) + d(b_n, y) : \\ &\qquad [a, b] \in f^{\times n}, n = 1, 2, ...\}. \\ m_d^f(x, y) &= \inf \{\max(d(x, a_1), d(b_1, a_2), \dots, d(b_{n-1}, a_n), d(b_n, y)) : \\ &\qquad [a, b] \in f^{\times n}, n = 1, 2, ...\}. \end{aligned} \tag{3.3}$$

The functions ℓ_d^f and m_d^f are the *barrier functions* for f. Clearly, $m_d^f \leq \ell_d^f$. Using $n = 1$, we see that for all $(a, b) \in f$

$$\begin{aligned} \ell_d^f(x, y) &\leq d(x, a) + d(b, y), \\ m_d^f(x, y) &\leq \max(d(x, a), d(b, y)). \end{aligned} \tag{3.4}$$

and so

$$(x,y) \in f \quad \Longrightarrow \quad m_d^f(x,y) \ = \ \ell_d^f(x,y) \ = \ 0. \tag{3.5}$$

by using $(a,b) = (x,y)$.

For the special case of $f = \emptyset$ we define

$$m_d^{\emptyset} \ = \ diam(X), \quad \text{and} \quad \ell_d^{\emptyset} \ = \ 2diam(X), \tag{3.6}$$

the constant functions.

By using equation (3.4) with $(a,b) = (y,y)$ and the triangle inequality in (3.3) we see that

$$\ell_d^{1_X}(x,y) \ = \ d(x,y). \tag{3.7}$$

Define for the pseudo-metric d

$$Z_d \ = \ \{(x,y) : d(x,y) = 0\}. \tag{3.8}$$

Thus, Z_d is a closed equivalence relation which equals 1_X exactly when d is a metric. Z_d is the d-closure in $X \times X$ of the diagonal 1_X.

Lemma 3.1. *Let f be a relation on (X,d) with $A = Dom(f) = f^{-1}(X)$. If $f \subset Z_d$, then*

$$\ell_d^f(x,y) \ = \ \inf\{d(x,a) + d(a,y) : a \in A\} \ \geq \ d(x,y) \tag{3.9}$$

with equality if either x or y is an element of A.

If d is a pseudo-ultrametric then

$$m_d^f(x,y) \ = \ \inf\{\max(d(x,a), d(a,y)) : a \in A\} \ \geq \ d(x,y) \tag{3.10}$$

with equality if either x or y is an element of A.

Proof: If $(a,b) \in f$ then $d(a,b) = 0$ and so the xy chain-length of $[(a,b)]$ is $d(x,a) + d(a,y)$. If $[a,b] \in f^{\times n}$ then $d(a_i,b_i) = 0$ for all i implies that with $a = a_1$ the xy chain-length of $[a,b]$ is at least $d(x,a) + d(a,y)$ by the triangle inequality.

If d is a pseudo-ultrametric then the xy chain-bound of $[(a,b)]$ is $\max(d(x,a), d(a,y)))$ and if $[a,b] \in f^{\times n}$, then with with $a = a_1$ the xy chain-bound of $[a,b]$ is at least $\max(d(x,a), d(a,y)))$ by the ultrametric version of the triangle inequality. $\square$

In particular, if A is a nonempty subset of X, then

$$\ell_d^{1_A}(x,y) \ = \ \inf\{d(x,a) + d(a,y) : a \in A\} \ \geq \ d(x,y) \tag{3.11}$$

with equality if either x or y is an element of A.

It is clear that $f \subset g$ implies $f^{\times n} \subset g^{\times n}$ and so

$$f \subset g \quad \Longrightarrow \quad \ell_d^g \ \leq \ \ell_d^f \quad \text{and} \quad m_d^g \ \leq \ m_d^f \quad \text{on } X \times X. \tag{3.12}$$

In particular, if A is a subset of X, then

$$\ell_d^f \ \leq \ \ell_d^{f|A} \quad \text{and} \quad m_d^f \ \leq \ m_d^{f|A}. \tag{3.13}$$

The relation f is reflexive when $1_X \subset f$. We see from (3.7)

$$1_X \subset f \quad \Longrightarrow \quad \ell_d^f \ \leq \ d \quad \text{on } X \times X. \tag{3.14}$$

If $[a,b] \in f^{\times n}$, then we let $[a,b]^{-1} \in (f^{-1})^{\times n}$ be $(b_n,a_n), (b_{n-1},a_{n-1}), ..., (b_1,a_1)$. Using these reverse sequences we see immediately that

$$\ell_d^f(x,y) \ = \ \ell_d^{f^{-1}}(y,x) \quad \text{and} \quad m_d^f(x,y) \ = \ m_d^{f^{-1}}(y,x) \tag{3.15}$$

for all $x,y \in X$.

A map h from (X_1, d_1) to (X_2, d_2) is Lipschitz with constant K if $d_2(h(x), h(y)) \leq K d_1(x, y)$ for all $x, y \in X_1$.

Proposition 3.2. *Let f be a relation on (X, d). Let $x, y, z, w \in X$.*

(a) *The directed triangle inequalities hold:*

$$\begin{aligned} \ell_d^f(x,y) &\leq \ell_d^f(x,z) + \ell_d^f(z,y), \\ m_d^f(x,y) &\leq m_d^f(x,z) + m_d^f(z,y). \end{aligned} \tag{3.16}$$

(b) *Related to the ultrametric inequalities, taking into account the fact that $m_d^f(z,z)$ can be strictly greater than zero, we have:*

$$m_d^f(x,y) \leq \max(m_d^f(x,z) + m_d^f(z,z), m_d^f(z,z) + m_d^f(z,y)). \tag{3.17}$$

(c) *From*

$$\begin{aligned} \ell_d^f(x,y) &\leq d(x,w) + \ell_d^f(w,z) + d(z,y) \quad \textit{for all } w,x,y,z \in X, \\ m_d^f(x,y) &\leq d(x,w) + m_d^f(w,z) + d(z,y) \quad \textit{for all } w,x,y,z \in X \end{aligned} \tag{3.18}$$

we obtain that the functions ℓ_d^f and m_d^f from $X \times X$ to $\mathbb{R}$ are Lipschitz with Lipschitz constant ≤ 2.

Proof: (a) For $x, y, z \in X$ and $[a, b] \in f^{\times n}, [c, d] \in f^{\times m}$, we note that $d(b_n, c_1) \leq d(b_n, z) + d(z, c_1)$. So the xz chain-length of $[a, b]$ plus the zy chain-length of $[c, d]$ is greater than or equal to the xy chain-length of $[a, b] \cdot [c, d]$. Furthermore, the xz chain-bound of $[a, b]$ plus the zy chain-bound of $[c, d]$ is greater than or equal to the xy chain-bound of $[a, b] \cdot [c, d]$. The directed triangle inequalities (3.16) follow.

(b) Let $[u, v] \in f^{\times p}$. We see that $d(b_n, u_1) \leq d(b_n, z) + d(z, u_1)$ and $d(v_p, c_1) \leq d(v_p, z) + d(z, c_1)$. Hence, the larger of the xz chain-bound of $[a, b]$ plus the zz chain-bound of $[u, v]$ and the zz chain-bound of $[u, v]$ plus the zy chain-bound of $[c, d]$ bounds the xy chain-bound of $[a, b] \cdot [u, v] \cdot [c, d]$. This implies (3.17).

(c) Similarly, $d(x, a_1) \leq d(x, w) + d(w, a_1)$ and $d(b_n, y) \leq d(b_n, z) + d(z, y)$ implies (3.18). Then $\ell_d^f(x, y) - \ell_d^f(w, z) \leq d(x, w) + d(y, z) \leq 2\max(d(x, w), d(y, z)) = 2(d \times d)((x, y), (w, z))$, and similarly for m_d^f. □

If h is a map from (X_1, d_1) to (X_2, d_2) then h is uniformly continuous if for every $\epsilon > 0$ there exists $\delta > 0$ such that $d_1(x, y) < \delta$ implies $d_2(h(x), h(y)) < \epsilon$ for all $x, y \in X_1$. We call δ an ϵ modulus of uniform continuity.

If f_1 is a relation on X_1 and f_2 is a relation on X_2 then we say that a function $h : X_1 \to X_2$ maps f_1 to f_2 if $(h \times h)(f_1) \subset f_2$, i.e. $(x, y) \in f_1$ implies $(h(x), h(y)) \in f_2$. Since h is a map, $1_{X_1} \subset h^{-1} \circ h$ and $h \circ h^{-1} \subset 1_{X_2}$. From these it easily follows that

$$(h \times h)(f_1) = h \circ f_1 \circ h^{-1}, \tag{3.19}$$

$$(h \times h)(f_1) \subset f_2 \iff h \circ f_1 \subset f_2 \circ h. \tag{3.20}$$

If h maps f_1 to f_2 then clearly h maps f_1^{-1} and f_2^{-1} and

$$h(|f_1|) \subset |f_2|. \tag{3.21}$$

Proposition 3.3. *Let f_1 and f_2 be relations on (X_1, d_1) and (X_2, d_2), respectively. Assume $h : X_1 \to X_2$ maps f_1 to f_2.*

(a) If h is uniformly continuous then for $\epsilon > 0$ with $\delta > 0$ an ϵ modulus of uniform continuity, $m_{d_1}^{f_1}(x, y) < \delta$ implies $m_{d_2}^{f_2}(h(x), h(y)) < \epsilon$ for all $x, y \in X_1$.

(b) If h is Lipschitz with constant K then $\ell^{f_1}_{d_1}(x,y) \leq K\ell^{f_2}_{d_2}(h(x),h(y))$ for all $x,y \in X_1$.

Proof: If $[a,b] \in f_1^{\times n}$ then $(h\times h)^{\times n}([a,b]) \in f_2^{\times n}$. If δ is an ϵ modulus of uniform continuity then if the xy chain-bound of $[a,b]$ is less than δ then the $h(x)h(y)$ chain-bound of $(h\times h)^{\times n}([a,b])$ is less than ϵ. If h is Lipschitz with constant K then the $h(x)h(y)$ chain-length is at most K times the xy chain-length. □

4. The Conley and Aubry-Mather chain-relations

For a relation f on (X,d), the *Conley chain relation* $\mathcal{C}_d f$ is defined by

$$\mathcal{C}_d f \;=\; \{(x,y) : m^d_f(x,y) = 0\}, \tag{4.1}$$

and the *Aubry-Mather chain relation* is defined by

$$\mathcal{A}_d f \;=\; \{(x,y) : \ell^d_f(x,y) = 0\}. \tag{4.2}$$

Because m^f_d and ℓ^f_d are continuous, it follows that $\mathcal{C}_d f$ and $\mathcal{A}_d f$ are closed in $(X\times X, d\times d)$. From the directed triangle inequalities (3.16), it follows that $\mathcal{C}_d f$ and $\mathcal{A}_d f$ are transitive, i.e.

$$\begin{aligned}\mathcal{C}_d f \circ \mathcal{C}_d f \;&\subset\; \mathcal{C}_d f,\\ \mathcal{A}_d f \circ \mathcal{A}_d f \;&\subset\; \mathcal{A}_d f.\end{aligned} \tag{4.3}$$

From (3.5) we see that,

$$f \;\subset\; \mathcal{A}_d f \;\subset\; \mathcal{C}_d f. \tag{4.4}$$

If $A \subset X$ with $f \subset A\times A$ we can regard f as a relation on (X,d) or as a relation on $(A, d|A\times A)$ where $d|(A\times A)$ is the restriction of the pseudo-metric d to $A\times A$. It is clear that if $f \subset A\times A$, then

$$m^f_d|(A\times A) \;=\; m^f_{d|(A\times A)} \quad\text{and}\quad \ell^f_d|(A\times A) \;=\; \ell^f_{d|(A\times A)}. \tag{4.5}$$

and so

$$(\mathcal{C}_d f)\cap(A\times A) \;=\; \mathcal{C}_{d|(A\times A)} f \quad\text{and}\quad (\mathcal{A}_d f)\cap(A\times A) \;=\; \mathcal{A}_{d|(A\times A)} f. \tag{4.6}$$

If A is closed and $x,y\in X$ with either $x \not\in A$ or $y \not\in A$, then $\ell^f_d(x,y) \geq m^f_d(x,y) > 0$ and so

$$\mathcal{C}_d f \;=\; \mathcal{C}_{d|(A\times A)} f \quad\text{and}\quad \mathcal{A}_d f \;=\; \mathcal{A}_{d|(A\times A)} f. \tag{4.7}$$

From (3.12) we get monotonicity

$$f \subset g \quad\Longrightarrow\quad \mathcal{C}_d f \subset \mathcal{C}_d g \quad\text{and}\quad \mathcal{A}_d f \subset \mathcal{A}_d g. \tag{4.8}$$

and from (3.15)

$$\mathcal{C}_d(f^{-1}) \;=\; (\mathcal{C}_d f)^{-1} \quad\text{and}\quad \mathcal{A}_d(f^{-1}) \;=\; (\mathcal{A}_d f)^{-1}, \tag{4.9}$$

and so we can omit the parentheses.

Proposition 4.1. *Let f,g be relations on X.*

$$m^{\mathcal{C}_d f}_d \;=\; m^f_d \quad\textit{and}\quad \ell^{\mathcal{A}_d f}_d \;=\; \ell^f_d \tag{4.10}$$

The operators $\mathcal{C}_d$ and $\mathcal{A}_d$ on relations are idempotent. That is,

$$\mathcal{C}_d(\mathcal{C}_d f) \;=\; \mathcal{C}_d f \quad\textit{and}\quad \mathcal{A}_d(\mathcal{A}_d f) \;=\; \mathcal{A}_d f \tag{4.11}$$

In addition,

$$\mathcal{C}_d(\mathcal{C}_d f \cap \mathcal{C}_d g) = \mathcal{C}_d f \cap \mathcal{C}_d g \text{ and } \mathcal{A}_d(\mathcal{A}_d f \cap \mathcal{A}_d g) = \mathcal{A}_d f \cap \mathcal{A}_d g, \tag{4.12}$$

Proof: Since $f \subset \mathcal{A}_d f \subset \mathcal{C}_d f$ it follows from (3.12) that $m_d^{\mathcal{C}_d f} \leq m_d^f$ and $\ell_d^{\mathcal{A}_d f} \leq \ell_d^f$.

For the reverse inequality fix $x, y \in X$ and let $t > \ell_d^{\mathcal{A}_d f}(x,y)$ be arbitrary. Choose t_1 with $t > t_1 > \ell_d^{\mathcal{A}_d f}(x,y)$. Suppose that $[a,b] \in (\mathcal{A}_d f)^{\times n}$ with xy chain-length less than t_1. Let $\epsilon = (t - t_1)/2n$. For $i = 1, ..., n$ we can choose an element of some $f^{\times n_i}$ whose $a_i b_i$ chain-length is less than ϵ. Concatenating these in order we obtain a sequence in $f^{\times m}$ with $m = \Sigma_{i=1}^n n_i$ whose xy chain-length is at most $t_1 + 2n\epsilon \leq t$. Hence, $\ell_d^f(x,y) \leq t$. Letting t approach $\ell_d^{\mathcal{A}_d f}(x,y)$ we obtain in the limit that $\ell_d^f(x,y) \leq \ell_d^{\mathcal{A}_d f}(x,y)$.

The argument to show $m_d^f(x,y) \leq m_d^{\mathcal{C}_d f}(x,y)$ is completely similar.

It is clear that (4.10) implies (4.11).

Finally, $\mathcal{C}_d f \cap \mathcal{C}_d g \subset \mathcal{C}_d(\mathcal{C}_d f \cap \mathcal{C}_d g) \subset \mathcal{C}_d(\mathcal{C}_d f) = \mathcal{C}_d f$ and similarly, $\mathcal{C}_d f \cap \mathcal{C}_d g \subset \mathcal{C}_d(\mathcal{C}_d f \cap \mathcal{C}_d g) \subset \mathcal{C}_d g$. Intersect to get (4.12) for $\mathcal{C}_d$ and the same argument yields the $\mathcal{A}_d$ result. □

Corollary 4.2. *For a relation f on (X,d) let $\bar{f}^d$ be the closure of f in $(X \times X, d \times d)$.*

$$\begin{aligned} m_d^{\bar{f}^d} &= m_d^f \quad \text{and} \quad \ell_d^{\bar{f}^d} = \ell_d^f, \\ \mathcal{C}_d(\bar{f}^d) &= \mathcal{C}_d f \quad \text{and} \quad \mathcal{A}_d(\bar{f}^d) = \mathcal{A}_d f \end{aligned} \tag{4.13}$$

Proof: This is clear from (3.12) and (4.10) because $f \subset \bar{f}^d \subset \mathcal{A}_d f \subset \mathcal{C}_d f$. □

The *Conley set* (or *d-chain recurrent set*) is the cyclic set $|\mathcal{C}_d f| = \{x : (x,x) \in \mathcal{C}_d f\}$. Since $|\mathcal{C}_d f|$ is the pre-image of the closed set $\mathcal{C}_d f \subset X \times X$ via the continuous map $x \mapsto (x,x)$ it follows that $|\mathcal{C}_d f| \subset X$ is closed. The *Aubry Set* (or *d-strong chain recurrent set*) is the cyclic set $|\mathcal{A}_d f| \subset X$ which is similarly closed.

From (4.4) we clearly have $|\mathcal{A}_d f| \subset |\mathcal{C}_d f|$.

On $|\mathcal{C}_d f|$ the relation $\mathcal{C}_d f \cap \mathcal{C}_d f^{-1}$ is a closed equivalence relation and on $|\mathcal{A}_d f|$ $\mathcal{A}_d f \cap \mathcal{A}_d f^{-1}$ is a closed equivalence relation.

Example 4.3. Consider the map g in Example 1.2. With the usual metric d, we have $|\mathcal{C}_d g| = |\mathcal{A}_d g| = \emptyset$. With the metric d_1, we have $|\mathcal{C}_{d_1} g| = (0,1) \times (0,1)$, with all points equivalent, while $|\mathcal{A}_{d_1} g| = (0,1) \times \{1/2\}$, also with all points equivalent. Finally, with the metric d_2, we again have $|\mathcal{C}_{d_2} g| = (0,1) \times (0,1)$, with all points equivalent, while $|\mathcal{A}_{d_2} g| = (0,1) \times (0,1)$ with (x_1, y_1) equivalent to (x_2, y_2) iff $y_1 = y_2$.

Define the symmetrized functions

$$\begin{aligned} sm_d^f(x,y) &= \max\{m_d^f(x,y), m_d^f(y,x)\}, \\ s\ell_d^f(x,y) &= \max\{\ell_d^f(x,y), \ell_d^f(y,x)\}. \end{aligned} \tag{4.14}$$

Proposition 4.4. *Let f be a relation on X. Let $x, y, z \in X$.*

(a) $sm_d^f(x,y) \leq s\ell_d^f(x,y)$

(b) *The functions sm_d^f and $s\ell_d^f$ are symmetric and satisfy the triangle inequality.*

(c) *The functions* $sm^f_d, s\ell^f_d : X \times X \to \mathbb{R}$ *are Lipschitz with Lipschitz constant less than or equal to 2.*

(d)

$$\begin{aligned} sm^f_d(x,y) = 0 \quad &\Longleftrightarrow \quad (x,y),(y,x) \in \mathcal{C}_d f \quad \textit{and so } x,y \in |\mathcal{C}_d f|, \\ s\ell^f_d(x,y) = 0 \quad &\Longleftrightarrow \quad (x,y),(y,x) \in \mathcal{A}_d f \quad \textit{and so } x,y \in |\mathcal{A}_d f|. \end{aligned} \tag{4.15}$$

(e)

$$\begin{aligned} y \in |\mathcal{C}_d f| \quad &\Longrightarrow \quad sm^f_d(x,y) \le d(x,y), \\ y \in |\mathcal{A}_d f| \quad &\Longrightarrow \quad s\ell^f_d(x,y) \le d(x,y). \end{aligned} \tag{4.16}$$

(f) *If* $z \in |\mathcal{C}_d f|$ *then* $m^f_d(x,y) \le max(m^f_d(x,z), m^f_d(z,y))$.

Proof: (a) is obvious as is symmetry in (b), i.e. $sm^f_d(x,y) = sm^f_d(y,x)$ and $s\ell^f_d(x,y) = s\ell^f_d(y,x)$. The triangle inequality for $s\ell^f_d$ follows from

$$\begin{aligned} s\ell^f_d(x,z) + s\ell^f_d(z,y) \ge \ell^f_d(x,z) + \ell^f_d(z,y) \ge \ell^f_d(x,y), \\ s\ell^f_d(x,z) + s\ell^f_d(z,y) \ge \ell^f_d(z,x) + \ell^f_d(y,z) \ge \ell^f_d(y,x), \end{aligned} \tag{4.17}$$

with a similar argument for sm^f_d.

By Proposition 3.2(c) m^f_d and ℓ^f_d are Lipschitz. Then (c) follows from Lemma 2.1.

The equivalences in (d) are obvious. By transitivity, $(x,y),(y,x) \in \mathcal{C}_d f$ implies $(x,x),(y,y) \in \mathcal{C}_d f$. Similarly, for $\mathcal{A}_d f$.

(e) If $m^f_d(y,y) = 0$, then by (3.18) (with $w = z = y$), $m^f_d(x,y) \le d(x,y)$, and, switching the roles of x and y, $m^f_d(y,x) \le d(y,x)$. Similarly, for $s\ell^f_d$.

(f) follows from Proposition 3.2(b). □

We immediately obtain the following.

Corollary 4.5. *The map* $s\ell^f_d$ *restricts to define a pseudo-metric on* $|\mathcal{A}_d f|$ *and induces a metric on the quotient space of* $\mathcal{A}_d f \cap \mathcal{A}_d f^{-1}$ *equivalence classes. Furthermore, the projection map from* $|\mathcal{A}_d f|$ *to the space of equivalence classes has Lipschitz constant at most 2 with respect to this metric.*

The map sm^f_d *restricts to define a pseudo-ultrametric on* $|\mathcal{C}_d f|$ *and induces an ultrametric on the quotient space of* $\mathcal{C}_d f \cap \mathcal{C}_d f^{-1}$ *equivalence classes. Furthermore, the projection map from* $|\mathcal{C}_d f|$ *to the space of equivalence classes has Lipschitz constant at most 2 with respect to this metric.*

□

Example 4.6. Consider again the map g on $(0,1) \times (0,1)$ with metric d_2 from Examples 1.2 and 4.3. The projection map from $|\mathcal{A}_{d_2} g|$ to the space of $\mathcal{A}_{d_2} g \cap \mathcal{A}_{d_2} g^{-1}$ equivalence classes is just projection onto the second coordinate.

Let f_1 and f_2 be relations on X_1 and X_2, respectively. Recall that $h : X_1 \to X_2$ maps f_1 to f_2 when $h \circ f_1 \circ h^{-1} = (h \times h)(f_1) \subset f_2$, i.e. if $(x,y) \in f_1$ implies $(h(x),h(y)) \in f_2$. It then follows that h maps f_1^{-1} to f_2^{-1}.

Proposition 4.7. *Let* f_1 *and* f_2 *be relations on* (X_1,d_1) *and* (X_2,d_2)*, respectively. Assume* $h : X_1 \to X_2$ *maps* f_1 *to* f_2.

(a) If h is uniformly continuous, then h maps $\mathcal{C}_{d_1}f_1$ to $\mathcal{C}_{d_2}f_2$ and $\mathcal{C}_{d_1}f_1 \cap \mathcal{C}_{d_1}f_1^{-1}$ to $\mathcal{C}_{d_2}f_2 \cap \mathcal{C}_{d_2}f_2^{-1}$. So h maps each $\mathcal{C}_{d_1}f_1 \cap \mathcal{C}_{d_1}f_1^{-1}$ equivalence class in $|\mathcal{C}_{d_1}f_1|$ into a $\mathcal{C}_{d_2}f_2 \cap \mathcal{C}_{d_2}f_2^{-1}$ equivalence class in $|\mathcal{C}_{d_2}f_2|$.

(b) If h is Lipschitz, then h maps $\mathcal{A}_{d_1}f_1$ to $\mathcal{A}_{d_2}f_2$ and $\mathcal{A}_{d_1}f_1 \cap \mathcal{A}_{d_1}f_1^{-1}$ to $\mathcal{A}_{d_2}f_2 \cap \mathcal{A}_{d_2}f_2^{-1}$. So h maps each $\mathcal{A}_{d_1}f_1 \cap \mathcal{A}_{d_1}f_1^{-1}$ equivalence class in $|\mathcal{A}_{d_1}f_1|$ into a $\mathcal{A}_{d_2}f_2 \cap \mathcal{A}_{d_2}f_2^{-1}$ equivalence class in $|\mathcal{A}_{d_2}f_2|$.

Proof: This obviously follows from Proposition 3.3. □

We conclude this section with some useful computations.

Recall that

$$Z_d = \{(x,y) : d(x,y) = 0\}. \tag{4.18}$$

Proposition 4.8. *Let f be a relation on X and A be a nonempty, closed subset of X*

(a) For $x, y \in X$

$$\begin{aligned} \ell_d^{1_A \cup f}(x,y) &= \min(\ell_d^f(x,y), \ell_d^{1_A}(x,y)), \\ \ell_d^{1_X \cup f}(x,y) &= \min(\ell_d^f(x,y), d(x,y)), \\ s\ell_d^{1_X \cup f}(x,y) &= \min(s\ell_d^f(x,y), d(x,y)), \\ s\ell_d^{1_X \cup f}(x,y) &= s\ell_d^f(x,y) \qquad \text{if } x \in |\mathcal{A}_d f|. \end{aligned} \tag{4.19}$$

(b)

$$\begin{aligned} \mathcal{A}_d(1_A \cup f) &= (Z_d \cap (A \times A)) \cup \mathcal{A}_d f, \\ \mathcal{A}_d(1_X \cup f) &= Z_d \cup \mathcal{A}_d f. \end{aligned} \tag{4.20}$$

$s\ell_d^{1_X \cup f}$ is a pseudo-metric on X whose associated metric space is the quotient space of X by the equivalence relation $Z_d \cup (\mathcal{A}_d f \cap \mathcal{A}_d f^{-1})$. The quotient map has Lipschitz constant at most 2.

Proof: (a) By (3.12) $\ell_d^{1_A \cup f} \leq \min(\ell_d^f, \ell_d^{1_A})$. By (3.7) $\ell_d^{1_X} = d$.

Let $[a,b] \in (1_A \cup f)^{\times n}$. If $(a_i, b_i) \in 1_A$ for all i then omit all but one of the pairs to obtain an element of $1_A^{\times 1}$. Otherwise, omit the pairs $(a_i, b_i) \in 1_A$ and renumber. We then obtain a sequence in $f^{\times m}$ for some m with $1 \leq m \leq n$. Furthermore, in either case the xy chain-length has not increased. For example, if $(a_i, b_i) \in 1_A$ for some $1 < i < n$ then since $a_i = b_i$ the triangle inequality implies $d(b_{i-1}, a_{i+1}) \leq d(b_{i-1}, a_i) + d(b_i, a_{i+1})$. It follows that $\ell_d^{1_X \cup f} \geq \min(\ell_d^f, \ell_d^{1_A})$.

$s\ell_d^{1_X \cup f}(x,y) = \max[\min(\ell_d^f(x,y), d(x,y)), \min(\ell_d^f(y,x), d(x,y))]$. This is $d(x,y)$ except when $d(x,y) > \ell_d^f(x,y)$ and $d(x,y) > \ell_d^f(y,x)$, i.e. $d(x,y) > s\ell_d^f(x,y)$ in which case it is $s\ell_d^f(x,y)$. If $x \in |\mathcal{A}_d f|$ then by (4.16) $\min(s\ell_d^f(x,y), d(x,y)) = s\ell_d^f(x,y)$.

(b) It follows that $(x,y) \in |\mathcal{A}_d(1_A \cup f)|$ iff $\ell_d^f(x,y) = 0$ or $\ell_d^{1_A}(x,y) = 0$. By (3.11) the latter is true iff $x, y \in A$ with $d(x,y) = 0$ since A is closed. Thus, (4.20) holds and the rest is obvious. □

If A, B are subsets of X then we can regard $A \times B$ as a relation on X. For any relation g on X we clearly have:

$$(A \times B) \circ g \circ (A \times B) \subset A \times B. \tag{4.21}$$

Lemma 4.9. *If A and B are nonempty subsets of (X,d) and $x,y \in X$, then*

$$\begin{aligned} m_d^{A\times B}(x,y) &= \max(d(x,A),d(y,B)) \quad \text{and} \\ \ell_d^{A\times B}(x,y) &= d(x,A)+d(y,B) \end{aligned} \tag{4.22}$$

where $d(x,A)=\inf\{d(x,z) : z \in A\}$.

Proof: If $[a,b] \in (A\times B)^{\times n}$ then $(a_1,b_n) \in A \times B$ with xy chain-length $d(x,a_1)+d(y,b_n)$ no larger than the xy chain-length for $[a,b]$ and with xy chain-bound $\max(d(x,a_1),d(y,b_n))$ no larger than the xy chain-bound for $[a,b]$. This proves (4.22). □

From Proposition 4.8 we immediately get

Corollary 4.10. *If A and B are nonempty subsets of X and $x,y \in X$ then*

$$\ell_d^{1_X\cup(A\times B)}(x,y) = \min[d(x,y),d(x,A)+d(y,B)]. \tag{4.23}$$

□

Remark: If $A=B$ then $s\ell_d^{1_X\cup(A\times A)} = \ell_d^{1_X\cup(A\times A)}$ is the pseudo-metric on X induced by the equivalence relation $1_X \cup (A\times A)$ corresponding to smashing A to a point.

Lemma 4.11. *For $x,y,z \in X$*

$$\begin{aligned} &m_d^{f\cup\{(z,z)\}}(x,y) = \\ &\min[\ m_d^f(x,y),\ \max[\min(m_d^f(x,z),d(x,z)),\ \min(m_d^f(z,y),d(z,y))\ ]. \end{aligned} \tag{4.24}$$

In particular, with $z=y$ or $z=x$

$$m_d^{f\cup\{(y,y)\}}(x,y) = m_d^{f\cup\{(x,x)\}}(x,y) = min[m_d^f(x,y),d(x,y)]. \tag{4.25}$$

If $(z,z) \in \mathcal{C}_d f$, i.e. $z \in |\mathcal{C}_d f|$, then $m_d^{f\cup\{(z,z)\}} = m_d^f$.

Proof: Since $f \subset f\cup\{(z,z)\}$ we have $m_d^{f\cup\{(z,z)\}} \leq \min(m_d^f, m_d^{\{(z,z)\}})$.

Let $[a,b] \in (f\cup\{(z,z)\})^{\times n}$. If (z,z) occurs more than once in $[a,b]$ we can eliminate the repeat and all of the terms between them without increasing the xy chain-bound. Thus, we may take the infimum over those $[a,b]$ in which (z,z) occurs at most once.

The infimum of the xy chain-bounds in $f^{\times n}$ is $m_d^f(x,y)$.

- The xy chain-bound of $(z,z) \in (f\cup\{(z,z)\})^{\times 1}$ is $\max(d(x,z),d(z,y))$.
- If $[a,b]$ varies in $(f\cup\{(z,z)\})^{\times n}$ with $n>1$ and $(a_i,b_i)=(z,z)$ only for $i=1$, then the infimum of the xy chain-bounds is $\max(d(x,z),m_d^f(z,y))$.
- If $[a,b]$ varies in $(f\cup\{(z,z)\})^{\times n}$ with $n>1$ and $(a_i,b_i)=(z,z)$ only for $i=n$, then the infimum of the xy chain-bounds is $\max(m_d^f(x,z),d(z,y))$.
- If $[a,b]$ varies in $(f\cup\{(z,z)\})^{\times n}$ with $n>2$ and $(a_i,b_i)=(z,z)$ only for some i with $1<i<n$, then the infimum of the xy chain-bounds is $\max(m_d^f(x,z),m_d^f(z,y))$.

Equation (4.24) then follows from Lemma 2.1.

If $(z,z) \in \mathcal{C}_d f$ then $f \subset f\cup\{(z,z)\} \subset \mathcal{C}_d f$. So $m_d^{\mathcal{C}_d f} \leq m_d^{f\cup\{(z,z)\}} \leq m_d^f$ by (3.12) and so they are equal by (4.10). □

5. Lyapunov functions

A Lyapunov function for a relation f on a pseudo-metric space (X, d) is a continuous map $L : X \to \mathbb{R}$ such that

$$(x, y) \in f \qquad \Longrightarrow \qquad L(x) \leq L(y). \tag{5.1}$$

We follow [**1**] in using functions increasing on orbits rather than decreasing. For the compact case, Conley constructed Lyapunov functions for the chain recurrent set for flows [**5**], and Franks extended those results for maps [**8**]. Yokoi constructed them for the strong chain recurrent set for maps on compact spaces [**14**].

The set of Lyapunov functions contains the constants and is closed under addition, multiplication by positive scalars, max, min and post composition with any continuous non-decreasing function on $\mathbb{R}$. A continuous function which is a pointwise limit of Lyapunov functions is itself a Lyapunov function.

We define for a real-valued function L the relation

$$\leq_L \; = \; \{(x, y) : L(x) \leq L(y)\}. \tag{5.2}$$

This is clearly reflexive and transitive. If L is continuous, the relation $\leq_L$ is closed and so contains Z_d.

The Lyapunov function condition (5.1) can be restated as:

$$f \;\subset\; \leq_L . \tag{5.3}$$

For a Lyapunov function L and $x \in X$ we have

$$L(z) \leq L(x) \leq L(w) \qquad \text{for } \; z \in f^{-1}(x), w \in f(x) \tag{5.4}$$

The point x is called an *f-regular point* for L when the inequalities are strict for all $z \in f^{-1}(x), w \in f(x)$. Otherwise x is called an *f-critical point* for L. Notice, for example, that if $f^{-1}(x) = f(x) = \emptyset$ then these conditions hold vacuously and so x is an f-regular point.

We denote by $|L|_f$ the set of f-critical points for L. Clearly,

$$|L|_f \; = \; \pi_1(A) \cup \pi_2(A) \qquad \text{where} \quad A \; = \; f \cap (L \times L)^{-1}(1_{\mathbb{R}}), \tag{5.5}$$

and $\pi_1, \pi_2 : X \times X \to X$ are the two coordinate projections.

Definition 5.1. Let F be a transitive relation on (X, d) and let $\mathcal{L}$ be a collection of Lyapunov functions for F. We define three conditions on $\mathcal{L}$.

- ALG If $L_1, L_2 \in \mathcal{L}$ and $c \geq 0$ then $L_1 + L_2, \max(L_1, L_2), \min(L_1, L_2), cL_1, c, -c \in \mathcal{L}$.
- CON For every sequence $\{L_k\}$ of elements of $\mathcal{L}$ there exists a summable sequence of positive real numbers $\{a_k\}$ such that $\Sigma_k a_k L_k$ converges uniformly to an element of $\mathcal{L}$.
- POIN If $(x, y) \not\in Z_d \cup F$ then there exists $L \in \mathcal{L}$ such that $L(y) < L(x)$, i.e. $Z_d \cup F \; = \; \bigcap_{L \in \mathcal{L}} \; \leq_L$.

Theorem 5.2. *Assume (X, d) is separable. Let F be a closed, transitive relation and $\mathcal{L}$ be a collection of Lyapunov functions for F which satisfies ALG, CON and POIN. There exists a sequence $\{L_k\}$ in $\mathcal{L}$ such that*

$$\bigcap_k \; \leq_{L_k} \quad = \quad Z_d \cup F. \tag{5.6}$$

If $\{a_k\}$ is a positive, summable sequence such that $L = \Sigma_k\ a_k L_k \in \mathcal{L}$ then L is a Lyapunov function for F such that $Z_d \cup F\ = \leq_L$ and

$$x \in F(y) \quad \Longrightarrow \quad L(y) < L(x) \quad \text{unless } y \in F(x). \tag{5.7}$$

In particular,

$$|L|_F \ = \ |F| \tag{5.8}$$

Proof: For each $(x, y) \in (X \times X) \setminus (Z_d \cup F)$ use POIN to choose $L_{xy} \in \mathcal{L}$ such that $L_{xy}(y) < L_{xy}(x)$ and then neighborhoods V_{xy} of y and U_{xy} of x such that $\sup L_{xy}|V_{xy} < \inf L_{xy}|U_{xy}$ and so $\leq_{L_{xy}}$ is disjoint from $U_{xy} \times V_{xy}$. Because (X, d) is separable, it is second countable and so $(X \times X) \setminus (Z_d \cup F)$ is Lindelöf. Choose a sequence of pairs (x_k, y_k) so that $\{U_{x_k y_k} \times V_{x_k y_k}\}$ covers $(X \times X) \setminus (Z_d \cup F)$ and let $L_k = L_{x_k y_k}$. Since $Z_d \cup F \subset \leq_L$ for any Lyapunov function L, (5.6) holds.

Now with $L = \Sigma_k\ a_k L_k$, (5.6) implies $Z_d \cup F\ = \leq_L$. If $x \in F(y)$ and $d(y, x) = d(x, y) = 0$ then $(y, x) \in F$ implies $(x, x), (y, y), (x, y) \in F$, because F is closed. Hence, $y \in F(x)$. Assume $(x, y) \not\in Z_d$. Since $x \in F(y)$, $L_k(y) \leq L_k(x)$ for all k. If equality holds for all k then $(x, y) \in \bigcap_k \leq_{L_k} = Z_d \cup F$. Since $(x, y) \not\in Z_d$ we have $y \in F(x)$. If, instead, the inequality is strict for some k then since $a_k > 0$, $L(y) < L(x)$, proving (5.7).

If $x \not\in |F|$ then for $z \in F^{-1}(x)$ and $w \in F(x)$ we have $x \in F(z)$ but not $z \in F(x)$ else by transitivity $x \in |F|$. Hence, $L(z) < L(x)$. Similarly, $L(x) < L(w)$. Thus, $x \not\in |L|_F$. □

Definition 5.3. For a relation f on (X, d) and $K > 0$, a function $L : X \to \mathbb{R}$ is called $K\ell^f_d$ *dominated* if for all $x, y \in X$

$$L(x) - L(y) \quad \leq \quad K\ell^f_d(x, y), \tag{5.9}$$

Km^f_d *dominated* if for all $x, y \in X$

$$L(x) - L(y) \quad \leq \quad Km^f_d(x, y). \tag{5.10}$$

Theorem 5.4. *Let f be a relation on (X, d).*

(a) If L is a $K\ell^f_d$ dominated function then it is a Lyapunov function for $\mathcal{A}_d f$ and so is a Lyapunov function for f. If L is a Km^f_d dominated function then it is a $K\ell^f_d$ dominated function and is a Lyapunov function for $\mathcal{C}_d f$.

(b) If L is a Lyapunov function for f which is Lipschitz with respect to d with Lipschitz constant at most K then it is a $K\ell^f_d$ dominated function and so is a $\mathcal{A}_d f$ Lyapunov function.

Proof: (a) If $(x, y) \in \mathcal{A}_d f$ then $\ell^f_d(x, y) = 0$ and so for a $K\ell^f_d$ dominated function $L(x) - L(y) \leq 0$, that is, L is a Lyapunov function for $\mathcal{A}_d f$. Since $f \subset \mathcal{A}_d f$, L is a Lyapunov function for f as well. Similarly, if $(x, y) \in \mathcal{C}_d f$ and L is Km^f_d dominated, then $L(x) - L(y) \leq 0$. Since $m^f_d \leq \ell^f_d$ a Km^f_d dominated function is a $K\ell^f_d$ dominated function.

(b) Assume L is an f Lyapunov function with Lipschitz constant K and $x, y \in X$. For any $[a, b] \in f^{\times n}$ we note that each $L(a_i) - L(b_i) \leq 0$ since $(a_i, b_i) \in f$ and L is a Lyapunov function for f. Hence,

$$\begin{aligned} L(x) - L(y) \quad &= \quad L(x) - L(a_1) + L(a_1) - L(b_1) + L(b_1) - L(a_2) + \\ &\quad \dots + L(a_n) - L(b_n) + L(b_n) - L(y) \quad \leq \\ &L(x) - L(a_1) + \Sigma_{i=1}^{n-1} L(b_i) - L(a_{i+1}) + L(b_n) - L(y) \quad \leq K\ell, \end{aligned} \tag{5.11}$$

where ℓ is the xy chain-length of $[a, b]$. Taking the infimum over the sequences $[a, b]$ we obtain (5.9). Hence, L is a $\mathcal{A}_d f$ Lyapunov function by part (a). □

Proposition 5.5. *Let $f \subset g$ be relations on (X, d). For any $z \in X$, the function defined by $x \mapsto \ell^g_d(x, z)$ is a bounded, $1\ell^f_d$ dominated function, and the function defined by $x \mapsto m^g_d(x, z)$ is a bounded, $1m^f_d$ dominated function.*

Proof: By the directed triangle inequalities for ℓ^g_d and m^g_d we have

$$\ell^g_d(x, z) - \ell^g_d(y, z) \leq \ell^g_d(x, y) \quad \text{and} \quad m^g_d(x, z) - m^g_d(y, z) \leq m^g_d(x, y) \tag{5.12}$$

Since $f \subset g$, $\ell^g_d(x, y) \leq \ell^f_d(x, y)$ and $m^g_d(x, y) \leq m^f_d(x, y)$ by (3.12). □

Theorem 5.6. *For f a relation on (X, d) let $\mathcal{L}_\ell$ be the set of bounded, continuous functions which are $K\ell^f_d$ dominated for some positive K. Each $L \in \mathcal{L}_\ell$ is a $\mathcal{A}_d f$ Lyapunov function and so satisfies*

$$\mathcal{A}_d f \subset \leq_L \qquad \textit{and} \qquad |\mathcal{A}_d f| \subset |L|_{\mathcal{A}_d f}. \tag{5.13}$$

The collection $\mathcal{L}_\ell$ satisfies the conditions ALG, CON, and POIN with respect to $F = \mathcal{A}_d f$.

Proof: Each L in $\mathcal{L}_\ell$ is a $\mathcal{A}_d f$ Lyapunov function by Theorem 5.4 and so the first inclusion of (5.13) follows by definition. Clearly, if $(x, x) \in \mathcal{A}_d f$ then x is a $\mathcal{A}_d f$ critical point.

For $\mathcal{L}_\ell$ ALG is easy to check, see, e.g. Lemma 2.1. For CON let $\{L_k\}$ be a sequence in $\mathcal{L}_\ell$ and choose for each k, $M_k \geq 1$ which bounds $|L_k(x)|$ for all $x \in X$ and so that L_k is $M_k \ell^d_f$ dominated. If $\{b_k\}$ is any positive, summable sequence with $\sum b_k = 1$, then $a_k = b_k/M_k > 0$ is summable and $\Sigma_k\, a_k L_k$ converges uniformly to a function which is $1\ell^d_f$ dominated. Thus, CON holds as well.

Now assume $(x, y) \notin Z_d \cup \mathcal{A}_d f$. Let $g = 1_X \cup f$. By Proposition 5.5 $L(w) = \ell^g_d(w, y)$ defines a $1\ell^f_d$ dominated function which is a $\mathcal{A}_d f$ Lyapunov function by Theorem 5.4(a).

By Proposition 4.8 $L(w) = \min(\ell^f_d(w, y), d(w, y))$. Hence, $L(y) = 0$. Since $(x, y) \notin Z_d \cup \mathcal{A}_d f$, $L(x) > 0$. This proves POIN. □

Theorem 5.7. *For f a relation on (X, d) let $\mathcal{L}_m$ be the set of bounded, continuous functions which are Km^f_d dominated for some positive K. Each $L \in \mathcal{L}_m$ is a $\mathcal{C}_d f$ Lyapunov function and so satisfies*

$$\mathcal{C}_d f \subset \leq_L \qquad \textit{and} \qquad |\mathcal{C}_d f| \subset |L|_{\mathcal{C}_d f}. \tag{5.14}$$

The collection $\mathcal{L}_m$ satisfies the conditions ALG, CON, POIN with respect to $F = \mathcal{C}_d f$.

Proof: Each L in $\mathcal{L}_m$ is a $\mathcal{C}_d f$ Lyapunov function by Theorem 5.4 and so the first inclusion of (5.14) follows by definition. Clearly, if $(x, x) \in \mathcal{C}_d f$ then x is a $\mathcal{C}_d f$ critical point.

For $\mathcal{L}_m$ ALG again follows from Lemma 2.1. For CON let $\{L_k\}$ be a sequence in $\mathcal{L}_m$ and choose for each k, $M_k \geq 1$ which bounds $|L_k(x)|$ for all $x \in X$ and such that L_k is $M_k m^f_d$ dominated. If $\{b_k\}$ is any positive, summable sequence with $\sum b_k = 1$, then $a_k = b_k/M_k > 0$ is summable and $\Sigma_k\, a_k L_k$ converges uniformly to a function which is $1m^f_d$. Thus, CON holds as well.

Now assume $(x, y) \notin Z_d \cup \mathcal{C}_d f$. Let $g = f \cup \{(y, y)\}$. By Proposition 5.5 $L(w) = m^g_d(w, y)$ defines a $1m^f_d$ dominated function. By Equation (4.25) $L(w) =$

$\min(\ell_d^f(w,y), d(w,y))$. Hence, $L(y) = 0$. Since $(x,y) \notin Z_d \cup \mathcal{C}_d f$, $L(x) > 0$. This proves POIN. $\square$

6. Conley and Aubry-Mather relations for uniform spaces

Let $\mathcal{U}$ be a uniformity on X with gage Γ, the set of all bounded pseudo-metrics d on X such that the uniformity $\mathcal{U}(d)$ is contained in $\mathcal{U}$.

For a relation f on X we define the Conley relation and Aubry-Mather relation associated with the uniformity.

$$\mathcal{C}_{\mathcal{U}} f \ = \ \bigcap_{d \in \Gamma} \mathcal{C}_d f, \qquad \text{and} \qquad \mathcal{A}_{\mathcal{U}} f \ = \ \bigcap_{d \in \Gamma} \mathcal{A}_d f \tag{6.1}$$

with $|\mathcal{C}_{\mathcal{U}} f|$ the Conley set and $|\mathcal{A}_{\mathcal{U}} f|$ the Aubry set.

Thus, $\mathcal{C}_{\mathcal{U}} f$ and $\mathcal{A}_{\mathcal{U}} f$ are closed, transitive relations on X which contain f. We define $\mathcal{G} f$ to be the intersection of all the closed, transitive relations which contain f. Thus, $\mathcal{G} f$ is the smallest closed, transitive relation which contains f. Clearly,

$$f \subset \mathcal{G} f \ \subset \ \mathcal{A}_{\mathcal{U}} f \ \subset \mathcal{C}_{\mathcal{U}} f. \tag{6.2}$$

Thus, $(x,y) \in \mathcal{C}_{\mathcal{U}} f$ if for every $d \in \Gamma$ and every $\epsilon > 0$ there exists $[a,b] \in f^{\times n}$ with $n \geq 1$ such that the xy chain-bound of $[a,b]$ with respect to d is less than ϵ.

If $[a,b] \in f^{\times n}$ with $n \geq 1$ and $U \in \mathcal{U}$ we say that $[a,b]$ is an *xy, U chain* for f if $(x, a_1), (b_1, a_2), \dots (b_{n-1}, a_n), (b_n, y) \in U$. Clearly, then $[a,b]^{-1}$ is a yx, U^{-1} chain for f^{-1}.

Since the V_d^ϵ's for $d \in \Gamma(\mathcal{U})$ and $\epsilon > 0$ generate the uniformity, it is clear that the pair $(x,y) \in \mathcal{C}_{\mathcal{U}} f$ iff for every $U \in \mathcal{U}$ there exists an xy, U chain for f. This provides a uniformity description of $\mathcal{C}_{\mathcal{U}} f$.

Similarly, $(x,y) \in \mathcal{A}_{\mathcal{U}} f$ if for every $d \in \Gamma$ and every $\epsilon > 0$ there exists $[a,b] \in f^{\times n}$ with $n \geq 1$ such that the xy chain-length of $[a,b]$ with respect to d is less than ϵ.

Following [**12**] we obtain a uniformity description of $\mathcal{A}_{\mathcal{U}} f$. We will need the following lemma.

Lemma 6.1. *Let $\phi : \mathbb{R} \to [0, \infty)$ be given by $\phi(0) = 0$ and $\phi(t) = e^{-1/t^2}$ for $t \neq 0$. Then ϕ is a C^∞ function such that*

(i) *For all $t > 0$, $\phi'(t) > 0$ and for all $\sqrt{2/3} > t > 0$, $\phi''(t) > 0$.*
(ii) *For $\epsilon = e^{-3/2}/2$, $\bar{d}(x,y) = \phi^{-1}(\min(d(x,y), \epsilon))$ defines a pseudo-metric on X with $\mathcal{U}(\bar{d}) = \mathcal{U}(d) \subset \mathcal{U}$ and so $\bar{d} \in \Gamma$.*
(iii) *If $\{\alpha_k\}$ is a finite or infinite, non-increasing sequence of non-negative numbers with $\sum_k \alpha_k < \phi^{-1}(\epsilon) < 1$ then $\bar{d}(x,y) \leq \alpha_k$ implies $d(x,y) < 2^{-k}$, for all $k \in \mathbb{N}$.*

Proof: (i) is an easy direct computation.

(ii) Observe that if $\psi : [0,a] \to \mathbb{R}$ is C^2 with $\psi(0) = 0, \psi'(t) > 0$ and $\psi''(t) < 0$ for $0 < t < a$ then for all $t, s \leq a/2$, $\psi(t) + \psi(s) - \psi(t+s) \geq 0$, because with t fixed it is true for $s = 0$ and the derivative with respect to s is positive for $a - t > s > 0$. It follows that if d is a pseudo-metric with $d \leq a/2$ then $\psi(d)$ is a pseudo-metric. Clearly, $\mathcal{U}(\psi(d)) = \mathcal{U}(d)$. For (ii) we apply this with $\psi = \phi^{-1}$.

(iii) Observe that for all $k \in \mathbb{N}$, $\phi(1/k) = e^{-k^2} < 2^{-k}$. Each $\alpha_k < \phi^{-1}(\epsilon)$ and so $\bar{d}(x,y) \leq \alpha_k$ iff $d(x,y) \leq \phi(\alpha_k)$. If $\phi(\alpha_k) \geq 2^{-k}$ then for $1 \leq j \leq k$

$\phi(\alpha_j) \geq \phi(\alpha_k) \geq 2^{-k} \geq \phi(1/k)$ and so $\alpha_j \geq 1/k$ for $j = 1, \dots, k$. Hence, $\sum_j \alpha_j \geq k(1/k) = 1 > \phi^{-1}(\epsilon)$, contradicting the assumption on the sum. □

If $\xi = \{U_k : k \in \mathbb{N}\}$ is a sequence of elements of $\mathcal{U}$ and $(x, y) \in X \times X$, we call $[a, b] \in f^{\times n}$ an ξ *sequence chain from* x *to* y if there is an injective map $\sigma : \{0, \dots, n\} \to \mathbb{N}$ such that $(b_i, a_{i+1}) \in U_{\sigma(i)}$ for $i = 0, \dots, n$ with $b_0 = x, a_{n+1} = y$.

Theorem 6.2. *For a relation* f *on a uniform space* $(X, \mathcal{U})$, $(x, y) \in \mathcal{A}_{\mathcal{U}} f$ *iff for every sequence* ξ *in* $\mathcal{U}$ *there is a* ξ *sequence chain from* x *to* y.

Proof: Assume (x, y) satisfies the sequence chain condition. If $d \in \Gamma(\mathcal{U})$ and $\epsilon > 0$ the chain-length with respect to d of any sequence chain with $\xi = \{V^d_{\epsilon/2^n}\}$ from x to y is less than ϵ. Hence, $(x, y) \in \mathcal{A}_d f$. As d was arbitrary, $(x, y) \in \bigcap_{d \in \Gamma} \mathcal{A}_d f = \mathcal{A}_{\mathcal{U}} f$.

Now let $(x, y) \in \mathcal{A}_{\mathcal{U}} f$ and $\xi = \{U_k : k \in \mathbb{N}\}$ be a sequence in $\mathcal{U}$. We must show that there is a ξ sequence chain from x to y.

Let $V_0 = X \times X$. For $k \in \mathbb{N}$, inductively choose $V_k = V_k^{-1} \in \mathcal{U}$ such that $V_k \circ V_k \circ V_k \subset V_{k-1} \cap U_k$. By the Metrization Lemma [**10**, Lemma 6.12], there exists a pseudo-metric $d \leq 1$ such that $V_k \subset V^d_{1/2^{k-1}} \subset V_{k-1}$ for $k \in \mathbb{N}$. It follows that $d \in \Gamma$ and since $V^d_{1/2^k} \subset U_k$ it follows that if $\xi' = \{V^d_{1/2^k}\}$ then a ξ' sequence chain is a ξ sequence chain. It suffices to show that there is a ξ' sequence chain from x to y.

Since $(x, y) \in \mathcal{A}_{\mathcal{U}} f$, there exists $[a, b] \in f^{\times n}$ for some $n \geq 1$ such that with respect to the metric $\bar{d}$, the xy chain-length of $[a, b]$ is less than $\phi^{-1}(\epsilon)$, where ϕ is the function from Lemma 6.1. Let $b_0 = x$ and $a_{n+1} = y$. Let $k \mapsto i(k)$ be a bijection on $\{1, \dots, n+1\}$ so that the sequence $\alpha_k = d(b_{i(k)-1}, a_{i(k)})$ is non-increasing. From (iii) it follows that $(b_{i(k)-1}, a_{i(k)}) \in V^d_{2^{-k}}$ for $k = 1, \dots, n+1$ and so $[a, b]$ is a ξ' sequence chain from x to y as required. □

It is clear that $(\mathcal{G} f)^{-1}$ is the smallest closed, transitive relation which contains f^{-1}. So from (4.9) we obtain:

$$\mathcal{G}(f^{-1}) = (\mathcal{G} f)^{-1}, \quad \mathcal{A}_{\mathcal{U}}(f^{-1}) = (\mathcal{A}_{\mathcal{U}} f)^{-1}, \quad \mathcal{C}_{\mathcal{U}}(f^{-1}) = (\mathcal{C}_{\mathcal{U}} f)^{-1}, \tag{6.3}$$

and so again we may omit the parentheses.

Proposition 6.3. *For a relation* f *on a uniform space* $(X, \mathcal{U})$, *the image* $f(X)$ *is dense in* $\mathcal{C}_{\mathcal{U}} f(X)$ *and the domain* $f^{-1}(X)$ *is dense in* $\mathcal{C}_{\mathcal{U}} f^{-1}(X)$.

Proof: Let $A = \overline{f(X)}$ and let $y \in \mathcal{C}_{\mathcal{U}} f(x)$. If $U \in \mathcal{U}$ and $[a, b] \in f^{\times n}$ is an xy, U chain, then $b_i \in A$ for all i and so $y \in U(A)$. Because A is closed, it equals the intersection $\bigcap_{U \in \mathcal{U}} U(A)$. Thus, $\mathcal{C}_{\mathcal{U}} f(X) \subset A$. Replacing f by f^{-1} we obtain the domain result. □

From (4.8) we obtain monotonicity: If $f \subset g$ are relations on $(X, \mathcal{U})$ then

$$\mathcal{G} f \subset \mathcal{G} g, \quad \mathcal{A}_{\mathcal{U}} f \subset \mathcal{A}_{\mathcal{U}} g, \quad \mathcal{C}_{\mathcal{U}} f \subset \mathcal{C}_{\mathcal{U}} g, \tag{6.4}$$

Again the operators are idempotent.

Proposition 6.4.

$$\begin{aligned} f \subset g \subset \mathcal{C}_{\mathcal{U}} f \quad &\Longrightarrow \quad \mathcal{C}_{\mathcal{U}} f = \mathcal{C}_{\mathcal{U}} g, \\ f \subset g \subset \mathcal{A}_{\mathcal{U}} f \quad &\Longrightarrow \quad \mathcal{A}_{\mathcal{U}} f = \mathcal{A}_{\mathcal{U}} g, \\ f \subset g \subset \mathcal{G} f \quad &\Longrightarrow \quad \mathcal{G} f = \mathcal{G} g. \end{aligned} \tag{6.5}$$

Proof: For any $d \in \Gamma$, $f \subset g \subset \mathcal{C}_{\mathcal{U}} f \subset \mathcal{C}_d f$ and so by (4.11) and montonicity, $\mathcal{C}_d f = \mathcal{C}_d g$. Intersect over $d \in \Gamma$. The proof for $\mathcal{A}_{\mathcal{U}}$ is similar.

Finally, if F is a closed, transitive relation then $F = \mathcal{G}F$. $\square$

Proceeding just as with (4.12) we see that for relations f and g on $(X, \mathcal{U})$

$$
\begin{aligned}
\mathcal{C}_{\mathcal{U}} f \cap \mathcal{C}_{\mathcal{U}} g &= \mathcal{C}_{\mathcal{U}}(\mathcal{C}_{\mathcal{U}} f \cap \mathcal{C}_{\mathcal{U}} g), \\
\mathcal{A}_{\mathcal{U}} f \cap \mathcal{A}_{\mathcal{U}} g &= \mathcal{A}_{\mathcal{U}}(\mathcal{A}_{\mathcal{U}} f \cap \mathcal{A}_{\mathcal{U}} g), \\
\mathcal{G} f \cap \mathcal{G} g &= \mathcal{G}(\mathcal{G} f \cap \mathcal{G} g).
\end{aligned} \tag{6.6}
$$

If $\mathcal{U}_1$ and $\mathcal{U}_2$ are uniformities on X then

$$
\mathcal{U}_1 \subset \mathcal{U}_2 \quad \Longrightarrow \quad \mathcal{C}_{\mathcal{U}_2} f \subset \mathcal{C}_{\mathcal{U}_1} f \quad \text{and} \quad \mathcal{A}_{\mathcal{U}_2} f \subset \mathcal{A}_{\mathcal{U}_1} f. \tag{6.7}
$$

More generally, we have

Proposition 6.5. *If $h : (X_1, \mathcal{U}_1) \to (X_2, \mathcal{U}_2)$ is a continuous map which maps the relation f_1 on X_1 to f_2 on X_2, then h maps $\mathcal{G} f_1$ to $\mathcal{G} f_2$. If, in addition, h is uniformly continuous, then h maps $\mathcal{C}_{\mathcal{U}_1} f_1$ to $\mathcal{C}_{\mathcal{U}_2} f_2$, and maps $\mathcal{A}_{\mathcal{U}_1} f_1$ to $\mathcal{A}_{\mathcal{U}_2} f_2$.*

Proof: If h is continuous then, $(h \times h)^{-1}(\mathcal{G} f_2)$ is a closed, transitive relation which contains f_1 and so contains $\mathcal{G} f_1$.

Now assume that h is uniformly continuous. Let $d_2 \in \Gamma(\mathcal{U}_2)$. By uniform continuity, $d_1 = h^* d_2 \in \Gamma(\mathcal{U}_1)$, where

$$
h^* d_2(x, y) = d_2(h(x), h(y)). \tag{6.8}
$$

Thus, $h : (X_1, d_1) \to (X_2, d_2)$ is Lipschitz. In fact, it is an isometry. By Proposition 4.7, h maps $\mathcal{A}_{\mathcal{U}_1} f_1 \subset \mathcal{A}_{d_1} f$ into $\mathcal{A}_{d_2} f$ and similarly for $\mathcal{C}$. Intersect over all $d_2 \in \Gamma(\mathcal{U}_2)$. $\square$

For a relation f on X let $f^{[1,k]} = \bigcup_{j=1}^{k} f^j$ for any positive integer k. Let $f^{[0,k]} = 1_X \cup f^{[1,k]}$. If d is a pseudo-metric on X and f is a map on X we let $d^k = \max_{j=0}^{k} (f^j)^* d$. Let $d^0 = d$.

Corollary 6.6. *Let $k \geq 2$ be an integer and f be a continuous map on a uniform space $(X, \mathcal{U})$.*

$$
\mathcal{G} f = f^{[1,k-1]} \cup \mathcal{G}(f^k) \circ f^{[0,k-1]}, \tag{6.9}
$$

and $|\mathcal{G}(f^k)| = |\mathcal{G} f|$.

If f is a uniformly continuous map, then

$$
\mathcal{A}_{\mathcal{U}} f = f^{[1,k-1]} \cup \mathcal{A}_{\mathcal{U}}(f^k) \circ f^{[0,k-1]}, \quad \mathcal{C}_{\mathcal{U}} f = f^{[1,k-1]} \cup \mathcal{C}_{\mathcal{U}}(f^k) \circ f^{[0,k-1]}, \tag{6.10}
$$

and $|\mathcal{A}_{\mathcal{U}}(f^k)| = |\mathcal{A}_{\mathcal{U}} f|$, $\quad |\mathcal{C}_{\mathcal{U}}(f^k)| = |\mathcal{C}_{\mathcal{U}} f|$.

Proof: If F is a closed relation on X and f is a continuous map on X then $F \circ f$ is a closed relation. For suppose $\{(x_i, y_i)\}$ is a net in $F \circ f$ converging to (x, y). Then $\{f(x_i)\}$ converges to $f(x)$ by continuity and $\{(f(x_i), y_i)\}$ is a net in F converging to $(f(x), y)$. Since F is closed, $(f(x), y) \in F$ and $(x, y) \in F \circ f$.

Hence, $f^{[1,k-1]} \cup \mathcal{G}(f^k) \circ f^{[0,k-1]}$ is a closed relation which contains f. Since $f \subset \mathcal{G} f$, transitivity of $\mathcal{G} f$ implies that $f^k \subset \mathcal{G} f$. Hence, $\mathcal{G}(f^k) \subset \mathcal{G} f$. Transitivity again implies $f^{[1,k-1]} \cup \mathcal{G}(f^k) \circ f^{[0,k-1]} \subset \mathcal{G} f$.

Because f maps f^k to f^k it follows from Proposition 6.5 that it maps $\mathcal{G}(f^k)$ to itself. Hence, $f^{[0,k-1]} \circ \mathcal{G}(f^k) \subset \mathcal{G}(f^k) \circ f^{[0,k-1]}$. Furthermore, $f^{[0,k-1]} \circ f^{[1,k-1]} \subset f^{[1,k-1]} \cup f^k \circ f^{[0,k-1]}$. It follows that $f^{[1,k-1]} \cup \mathcal{G}(f^k) \circ f^{[1,k-1]}$ is transitive and so contains $\mathcal{G} f$ since it is closed and contains f.

It clearly follows that $|\mathcal{G}(f^k)| \subset |\mathcal{G}f|$. Assume that $x \in |\mathcal{G}f|$. From (6.9) it follows that either $x \in f^j(x)$ for some $j \in [1, k-1]$ or $x \in \mathcal{G}(f^k) \circ f^j(x)$ for some $j \in [0, k-1]$. If $x = f^j(x)$ then $x = (f^j)^k(x) = (f^k)^j(x)$ and so $x \in \mathcal{G}(f^k)(x)$. Similarly, since f^j maps $\mathcal{G}(f^k)$ to itself,

$$(\mathcal{G}(f^k) \circ f^j)^k \subset (\mathcal{G}(f^k))^k \circ (f^j)^k \subset \mathcal{G}(f^k)$$

and so $x \in |\mathcal{G}(f^k)|$ if $x \in \mathcal{G}(f^k) \circ f^j(x)$.

Transitivity again implies $f^k \subset \mathcal{A}_{\mathcal{U}} f \subset \mathcal{C}_{\mathcal{U}} f$, and so monotonicity and transitivity imply

$$\begin{aligned} \mathcal{A}_{\mathcal{U}} f &\supset f^{[1,k-1]} \cup \mathcal{A}_{\mathcal{U}}(f^k) \circ f^{[0,k-1]}, \\ \mathcal{C}_{\mathcal{U}} f &\supset f^{[1,k-1]} \cup \mathcal{C}_{\mathcal{U}}(f^k) \circ f^{[0,k-1]}. \end{aligned} \tag{6.11}$$

Now assume that f is a uniformly continuous map. Notice that if $[a, b] \in f^{\times n}$ then $b_i = f(a_i)$ for $i = 1, \ldots, n$. Observe that if $x \in X$ and $j \leq k$

$$\begin{aligned} d(f^j(a_1), a_{j+1}) &\leq \Sigma_{i=1}^{j} d(f^{j-i+1}(a_i), f^{j-i}(a_{i+1})) \\ &\leq \Sigma_{i=1}^{j-1} d^k(f(a_i), a_{i+1}), \\ \text{and} \quad d(f^j(x), f^j(a_1)) &\leq d^k(x, a_1). \end{aligned} \tag{6.12}$$

Let $(x, y) \in \mathcal{A}_{\mathcal{U}}$. For $\alpha = (d, \epsilon) \in \Gamma \times (0, \infty)$ there exists $[a, b]_\alpha \in f^{n_\alpha}$ with xy chain-length with respect to d less than ϵ.

If $n_\alpha < k$ frequently then for some $j \in [1, k-1]$ frequently $n_\alpha = j$ and it follows from continuity of f that $y = f^j(x)$.

Instead assume that eventually $n_\alpha \geq k$. If $\epsilon > 0$ and $d_1 \in \Gamma(\mathcal{U})$, there exists $d \geq d_1$ and $[a, b] \in f^{\times n}$ with $n \geq k$ so that the xy chain-length of $[a, b]$ with respect to d^k is less than ϵ. Let $n = j + qk$ with $j \in [0, k-1]$ and $q \geq 1$. The sequence

$$\begin{aligned} [a, b]^k = (a_{j+1}, f^k(a_{j+1})), (a_{j+k+1}, f^k(a_{j+k+1})) \ldots \\ (a_{j+(q-1)k+1}, f^k(a_{j+(q-1)k+1})) \in (f^k)^{\times q}, \end{aligned} \tag{6.13}$$

and with $y = a_{n_\alpha+1}$, (6.12) implies that the $f^j(x)y$ chain-length with respect to d and so with respect to d_1 is less than ϵ. Since d_1 was arbitrary it follows that $y \in \mathcal{G}(f^k) \circ f^{[0,k-1]}(x)$.

For $\mathcal{C}_{\mathcal{U}} f$ we proceed as before, but use chain-bound less than ϵ/k.

For $|\mathcal{A}_{\mathcal{U}}(f^k)|$ and $|\mathcal{C}_{\mathcal{U}}(f^k)|$ we use the same argument as for $|\mathcal{G}(f^k)|$ above. □

If a real-valued function on X is uniformly continuous with respect to some $d \in \Gamma(\mathcal{U})$ then it is uniformly continuous from $(X, \mathcal{U})$. In particular, for every $d \in \Gamma(\mathcal{U})$ and $f \subset X \times X$, the functions ℓ_d^f and m_d^f are uniformly continuous from $(X \times X, \mathcal{U} \times \mathcal{U})$. It follows that the sets $\mathcal{C}_{\mathcal{U}} f, \mathcal{A}_{\mathcal{U}} f \subset X \times X$ and $|\mathcal{C}_{\mathcal{U}} f|, |\mathcal{A}_{\mathcal{U}} f| \subset X$ are closed.

As before, a Lyapunov function for a relation f on a uniform space $(X, \mathcal{U})$ is a continuous map $L : X \to \mathbb{R}$ such that $(x, y) \in f$ implies $L(x) \leq L(y)$. Hence, the relation $\leq_L \subset X \times X$ is closed.

As in Definition 5.1:

Definition 6.7. Let F be a closed, transitive relation on a Hausdorff uniform space $(X, \mathcal{U})$ and let $\mathcal{L}$ be a collection of Lyapunov functions for F. We define three conditions on $\mathcal{L}$.

- ALG If $L_1, L_2 \in \mathcal{L}$ and $c \geq 0$ then $L_1 + L_2, \max(L_1, L_2), \min(L_1, L_2), cL_1, c, -c \in \mathcal{L}$.
- CON For every sequence $\{L_k\}$ of elements of $\mathcal{L}$ there exists a summable sequence of positive real numbers $\{a_k\}$ such that $\Sigma_k a_k L_k$ converges uniformly to an element of $\mathcal{L}$.
- POIN If $(x, y) \notin 1_X \cup F$ then there exists $L \in \mathcal{L}$ such that $L(y) < L(x)$, i.e. $1_X \cup F = \bigcap_{L\in\mathcal{L}} \leq_L$.

Theorem 6.8. *Let f be a relation on a Hausdorff uniform space $(X, \mathcal{U})$ with gage Γ.*

(a) Let $\mathcal{L}_\ell$ be the set of bounded, uniformly continuous functions which are $K\ell_d^f$ dominated for some $d \in \Gamma$ and some positive K. Each $L \in \mathcal{L}_\ell$ is a $\mathcal{A}_\mathcal{U} f$ Lyapunov function and so satisfies

$$\mathcal{A}_\mathcal{U} f \ \subset\ \leq_L \qquad \text{and} \qquad |\mathcal{A}_\mathcal{U} f| \ \subset\ |L|_{\mathcal{A}_\mathcal{U} f}. \tag{6.14}$$

The collection $\mathcal{L}_\ell$ satisfies the conditions ALG, CON, and POIN with respect to $F = \mathcal{A}_\mathcal{U} f$.

(b) Let $\mathcal{L}_m$ be the set of bounded, uniformly continuous functions which are Km_d^f dominated for some $d \in \Gamma$ and some positive K. Each $L \in \mathcal{L}_m$ is a $\mathcal{C}_\mathcal{U} f$ Lyapunov function and so satisfies

$$\mathcal{C}_\mathcal{U} f \ \subset\ \leq_L \qquad \text{and} \qquad |\mathcal{C}_\mathcal{U} f| \ \subset\ |L|_{\mathcal{C}_\mathcal{U} f}. \tag{6.15}$$

The collection $\mathcal{L}_m$ satisfies the conditions ALG, CON, and POIN with respect to $F = \mathcal{C}_\mathcal{U} f$.

Proof: If $\{d_k\}$ is a sequence in Γ and $K_k \geq 1$ so that $d_k \leq K_k$ and $\{K_k a_k\}$ is a summable sequence of positive reals, then by Lemma 2.2 $d = \Sigma_k\ (a_k)d_k \in \Gamma$. Furthermore,

$$a_k \ell_f^{d_k} \ =\ \ell_f^{a_k d_k} \ \leq\ \ell_f^d \quad \text{and} \quad a_k m_f^{d_k} \ =\ m_f^{a_k d_k} \ \leq\ m_f^d. \tag{6.16}$$

So if L is $K\ell_f^{d_k}$ dominated then it is $(K/a_k)\ell_f^d$ dominated. Thus, if $\{L_k\}$ is a sequence in $\mathcal{L}_\ell$ we can choose $d \in \Gamma$ such that each L_k is $K_k \ell_f^d$ dominated for some K_k. Then ALG and CON follow for L_ℓ from Theorem 5.6 for (X, d).

Now assume that $(x, y) \notin 1_X \cup \mathcal{A}_\mathcal{U} f$. Because X is Hausdorff there exists $d_1 \in \Gamma$ such that $d_1(x, y) > 0$. There exists $d_2 \in \Gamma$ such that $(x, y) \notin \mathcal{A}_{d_2} f$. Let $d = d_1 + d_2$. Since $\ell_f^{d_2} \leq \ell_f^d$ it follows that $(x, y) \notin Z_d \cup \mathcal{A}_d f$. From Theorem 5.6 again there exists a function L which is d uniformly continuous, $K\ell_f^d$ dominated for some K and satisfied $L(x) > L(y)$. Hence, $L \in \mathcal{L}_\ell$ with $L(x) > L(y)$, proving POIN.

The results in (b) for $\mathcal{L}_m$ are proved exactly the same way with Theorem 5.6 replaced by Theorem 5.7. □

Theorem 6.9. *Let f be a relation on a uniform space $(X, \mathcal{U})$.*

If L is a Lyapunov function for f, then L is a Lyapunov function for $\mathcal{G}f$.

If L is a uniformly continuous Lyapunov function for f, then L is a Lyapunov function for $\mathcal{A}_\mathcal{U} f$.

Proof: If L is a Lyapunov function for f then, by continuity of L, $\leq_L$ is a closed, transitive relation which contains f and so contains $\mathcal{G}f$.

If L is bounded and uniformly continuous, then $d_L(x, y) = |L(x) - L(y)|$ is a pseudo-metric in $\Gamma(\mathcal{U})$. Let $(x, y) \in \mathcal{A}_\mathcal{U} f$ and $\epsilon \in (0, 1)$. There exists $[a, b] \in f^{\times n}$

such that the xy chain-length of $[a,b]$ with respect to d_L is less than ϵ. Since L is a Lyapunov function for f, we have that $L(a_i) \leq L(b_i)$ for $i = 1, \dots, n$.

$$\begin{aligned} L(y) - L(x) &= L(y) - L(b_n) + \Sigma_{i=1}^{n} L(b_i) - L(a_i) + \\ &\qquad \Sigma_{i=1}^{n-1} L(a_i) - L(b_{i+1}) + L(a_1) - L(x). \end{aligned} \tag{6.17}$$

The first sum is non-negative and the rest has absolute value at most the chain-length. Hence, $L(y) - L(x) \geq -\epsilon$. Since ϵ was arbitrary, $L(y) - L(x) \geq 0$.

If L is unbounded then for each positive K, $L_K = \max(\min(L,K), -K)$ is a bounded, uniformly continuous Lyapunov function and so is an $\mathcal{A}_{\mathfrak{U}} f$ Lyapunov function. If $(x,y) \in \mathcal{A}_{\mathfrak{U}} f$ then by choosing K large enough we have $L_K(x) = L(x)$ and $L_K(y) = L(y)$. So $L(y) - L(x) = L_K(y) - L_K(x) \geq 0$. □

Corollary 6.10. *Let f be a relation on a Tychonoff space X and let $\mathfrak{U}_M$ be the maximum uniformity compatible with the topology. Let $\mathcal{L}$ be the set of all bounded Lyapunov functions for f. Each $L \in \mathcal{L}$ is a Lyapunov function for $\mathcal{A}_{\mathfrak{U}_M} f$ and*

$$1_X \cup \mathcal{A}_{\mathfrak{U}_M} f \;=\; \bigcap_{L \in \mathcal{L}} \leq_L \tag{6.18}$$

Proof: With respect to the maximum uniformity every continuous real-valued function is uniformly continuous. So every $L \in \mathcal{L}$ is a Lyapunov function for $\mathcal{A}_{\mathfrak{U}_M} f$ by Theorem 6.9. Hence $1_X \cup \mathcal{A}_{\mathfrak{U}_M} f \;\subset\; \bigcap_{L \in \mathcal{L}} \leq_L$. The reverse inclusion follows from POIN in Theorem 6.8 (a). □

Theorem 6.11. *Let F be a closed, transitive relation on a Hausdorff uniform space $(X, \mathfrak{U})$ whose topology is second countable. Let $\mathcal{L}$ be a collection of Lyapunov functions for F which satisfies ALG, CON and POIN. There exists a sequence $\{L_k\}$ in $\mathcal{L}$ such that*

$$\bigcap_k \leq_{L_k} \;=\; 1_X \cup F. \tag{6.19}$$

If $\{a_k\}$ is a positive, summable sequence such that $L = \Sigma_k\ a_k L_k \in \mathcal{L}$ then L is a Lyapunov function for F such that $1_X \cup F \;=\; \leq_L$ and

$$x \in F(y) \quad\Longrightarrow\quad L(y) < L(x) \quad \text{unless } y \in F(x). \tag{6.20}$$

In particular,

$$|L|_F \;=\; |F| \tag{6.21}$$

Proof: Proceed just as in the proof of Theorem 5.2 using the fact that $(X \times X) \setminus (1_X \cup F)$ is Lindelöf. □

For a metrizable space X we let $\Gamma_m(X)$ be the set of metrics compatible with the topology on X.

Theorem 6.12. *Let f be a relation on a Hausdorff uniform space $(X, \mathfrak{U})$ whose topology is second countable. There exist bounded, uniformly continuous Lyapunov functions L_ℓ, L_m for f such that*

$$\begin{aligned} 1_X \cup \mathcal{A}_{\mathfrak{U}} f \;&=\; \leq_{L_\ell}, \quad 1_X \cup \mathcal{C}_{\mathfrak{U}} f \;=\; \leq_{L_m} \qquad \text{and,} \\ x \in \mathcal{A}_{\mathfrak{U}} f(y) \quad&\Longrightarrow\quad L_\ell(y) < L_\ell(x) \quad \text{unless } y \in \mathcal{A}_{\mathfrak{U}} f(x), \\ x \in \mathcal{C}_{\mathfrak{U}} f(y) \quad&\Longrightarrow\quad L_m(y) < L_m(x) \quad \text{unless } y \in \mathcal{C}_{\mathfrak{U}} f(x) \end{aligned} \tag{6.22}$$

In particular,

$$|L_\ell|_{\mathcal{A}_\mathcal{U} f} = |\mathcal{A}_\mathcal{U} f|, \quad \text{and} \quad |L_m|_{\mathcal{C}_\mathcal{U} f} = |\mathcal{C}_\mathcal{U} f| \tag{6.23}$$

Furthermore, there exists a metric $d \in \Gamma_m(X) \cap \Gamma(\mathcal{U})$ *such that* L_ℓ *and* L_m *are Lipschitz functions on* (X, d) *and*

$$\mathcal{A}_\mathcal{U} f = \mathcal{A}_d f \quad \text{and} \quad \mathcal{C}_\mathcal{U} f = \mathcal{C}_d f. \tag{6.24}$$

Proof: The pseudo-metrics chosen below are all assumed bounded by 1. We can always replace d by $\min(d, 1)$.

We apply Theorem 6.11 to $\mathcal{L}_\ell$ and $\mathcal{A}_\mathcal{U} f$ and to $\mathcal{L}_m$ and $\mathcal{C}_\mathcal{U} f$ and obtain $L_\ell \in \mathcal{L}_\ell$ and $L_m \in \mathcal{L}_m$ which satisfy (6.22) and (6.23). We may assume that each maps to $[0, 1]$. In particular, there exist $d_1, d_2 \in \Gamma(\mathcal{U})$ and positive K_1, K_2 so that L_ℓ is $K_1 \ell_f^{d_1}$ dominated and L_m is $K_2 m_f^{d_2}$ dominated.

Let $\mathcal{B}$ be a countable base and D be a countable dense subset of X. For each pair (x, U) with $U \in \mathcal{B}$ and $x \in U \cap D$ there exists $d = d_{(x,U)} \in \Gamma(\mathcal{U})$ and a rational $\epsilon > 0$ such that the ball $V_\epsilon^d(x) \subset U$

For each $x \notin |\mathcal{A}_\mathcal{U} f|$ there exists $d_{x,1} \in \Gamma(\mathcal{U})$ such that $\ell_f^{d_{x,1}}(x, x) > 0$ and for each $x \notin |\mathcal{C}_\mathcal{U} f|$ there exists $d_{x,2} \in \Gamma(\mathcal{U})$ such that $m_f^{d_{x,2}}(x, x) > 0$. These are open conditions and so we can choose a sequence $\{d_3, d_4, \dots\}$ in $\mathcal{G}$ and a positive sequence $\{a_1, a_2, \dots\}$ with sum $= 1$ so that d defined by $d(x, y) = \frac{1}{3}[|L_\ell(x) - L_\ell(y)| + |L_m(x) - L_m(y)| + \Sigma_{i=1}^\infty a_i d_i]$ satisfies

(i) $d \in \Gamma(\mathcal{U})$.
(ii) The $\mathcal{U}(d)$ topology is that of X, i.e. $d \in \Gamma_m(X)$.
(iii) $x \notin |\mathcal{A}_\mathcal{U} f|$ implies $\ell_f^d(x, x) > 0$, and $x \notin |\mathcal{C}_\mathcal{U} f|$ implies $m_f^d(x, x) > 0$.
(iv) There exist positive K_ℓ and K_m so that L_ℓ is $K_\ell \ell_f^d$ dominated and L_m is $K_m m_f^d$ dominated.

Condition (i) follows from Lemma 2.2. Condition (ii) implies that d is a metric since X is Hausdorff. From condition (iv) and (6.22) we obtain

$$\begin{aligned} 1_X \cup \mathcal{A}_d f &\subset \leq_{L_\ell} = 1_X \cup \mathcal{A}_\mathcal{U} f, \\ 1_X \cup \mathcal{C}_d f &\subset \leq_{L_m} = 1_X \cup \mathcal{C}_\mathcal{U} f. \end{aligned} \tag{6.25}$$

On the other hand, $d \in \Gamma(\mathcal{U})$ implies $\mathcal{A}_\mathcal{U} f \subset \mathcal{A}_d f$ and $\mathcal{C}_\mathcal{U} f \subset \mathcal{C}_d f$. Hence, if $(x, y) \in \mathcal{A}_d f \setminus \mathcal{A}_\mathcal{U} f$ then $(x, y) \in 1_X$ and so $\ell_f^d(x, x) = 0$. By condition (iii) this implies $x \in |\mathcal{A}_\mathcal{U} f|$ and so $(x, y) = (x, x) \in \mathcal{A}_\mathcal{U} f$. This contradiction proves the first equation in (6.24). The second follows similarly.

Clearly, L_ℓ and L_m are Lipschitz with Lipschitz constant at most 3. □

If $\mathcal{U}_M$ is the maximum uniformity compatible with the topology for a metrizable space X, then since such a space is paracompact, $\mathcal{U}_M$ consists of all neighborhoods of the diagonal. The gage $\Gamma(\mathcal{U}_M)$ consists of all pseudo-metrics which are continuous on X. In particular, $\Gamma_m(X) \subset \Gamma(\mathcal{U}_M)$.

Corollary 6.13. *Let* f *be a relation on a second countable Tychonoff space* X *and let* $\mathcal{U}_M$ *be the maximum uniformity compatible with the topology. There exists a metric* $d_0 \in \Gamma_m(X)$ *such that*

$$\mathcal{A}_{\mathcal{U}_M} f = \mathcal{A}_{d_0} f \quad \text{and} \quad \mathcal{C}_{\mathcal{U}_M} f = \mathcal{C}_{d_0} f. \tag{6.26}$$

Furthermore,

$$\mathcal{A}_{\mathcal{U}_M} f = \bigcap_{d \in \Gamma_m(X)} \mathcal{A}_d f \quad \text{and} \quad \mathcal{C}_{\mathcal{U}_M} f = \bigcap_{d \in \Gamma_m(X)} \mathcal{C}_d f. \tag{6.27}$$

Proof: A second countable Hausdorff space is metrizable, i.e. there exists a metric $\bar{d}$ with the $\mathcal{U}(\bar{d})$ topology that of X. Thus, $\bar{d} \in \Gamma_m(X) \subset \Gamma(\mathcal{U}_M)$. If $d_0 \in \Gamma(\mathcal{U}_M)$, then $d = \bar{d} + d_0$ is a metric in $\Gamma(\mathcal{U}_M)$ and so is continuous. Since $d \geq \bar{d}$ it follows that the $\mathcal{U}(d)$ topology is that of X as well, i.e. $d \in \Gamma_m(X)$. Furthermore,

$$\begin{aligned} \mathcal{A}_{\mathcal{U}_M} f \; &\subset \; \mathcal{A}_d f \; \subset \; \mathcal{A}_{d_0} f \\ \mathcal{C}_{\mathcal{U}_M} f \; &\subset \; \mathcal{C}_d f \; \subset \; \mathcal{C}_{d_0} f. \end{aligned} \tag{6.28}$$

Hence, the intersection over $\Gamma_m(X)$ yields the same result as intersecting over the entire gage, $\Gamma(\mathcal{U}_M)$. Furthermore, if d_0 is a metric in $\Gamma(\mathcal{U})$ satisfying (6.24) then (6.24) together with (6.28) implies (6.26). □

For d a metric on X, $\mathcal{U}(d)$ is the uniformity generated by V^d_ϵ for all $\epsilon > 0$. We say that d generates the uniformity $\mathcal{U}(d)$ and that $\mathcal{U}$ is metrizable if $\mathcal{U} = \mathcal{U}(d)$ for some metric d. The Metrization Theorem, Lemma 6.12 of [**10**], implies that a Hausdorff uniformity is metrizable iff it is countably generated. Two metrics d_1 and d_2 generate the same uniformity exactly when they are uniformly equivalent. That is, the identity maps between (X, d_1) and (X, d_2) are uniformly continuous. For a metrizable uniformity $\mathcal{U}$ we let $\Gamma_m(\mathcal{U}) = \{d : d \text{ is a metric with } \mathcal{U}(d) = \mathcal{U}\}$.

If (X, d) is a metric space and the set of non-isolated points is not compact, then the maximum uniformity $\mathcal{U}_M$ is not metrizable even if X is second countable. Since a metric space is paracompact, $\mathcal{U}_M$ consists of all neighborhoods of the diagonal. By hypothesis there is a sequence $\{x_1, x_2, \dots\}$ of distinct non-isolated points with no convergent subsequence and so we can choose open sets G_i pairwise disjoint and with $x_i \in G_i$. We can choose $y_i \in G_i \setminus \{x_i\}$ such that $\epsilon_i = d(x_i, y_i) \to 0$ as $i \to \infty$ and let $\epsilon_0 = 1$. Let G_0 be the complement of a closed neighborhood of $\{x_i\}$ in $\bigcup_{i=1}^{\infty} G_i$. Thus, $\{G_i\}$ is a locally finite open cover. Choose $\{\phi_i\}$ a partition of unity, i.e. each ϕ_i is a continuous real-valued function with support in G_i and with $\Sigma_i \phi_i = 1$. Define $\psi(x) = \Sigma_i \epsilon_i \phi_i(x)/2$. In particular, $\psi(x_i) = \epsilon_i/2$ for $i = 1, 2, \dots$. Thus, ψ is a continuous, positive function with infimum 0. So $U = \{(x, y) : d(x, y) < \psi(x)\}$ is a neighborhood of the diagonal disjoint from $\{(x_i, y_i) : i = 1, 2, \dots\}$. But if $\epsilon_i < \epsilon$ then $(x_i, y_i) \in V^d_\epsilon$. It follows that for any metric d compatible with the topology of X there exists a neighborhood of the diagonal, and so an element of $\mathcal{U}_M$, which is not in $\mathcal{U}(d)$.

Theorem 6.14. *Let $(X, \mathcal{U})$ be a uniform space with $\mathcal{U}$ metrizable and let f be a relation on X.*

(a) For every $d \in \Gamma_m(\mathcal{U})$, $\mathcal{C}_{\mathcal{U}} f = \mathcal{C}_d f$.

(b) $\mathcal{A}_{\mathcal{U}} f = \bigcap_{d \in \Gamma_m(\mathcal{U})} \mathcal{A}_d f$.

Proof: If $\bar{d} \in \Gamma(\mathcal{U})$ and $d_1 \in \Gamma_m(\mathcal{U})$ then $d = \bar{d} + d_1 \in \Gamma_m(\mathcal{U})$ and $\mathcal{C}_d f \subset \mathcal{C}_{\bar{d}} f$. Thus, we need only intersect over $\Gamma_m(\mathcal{U})$ to get $\mathcal{C}_{\mathcal{U}} f$. Similarly, for $\mathcal{A}_{\mathcal{U}} f$.

On the other hand, if $d_1, d_2 \in \Gamma_m(\mathcal{U})$ then d_1 and d_2 are uniformly equivalent metrics and so Proposition 4.7 implies that $\mathcal{C}_{d_1} f = \mathcal{C}_{d_2} f$. Hence, the intersection $\mathcal{C}_{\mathcal{U}} f$ is this common set. □

There are special constructions for the Conley relations. For the compact case, Conley [**5**] gives a construction of the chain recurrent set involving attractors, and Bernardi and Florio [**3**] do so for the strong chain recurrent set.

Definition 6.15. Let f be a relation on a uniform space $(X, \mathcal{U})$.

(a) A set $A \subset X$ is called $\mathcal{U}$ *inward* if there exists $U \in \mathcal{U}$ such that $U(f(A)) \subset A$, or, equivalently, if there exist $d \in \Gamma(\mathcal{U})$ and $\epsilon > 0$ such that A is $(V^d_\epsilon \circ f)$ $^+$invariant.

(b) A $\mathcal{U}$ uniformly continuous function $L : X \to [0,1]$ is called a $\mathcal{U}$ *elementary Lyapunov function* for f if $(x,y) \in f$ and $L(x) > 0$ imply $L(y) = 1$.

If $\mathcal{U} = \mathcal{U}_M$ for the space X, then a $\mathcal{U}$ inward set A for f is just called an inward set for f. For a paracompact Hausdorff space any neighborhood of a closed set is a $\mathcal{U}_M$ uniform neighborhood and so a set A is inward for a relation f on such a space iff $\overline{f(A)} \subset A^\circ$. A continuous function $L : X \to [0,1]$ is $\mathcal{U}_M$ uniformly continuous and we will call a $\mathcal{U}_M$ elementary Lyapunov function just an elementary Lyapunov function.

Observe for $L : X \to [0,1]$ that if $L(x) = 0$ or $L(y) = 1$ then $L(y) \geq L(x)$. So an elementary Lyapunov function is a Lyapunov function. In addition, the points of $G_L = \{x : 1 > L(x) > 0\}$ are regular points for L and so $|L|_f \subset L^{-1}(0) \cup L^{-1}(1)$ with equality if f is a surjective relation.

If $u : X \to \mathbb{R}$ is a bounded real-valued function we define the pseudometric d_u on X by $d_u(x,y) = |u(x) - u(y)|$. If u is uniformly continuous on $(X, \mathcal{U})$ then $d_u \in \Gamma(\mathcal{U})$.

Theorem 6.16. *Let f be a relation on a uniform space $(X, \mathcal{U})$.*

(a) *If A is a $\mathcal{U}$ inward subset for f then there exist $d \in \Gamma(\mathcal{U})$ and $\epsilon > 0$ such that $\overline{V^d_\epsilon(f(A))} \subset A^\circ$. In particular, $A_1 = A^\circ$ and $A_2 = \overline{V^d_\epsilon(f(A))}$ are $\mathcal{U}$ inward with A_1 open, A_2 closed and $\overline{f(A)} \subset A_2 \subset A_1 \subset A$.*

(b) *Let A be an open $\mathcal{U}$ inward subset for f. If for some $d \in \Gamma(\mathcal{U})$ and $\epsilon > 0$ $V^d_\epsilon(f(A)) \subset A$, then $V^d_\epsilon(\mathcal{C}_\mathcal{U} f(A)) \subset A$. In particular, A is a $\mathcal{U}$ inward subset of X for $\mathcal{C}_\mathcal{U} f$ and is $(V^d_\epsilon \circ \mathcal{C}_\mathcal{U} f)$ $^+$invariant.*

(c) *If A is a $\mathcal{U}$ inward subset for f, then there exists B a closed $\mathcal{U}$ inward subset for f^{-1} such that $A^\circ \cup B^\circ = X$ and $B \cap f(A) = \emptyset = A \cap f^{-1}(B)$.*

(d) *If A is a $\mathcal{U}$ inward subset of X, then there exists a $\mathcal{U}$ uniformly continuous elementary Lyapunov function L for f such that $L^{-1}(0) \cup A = X$ and $f(A) \subset L^{-1}(1)$.*

(e) *If L is a $\mathcal{U}$ elementary Lyapunov function for f and $1 \geq \epsilon > 0$, then $A = \{x : L(x) > 1 - \epsilon\}$ is an open set such that*

$$\begin{aligned} &f(A) \subset \mathcal{C}_\mathcal{U} f(A) \subset \mathcal{C}_{d_L} f(A) \subset L^{-1}(1), \\ &V^{d_L}_\epsilon(f(A)) \subset V^{d_L}_\epsilon(\mathcal{C}_\mathcal{U} f(A)) \subset V^{d_L}_\epsilon(\mathcal{C}_{d_L} f(A)) \subset A. \end{aligned} \tag{6.29}$$

In particular, L is a $\mathcal{U}(d_L)$ elementary Lyapunov function for $\mathcal{C}_{d_L} f$ and hence is a $\mathcal{U}$ elementary Lyapunov function for $\mathcal{C}_\mathcal{U} f$ and for $\overline{f}$.

(f) *If L is a $\mathcal{U}$ elementary Lyapunov function for f, then $1 - L$ is a $\mathcal{U}$ elementary Lyapunov function for f^{-1}.*

Proof: (a) There exist $d \in \Gamma$ and $\epsilon > 0$ such that $V^d_{2\epsilon}(f(A))$ is contained in A and so is contained in A°. For a subset B of X, $x \in \overline{B}$ implies $d(x,B) = 0$ and so $\overline{V^d_\epsilon(f(A))} \subset V^d_{2\epsilon}(f(A))$ and $\overline{f(A)} \subset V^d_\epsilon(f(A))$.

(b) Assume that $x \in A$ and $z \in V^d_\epsilon(\mathcal{C}_\mathcal{U} f(x))$. So there exist $z_1 \in \mathcal{C}_\mathcal{U} f(x)$ and $\epsilon_1 > 0$ such that $d(z_1, z) < \epsilon$. There exist $d_1 \in \Gamma$ and $\epsilon_1 > 0$ such that $V^{d_1}\epsilon_1(x) \subset A$ and $d(z_1, z) + \epsilon_1 < \epsilon$. Let $\bar{d} = d + d_1$. There exists $[a, b] \in f^{\times n}$ such that the xz_1 chain-bound of $[a, b]$ with respect to $\bar{d}$ is less than ϵ_1. Because $d_1(x, a_1) < \bar{d}(x, z_1) < \epsilon_1$, $a_1 \in A$. Since $b_1 \in f(A)$ and $d(b_1, a_2) < \bar{d}(b_1, a_2) < \epsilon$, $a_2 \in A$. Inductively, we obtain $a_i \in A$ and $b_i \in f(A)$ for $i = 1, \dots, n$. Finally, $d(b_n, z) \leq \bar{d}(b_n, z_1) + d(z_1, z) < \epsilon$. So $z \in A$.

(c) Let $d \in \Gamma$ and $\epsilon > 0$ be such that $V^d_{2\epsilon}(f(A))$ is contained in A and so is contained in A°. Let $B = X \setminus V^d_\epsilon(f(A))$ so that $B^\circ = X \setminus \overline{V^d_\epsilon(f(A))}$. Thus, B is closed, $A^\circ \cup B^\circ = X$ and $B \cap f(A) = \emptyset$. Assume that $(x, y) \in f$ and $z \in V_\epsilon(x)$. If $y \in B$ then $x \notin A$ and so $x \notin V^d_{2\epsilon}(f(A))$ and $z \notin V^d_\epsilon(f(A))$. That is, $z \in B$. Thus, $V^d_\epsilon(f^{-1}(B)) \subset B$. Finally, if $y \in B$ then $y \notin f(A)$ and so $x \notin A$. That is, $f^{-1}(B) \cap A = \emptyset$.

(d) Assume that $V^d_\epsilon(f(A)) \subset A$. Let $L(x) = \max(\epsilon - d(x, f(A)), 0)/\epsilon$. If $(x, y) \in f$ and $L(x) > 0$ then $d(x, f(A)) < \epsilon$ and so $x \in A$. Then $y \in f(A)$ implies $L(y) = 1$.

(e) Clearly, $f(A) \subset L^{-1}(1)$. Let $\epsilon > \epsilon_1 > 0$. We show that $V^{d_L}_{\epsilon_1}(\mathcal{C}_{d_L} f(A)) \subset \{y : L(y) > 1 - \epsilon_1\}$. Assume $x \in A, y \in V^{d_L}_{\epsilon_1}(\mathcal{C}_{d_L} f(x)$. So there exists $z \in \mathcal{C}_{d_L} f(x)$ with $d_L(z, y) < \epsilon_1$. Choose $\epsilon_2 > 0$ so that $d_L(z, y) + \epsilon_2 < \epsilon_1$ and $L(x) > 1 - \epsilon + \epsilon_2$. Since L is uniformly continuous, $d_L \in \Gamma(\mathcal{U})$ and so there exists $[a, b] \in f^{\times n}$ such that the xz chain-bound of $[a, b]$ with respect to d_L is less than ϵ_2. Since $d_L(x, a_1) < \epsilon_2$, $a_1 \in A$. Hence, $b_1 \in L^{-1}(1)$. Inductively, $a_i \in A$ and $b_i \in L^{-1}$ for all $i = 1, \dots, n$. Finally, $d_L(b_n, y) \leq d_L(b_n, z) + d_L(z, y) < \epsilon_1$. Since $L(b_n) = 1$, $L(y) > 1 - \epsilon_1$. Letting $\epsilon_1 \to 0$ we obtain $\mathcal{C}_{d_L} f(A) \subset L^{-1}(1)$. Letting $\epsilon_1 \to \epsilon$ we obtain $V^{d_L}_\epsilon(\mathcal{C}_{d_L} f(A)) \subset \{y : L(y) > 1 - \epsilon\} = A$.

(f) The contrapositive of the definition of an elementary Lyapunov function says that if $(x, y) \in f$ with $L(y) < 1$ then $L(x) = 0$. It follows that $1 - L$ is an elementary Lyapunov function for f^{-1}. □

Proposition 6.17. *Let f be a relation on a uniform space $(X, \mathcal{U})$, $\epsilon > 0$ and $d \in \Gamma(\mathcal{U})$. Let $K \subset X$ be closed and compact.*

(a) For $x \in X$, the set $\{y : \ell^f_d(x, y) < \epsilon\}$ is an open subset of X containing $\mathcal{A}_d f(x) \supset \mathcal{A}_\mathcal{U} f(x)$. It is $\mathcal{A}_d f$ $^+$invariant and so is $\mathcal{A}_\mathcal{U} f$ $^+$invariant.

$$
\begin{aligned}
\mathcal{A}_\mathcal{U} f(K) &= \bigcap_{d \in \Gamma, \epsilon > 0} \bigcup_{x \in K} \{y : \ell^f_d(x, y) < \epsilon\}, \\
K \cup \mathcal{A}_\mathcal{U} f(K) &= \bigcap_{d \in \Gamma, \epsilon > 0} \bigcup_{x \in K} \{y : \min(\ell^f_d(x, y), d(x, y)) < \epsilon\}
\end{aligned} \tag{6.30}
$$

(b) For $x \in X$, the set $\{y : m^f_d(x, y) < \epsilon\}$ is an open subset of X containing $V^d_\epsilon \circ \mathcal{C}_d f \circ V^d_\epsilon(x) \supset V^d_\epsilon \circ \mathcal{C}_\mathcal{U} f \circ V^d_\epsilon(x)$. It is $V^d_\epsilon \circ \mathcal{C}_d f$ $^+$invariant and so is $V^d_\epsilon \circ \mathcal{C}_\mathcal{U} f$ and $V^d_\epsilon \circ f$ $^+$invariant. In particular, $\{(x, y) : m^f_d(x, y) < \epsilon\}$ is a $\mathcal{U}$ inward set for f.

$$
\begin{aligned}
\mathcal{C}_\mathcal{U} f(K) &= \bigcap_{d \in \Gamma, \epsilon > 0} \bigcup_{x \in K} \{y : m^f_d(x, y) < \epsilon\}, \\
K \cup \mathcal{C}_\mathcal{U} f(K) &= \bigcap_{d \in \Gamma, \epsilon > 0} \bigcup_{x \in K} \{y : \min(m^f_d(x, y), d(x, y)) < \epsilon\}
\end{aligned} \tag{6.31}
$$

Proof: The sets are open because ℓ_f^d and m_f^d are continuous. The set in (a) clearly contains $\mathcal{A}_d f(x) = \{y : \ell_d^f(x,y) = 0\}$. If $(y,z) \in \mathcal{A}_d f$ then by Proposition 3.2 $\ell_d^f(x,z) \leq \ell_d^f(x,y) + \ell_d^f(y,z) = \ell_d^f(x,y) < \epsilon$.

If $y \in V_\epsilon^d \circ \mathcal{C}_d f(z)$ with $m_d^f(x,z) < \epsilon$ then there exists $z_1 \in \mathcal{C}_d f(z)$ with $d(z_1,y) < \epsilon$. Let $\epsilon_1 > 0$ and such that $d(z_1,y) + \epsilon_1, m_d^f(x,z) + 2\epsilon_1 < \epsilon$. There exist $[a,b] \in f^{\times n}$ and $[c,d] \in f^{\times m}$ such that with respect to d the xz chain-bound of $[a,b]$ is less than $m_d^f(x,z) + \epsilon_1$ and the zz_1 chain-bound of $[c,d]$ is less than ϵ_1. Notice that $d(b_n,c_1) \leq d(b_n,z) + d(z,c_1) < \epsilon$ and $d(c_m,y) \leq d(c_m,z_1) + d(z_1,y) < \epsilon$. Hence, the xy chain-bound of the concatenation $[a,b] \cdot [c,d]$ is less than ϵ. Thus, $\{y : m_d^f(x,y) < \epsilon\}$ is $(V_\epsilon^d \circ \mathcal{C}_d f)$ $^+$invariant.

Similarly, if $y \in V_\epsilon \circ \mathcal{C}_d f(z)$ with $d(x,z) < \epsilon$ then there exists $z_1 \in \mathcal{C}_d f(z)$ with $d(z_1,y) < \epsilon$. Let $\epsilon_1 > 0$ and such that $d(z_1,y) + 2\epsilon_1, d(x,z) + 2\epsilon_1 < \epsilon$. There exists $[c,d] \in f^{\times m}$ such that with respect to d the zz_1 chain-bound of $[c,d]$ is less than ϵ_1. Notice that $d(x,c_1) \leq d(x,z) + d(z,c_1) < \epsilon$ and $d(c_m,y) \leq d(c_m,z_1) + d(z_1,y) < \epsilon$. Hence, the xy chain-bound of the concatenation $[c,d]$ is less than ϵ. Thus, $\{y : m_d^f(x,y) < \epsilon\}$ contains $V_\epsilon^d \circ \mathcal{C}_d f \circ V_\epsilon^d(x)$.

If $Q : X \times X \to \mathbb{R}$ is a continuous function with $Q \geq 0$, then we let $Q(K,y) = \inf\{Q(x,y) : x \in K\}$. Clearly, $Q(K,y) \leq \epsilon$ iff there exists $x \in K$ such that $Q(x,y) < \epsilon$. Also,

$$\{x : Q(K,y) = 0\} = \bigcap_{\epsilon > 0} \{x : Q(K,y) < \epsilon\}. \tag{6.32}$$

Furthermore, if K is compact then $Q(K,y) = 0$ iff there exists $x \in K$ such that $Q(x,y) = 0$.

Recall from (4.19) that $\ell_d^{f \cup 1_X}(x,y) = \min(\ell_d^f(x,y), d(x,y))$ and from (4.25) that $m_d^{f \cup \{(x,x)\}}(x,y) = \min(m_d^f(x,y), d(x,y))$.

Let $Q_d(x,y) = m_d^{f \cup \{(x,x)\}}(x,y)$ so that $Q_d(K,y) = \min(m_d^f(K,y), d(K,y))$. Observe that if $d_1, d_2 \in \Gamma(\mathcal{U})$ and $\epsilon_1, \epsilon_2 \geq 0$ then with $d = d_1 + d_2$ and $\epsilon = \min(\epsilon_1, \epsilon_2)$,

$$\{(x,y) : Q_d(x,y) \leq \epsilon\} \subset \{(x,y) : Q_{d_1}(x,y) \leq \epsilon_1\} \cap \{(x,y) : Q_{d_2}(x,y) \leq \epsilon_2\}. \tag{6.33}$$

So if K is compact, and $y \in \bigcap_{d \in \Gamma, \epsilon > 0} \bigcup_{x \in K} \{y : \min(m_d^f(x,y), d(x,y)) < \epsilon\}$ the collection of closed subsets $\{\{x \in K : Q_d(x,y) = 0\} : d \in \Gamma(\mathcal{U})\}$ satisfies the finite intersection property and so has a nonempty intersection. If $x \in K$ is a point of the intersection, then $y \in K \cup \mathcal{C}_\mathcal{U} f(x)$. This proves the second equation in (6.31). The three remaining equations in (6.30) and (6.31) follow from a similar argument with Q_d equal to ℓ_d^f, $\ell_d^{f \cup 1_X}$ and m_d^f.

Notice that by (3.18) (using $w = x$), $\ell_d^f(x,y) - \ell_d^f(x,z) \leq d(y,z)$, and similarly for m_d^f. Thus, as functions of y, $\ell_d^f(x,y)$ and $m_d^f(x,y)$ are d Lipschitz with Lipschitz constant at most 1. Hence, for any $K \subset X$, as functions of y, $\ell_d^f(K,y)$ and $m_d^f(K,y)$ are d Lipschitz with Lipschitz constant at most 1 as are $\min(\ell_d^f(K,y), d(K,y))$ and $\min(m_d^f(K,y), d(K,y))$. □

Theorem 6.18. *Let f be a relation on a uniform space $(X, \mathcal{U})$.*

(a) *If $(x,y) \notin 1_X \cup \mathcal{C}_\mathcal{U} f$, then there exists a $\mathcal{U}$ elementary Lyapunov function L such that $L(y) = 0$ and $L(x) = 1$.*

(b) *If* $x \not\in |\mathcal{C}_{\mathcal{U}} f|$, *then there exists a* $\mathcal{U}$ *elementary Lyapunov function* L *such that* $1 > L(x) > 0$.

Proof: (a) With $g = f \cup \{(x,x)\}$, $m^g_d(y) = \min(m^f_d(x,y), d(x,y))$ by Lemma 4.11. By hypothesis, there exist $d \in \Gamma$ and $\epsilon > 0$ so that $m^g_d(x,y) > \epsilon$. By Proposition 6.17 (b), the set $A = \{y : m^g_d(x,y) < \epsilon\}$ is a $\mathcal{U}$ inward set for g. By Proposition 6.16 (d) there is a $\mathcal{U}$ uniformly continuous elementary Lyapunov function L for g (and hence for f) so that $L^{-1}(0) \cup A = X$ and $g(A) \subset L^{-1}(1)$. Since $x \in A$ and $(x,x) \in g$, $x \in g(A)$ and so $L(x) = 1$. Since $y \not\in A$, $L(y) = 0$.

(b) By hypothesis, there exist $d \in \Gamma$ and $1 > \epsilon > 0$ so that $m^f_d(x,x) > 2\epsilon$. Let $A_0 = V^d_\epsilon(x)$ and $A_1 = \{y : m^f_d(x,y) < \epsilon\}$. Since $m^f(x,x) \leq m^f_d(x,y) + d(y,x)$, it follows that A_0 and A_1 are disjoint.

By Proposition 6.17 (b) $V^d_\epsilon(f(A_0 \cup A_1)) \subset A_1$. Let $B = f(A_0 \cup A_1)$. Define $L(y) = \max([\epsilon - d(y,B)]/\epsilon, \epsilon - d(y,x), 0)$. If $(y_1, y_2) \in f$ and $L(y_1) > 0$ then $y_1 \in A_0 \cup A_1$ and so $y_2 \in B$. Thus, $L(y_2) = 1$. Thus, L is a $\mathcal{U}$ elementary Lyapunov function. Since $x \in A_0$, $d(x,B) > \epsilon$. Hence, $L(x) = \epsilon$. □

Definition 6.19. Let f be a relation on a uniform space $(X, \mathcal{U})$. We denote by $\mathcal{L}_e$ the set of $\mathcal{U}$ elementary Lyapunov functions for f. We say that a set $\mathcal{L} \subset \mathcal{L}_e$ satisfies the condition POIN-E for $\mathcal{C}_{\mathcal{U}} f$ if it satisfies POIN for $\mathcal{C}_{\mathcal{U}} f$ and, in addition,

- If $x \not\in |\mathcal{C}_{\mathcal{U}} f|$, then there exists $L \in \mathcal{L}_e$ such that $1 > L(x) > 0$.

By Proposition 6.18, the set $\mathcal{L}_e$ satisfies POIN-E for $\mathcal{C}_{\mathcal{U}} f$.

Theorem 6.20. *Let* f *be a relation on a uniform space* $(X, \mathcal{U})$. *If* $\mathcal{L} \subset \mathcal{L}_e$ *satisfies POIN-E for* $\mathcal{C}_{\mathcal{U}} f$ *then*

$$\begin{aligned} \mathcal{C}_{\mathcal{U}} f &= \bigcap_{L \in \mathcal{L}} \leq_L, \\ |\mathcal{C}_{\mathcal{U}} f| &= \bigcap_{L \in \mathcal{L}} [L^{-1}(0) \cup L^{-1}(1)] = \bigcap_{L \in \mathcal{L}} |L|_f. \end{aligned} \tag{6.34}$$

Proof: The first equation follows from POIN for $\mathcal{C}_{\mathcal{U}} f$.

If $L \in \mathcal{L}_e$ then it is an elementary Lyapunov function for $\mathcal{C}_{\mathcal{U}} f$ by Proposition 6.16 (e) and $1 - L$ is an elementary Lyapunov function for $\mathcal{C}_{\mathcal{U}} f^{-1}$ by Proposition 6.16 (f). So with $G_L = \{x : 1 > L(x) > 0\}$,

$$\mathcal{C}_{\mathcal{U}} f(G_L) \subset L^{-1}(1) \quad \text{and} \quad \mathcal{C}_{\mathcal{U}} f^{-1}(G_L) \subset L^{-1}(0). \tag{6.35}$$

Hence, $G_L \cap |\mathcal{C}_{\mathcal{U}} f| = \emptyset$, i.e. $|\mathcal{C}_{\mathcal{U}} f| \subset |L|_f$.

On the other hand, if $x \not\in |\mathcal{C}_{\mathcal{U}} f|$ then by POIN-E there exists $L \in \mathcal{L}_e$ such that $x \in G_L$. □

If A is a $^+$invariant subset for a relation f we denote by $f^\infty(A)$ the (possibly empty) maximum invariant subset of A, i.e. the union of all f invariant subsets of A. We can obtain it by a transfinite construction

$$A_0 = A, \quad A_{\alpha+1} = f(A_\alpha), \quad A_\alpha = \bigcap_{\beta < \alpha} A_\beta \text{ for } \alpha \text{ a limit ordinal.} \tag{6.36}$$

The process stabilizes at α when $A_{\alpha+1} = A_\alpha$ which then equals $f^\infty(A)$.

Definition 6.21. If A is a $\mathcal{U}$ inward set for a relation f then $(\mathcal{C}_{\mathcal{U}} f)^\infty(A)$ is called the $\mathcal{U}$ *attractor associated with* A. A $\mathcal{U}$ attractor for f^{-1} is called a $\mathcal{U}$ *repellor* for f. If A is a $\mathcal{U}$ inward set for f and B is a $\mathcal{U}$ inward set for f^{-1}

such that $A \cup B = X, f(A) \cap B = \emptyset = f^{-1}(B) \cap A$ then the pair $(A_\infty, B_\infty) = ((\mathcal{C}_\mathcal{U} f)^\infty(A), (\mathcal{C}_\mathcal{U} f^{-1})^\infty(B))$ is called a $\mathcal{U}$ attractor-repellor pair with

$$B_\infty = (\mathcal{C}_\mathcal{U} f^{-1})^\infty(B)$$

the repellor dual to $A_\infty = (\mathcal{C}_\mathcal{U} f)^\infty(A)$ and vice-versa.

Again, if $\mathcal{U} = \mathcal{U}_M$ we will drop the label $\mathcal{U}$.

Proposition 6.22. *Let f be a relation on a uniform space $(X, \mathcal{U})$ and let $x, y \in X$ with $y \not\in \overline{\{x\}}$. The following are equivalent.*

(i) $y \in \mathcal{C}_\mathcal{U} f(x)$.
(ii) *For every $\mathcal{U}$ elementary Lyapunov function L for f, $L(x) > 0$ implies $L(y) = 1$.*
(iii) *For every open $\mathcal{U}$ inward set A for f, $x \in A$ implies $y \in A$.*

If $x \in |\mathcal{C}_\mathcal{U} f|$, then these conditions are further equivalent to

(iv) *For every $\mathcal{U}$ attractor A_∞ for f, $x \in A_\infty$ implies $y \in A_\infty$.*

Proof: (i) $\Rightarrow$ (ii): A $\mathcal{U}$ elementary Lyapunov function for f is a $\mathcal{U}$ elementary Lyapunov function for $\mathcal{C}_\mathcal{U} f$ by Theorem 6.16(e).

(i) $\Rightarrow$ (iii): A $\mathcal{U}$ inward set for f is $\mathcal{C}_\mathcal{U} f$ $^+$invariant by Theorem 6.16(b).

(ii) $\Rightarrow$ (i): Apply Theorem 6.18 (a).

(iii) $\Rightarrow$ (i): By Proposition 6.17 (b), with $g = f \cup \{(x, x)\}$, $\{y : m_d^g(x, y) < \epsilon\} = \{y : \min(m_d^f(x, y), d(x, y)) < \epsilon\}$ is a $\mathcal{U}$ inward set for g and hence for f. So (6.31) implies that $\overline{\{x\}} \cup \mathcal{C}_\mathcal{U} f(x)$ is the intersection of $\mathcal{U}$ inward sets.

If $x \in |\mathcal{C}_\mathcal{U} f|$, then $\mathcal{C}_\mathcal{U} f(x)$ is $\mathcal{C}_\mathcal{U} f$ invariant and so x is contained in an inward set A iff it is contained in the associated attractor. Hence (iii) $\Leftrightarrow$ (iv) in this case.

Notice that if $x \in |\mathcal{C}_\mathcal{U} f|$ then $\overline{\{x\}}$ is contained in the closed set $\mathcal{C}_\mathcal{U} f(x)$. □

Proposition 6.23. *If A_∞ is the $\mathcal{U}$ attractor associated with the $\mathcal{U}$ inward set A, then $A \cap |\mathcal{C}_\mathcal{U} f| \subset A_\infty$. Furthermore,*

$$|\mathcal{C}_\mathcal{U} f| \;=\; \bigcap\{A_\infty \cup B_\infty : \ (A, B) \textit{ a } \mathcal{U} \textit{ attractor-repellor pair for } f\}. \tag{6.37}$$

If $x \in |\mathcal{C}_\mathcal{U} f|$ then the $\mathcal{C}_\mathcal{U} f \cap \mathcal{C}_\mathcal{U} f^{-1}$ equivalence class of x in $|\mathcal{C}_\mathcal{U} f|$ is given by

$$(\mathcal{C}_\mathcal{U} f \cap \mathcal{C}_\mathcal{U} f^{-1})(x) \;=\; \bigcap\ \{B : B \ \textit{ a } \ \mathcal{U} \textit{ attractor or repellor with } \ x \in B\}. \tag{6.38}$$

Proof: For any $\mathcal{C}_\mathcal{U} f$ $^+$invariant set A, if $x \in |\mathcal{C}_\mathcal{U} f|$ then $\mathcal{C}_\mathcal{U} f(x)$ is a $\mathcal{C}_\mathcal{U} f$ invariant subset of A and so is contained $(\mathcal{C}_\mathcal{U} f)^\infty(A)$. So if (A, B) is an attractor-repellor pair then $|\mathcal{C}_\mathcal{U} f| = |\mathcal{C}_\mathcal{U} f| \cap (A \cup B) \subset |\mathcal{C}_\mathcal{U} f| \cap (A_\infty \cup B_\infty)$.

In particular, if L is a $\mathcal{U}$ elementary Lyapunov function then with $A = \{x : L(x) > 0\}$ and $B = \{x : L(x) < 1\}$, the associated attractor-repellor pair (A_∞, B_∞) satisfies $A_\infty \subset L^{-1}(1)$, $B_\infty \subset L^{-1}(0)$ and so $|\mathcal{C}_\mathcal{U} f| \cap L^{-1}(1) = |\mathcal{C}_\mathcal{U} f| \cap A_\infty$ and $|\mathcal{C}_\mathcal{U} f| \cap L^{-1}(0) = |\mathcal{C}_\mathcal{U} f| \cap B_\infty$. Hence, (6.37) follows from (6.34).

Finally, $(\mathcal{C}_\mathcal{U} f \cap \mathcal{C}_\mathcal{U} f^{-1})(x) = \mathcal{C}_\mathcal{U} f(x) \cap \mathcal{C}_\mathcal{U} f^{-1}(x)$. By Proposition 6.22 $\mathcal{C}_\mathcal{U} f(x)$ is the intersection of the attractors containing x and $\mathcal{C}_\mathcal{U} f^{-1}(x)$ is the intersection of the repellors containing x. □

References

[1] E. Akin, *The general topology of dynamical systems*, Graduate Studies in Mathematics, vol. 1, American Mathematical Society, Providence, RI, 1993. MR1219737

[2] E. Akin and J. Wiseman, *Chain recurrence for general spaces*, arXiv:1707.09601v1.

[3] O. Bernardi and A. Florio, *A Conley-type decomposition of the strong chain recurrent set*, Ergod. Theo. & Dyn. Sys., to appear, DOI:10.1017/etds.2017.70.

[4] N. Bourbaki, *General topology. Chapters 1–4*, Elements of Mathematics (Berlin), Springer-Verlag, Berlin, 1989. Translated from the French; Reprint of the 1966 edition. MR979294

[5] C. Conley, *Isolated invariant sets and the Morse index*, CBMS Regional Conference Series in Mathematics, vol. 38, American Mathematical Society, Providence, R.I., 1978. MR511133

[6] R. Easton, *Chain transitivity and the domain of influence of an invariant set*, The structure of attractors in dynamical systems (Proc. Conf., North Dakota State Univ., Fargo, N.D., 1977), Lecture Notes in Math., vol. 668, Springer, Berlin, 1978, pp. 95–102. MR518550

[7] A. Fathi and P. Pageault, *Aubry-Mather theory for homeomorphisms*, Ergodic Theory Dynam. Systems **35** (2015), no. 4, 1187–1207, DOI 10.1017/etds.2013.107. MR3345168

[8] J. Franks, *A variation on the Poincaré-Birkhoff theorem*, Hamiltonian dynamical systems (Boulder, CO, 1987), Contemp. Math., vol. 81, Amer. Math. Soc., Providence, RI, 1988, pp. 111–117, DOI 10.1090/conm/081/986260. MR986260

[9] M. Hurley, *Noncompact chain recurrence and attraction*, Proc. Amer. Math. Soc. **115** (1992), no. 4, 1139–1148, DOI 10.2307/2159367. MR1098401

[10] J. L. Kelley, *General topology*, D. Van Nostrand Company, Inc., Toronto-New York-London, 1955. MR0070144

[11] P. Pageault, *Conley barriers and their applications: chain-recurrence and Lyapunov functions*, Topology Appl. **156** (2009), no. 15, 2426–2442, DOI 10.1016/j.topol.2009.06.013. MR2546945

[12] J. Wiseman, *The generalized recurrent set and strong chain recurrence*, Ergodic Theory Dynam. Systems **38** (2018), no. 2, 788–800, DOI 10.1017/etds.2016.35. MR3774842

[13] J. Wiseman, *Generalized recurrence and the nonwandering set for products*, Topology Appl. **219** (2017), 111–121, DOI 10.1016/j.topol.2017.01.010. MR3606287

[14] K. Yokoi, *On strong chain recurrence for maps*, Ann. Polon. Math. **114** (2015), no. 2, 165–177, DOI 10.4064/ap114-2-6. MR3361230

Mathematics Department, The City College, 137 Street and Convent Avenue, New York City, NY 10031, USA

Email address: ethanakin@earthlink.net

Department of Mathematics, Agnes Scott College, 141 East College Avenue, Decatur, GA 30030, USA

Email address: jwiseman@agnesscott.edu

Contemporary Mathematics
Volume **736**, 2019
https://doi.org/10.1090/conm/736/14846

Towards the prediction of critical transitions in spatially extended populations with cubical homology

Laura S. Storch and Sarah L. Day

Abstract. The prediction of critical transitions, such as extinction events, is vitally important to preserving vulnerable populations in the face of a rapidly changing climate and continuously increasing human resource usage. Predicting such events in spatially distributed populations is challenging because of the high dimensionality of the system and the complexity of the system dynamics. Here, we reduce the dimensionality of the problem by quantifying spatial patterns via Betti numbers (β_0 and β_1), which count particular topological features in a topological space. Spatial patterns representing regions occupied by the population are analyzed in a coupled patch population model with Ricker map growth and nearest-neighbors dispersal on a two-dimensional lattice. We illustrate how Betti numbers can be used to characterize spatial patterns by type, which in turn may be used to track spatiotemporal changes via Betti number time series and characterize asymptotic dynamics of the model parameter space. En route to a global extinction event, we find that the Betti number time series of a population exhibits characteristic changes. We hope these preliminary results will be used to aide in the prediction of critical transitions in spatially extended systems. Additional applications of this technique include analysis of spatial data (e.g., GIS) and model validation.

1. Introduction

The dynamics of spatially distributed populations are difficult to understand and predict, both due to their complex behavior and high dimensionality. Improving our understanding of these systems, and attempting to predict critical changes in their behavior, become vitally important in the context of a globally changing climate, habitat destruction, and exploitation pressures, all of which contribute to increased dynamical volatility and extinction risk (e.g., [**20**], [**13**], [**22**]). Here, we aim to understand and predict the dynamics of a spatially distributed population by examining spatial patterns and how they change over time. We accomplish this by analyzing the topological features of spatial patterns representing the regions of space that are occupied by members of the population. More specifically, using cubical homology to calculate the first and second Betti numbers (β_0 and β_1) of the union of occupied patches in a two-dimensional lattice model allows us to quantify properties of these population patterns. These numbers, β_0 and β_1, count the number of connected components and one-dimensional holes, respectively, in the population pattern. Population time series are generated via a coupled patch

2010 *Mathematics Subject Classification.* 92-XX (Primary) and 54-XX (Secondary).

model with Ricker map growth ([**23**]), symmetric nearest-neighbors dispersal, and absorbing boundaries. We find that β_0 and β_1 can be used to characterize spatial patterns by type, and to track spatial pattern changes over time. We track population global extinction events via Betti number time series and find characteristic changes in β_0 and β_1 en route to global extinction, suggesting that Betti numbers may be useful in the prediction of critical transitions.

Coupled patch lattice models and tools from computational topology have a separate, established history in the literature. Coupled patch models are relatively simple, yet can exhibit complicated dynamics in both space and time. Studies have experimented with the coupling of two patches (e.g., [**26**], [**11**], [**32**]), the coupling of multiple patches along a line (e.g., [**3**], [**14**], [**16**], [**17**], [**19**], [**31**], [**30**]) and the coupling of patches in a two-dimensional spatial lattice (e.g., [**15**]). Here, we focus on the two-dimensional lattice models. Likewise, computational topology has been used to measure structure in many systems, including time-varying models and data. These studies range from using cubical homology to study patterns in models ([**18**], [**5**], [**7**]) to using simplicial persistent homology to study time evolving patterns of population point cloud data ([**6**], [**29**], [**1**], [**2**], [**27**]). Our focus here is to use cubical homology to quantify and track population patterns representing the occupied regions of space, forming the foundation for later studies using the more sophisticated tool of cubical persistent homology to study the additional information offered by patch-wise abundance data.

Our goal is to use topology to study the complex patterns that arise in coupled patch lattice models. In a similar model to our own, Kaneko ([**15**]) finds that a population distributed across two-dimensional space can fall into several typical patterns, depending on the parameter combination. Such patterns include checkerboard, fully-developed spatiotemporal chaos, and the frozen random pattern, which consists of a population pattern that appears largely periodic in space and time, with small sections of sustained non-periodic behavior. The Kaneko model consists of a coupled patch model with logistic map growth, symmetric nearest-neighbors dispersal, and periodic boundary conditions. Our model employs a different growth map and boundary conditions, but we observe qualitatively similar dynamics. While previous work has also focused on characterizing system dynamics over the model parameter space, here, we focus specifically on the topological features of the spatial patterns produced by the model. The topological features allow for the characterization of spatial patterns by type, allow for observation of changes in spatial patterns over time, and provide an additional tool for labeling of the parameter space.

One goal of our spatial pattern analysis is to determine if there is a characteristic change in spatial patterns that occurs en route to a critical dynamical transition, and if we can use this information to predict critical transitions. More specifically, we focus on the changes in spatial pattern en route to a global extinction event. To our knowledge, this is a novel approach to the prediction of critical transitions, which currently focuses on dynamical or statistical markers of impending change.

Scheffer et al. ([**25**]) outline several potential symptoms of critical dynamical transitions, which manifest as critical slowing down events. The symptoms of an impending transition can include slowing of recovery from perturbation, increased variance, and increased autocorrelation. Such symptoms have been observed in real-world systems (e.g., [**13**]) but direct applications remain limited. In spatially

FIGURE 1. Example population patterns on a 10×10 lattice, displaying gaps, labyrinth, and spots, respectively, moving from left to right. Black color indicates that the patch is occupied, white color indicates that the patch is unoccupied. We compute Betti numbers for the pattern given by the union of the occupied (black) patches. Please note that each patch is a closed square so that two edge or corner adjacent patches form a connected pattern. Patch boundaries are displayed for visualization purposes. The leftmost image has $\beta_0 = 1$ and $\beta_1 = 6$, the middle image has Betti numbers $\beta_0 = 3$ and $\beta_1 = 3$, the rightmost image has $\beta_0 = 5$ and $\beta_1 = 0$.

extended systems, the distribution of the population may also provide early warning signals, e.g., increase in spatial coherence preceding an event ([**25**]), or self-organized patchiness ([**24**]).

For spatially extended systems specifically, increased attention has been given to the prediction of critical transitions in semiarid grassland environments. These tend to self-organize into characteristic patterns, dependent on rainfall (e.g., [**25**], [**24**], [**10**], [**9**]). The characteristic pattern of the system corresponds with the dynamical state, and so tracking changes in pattern may provide early warning signals. Grassland models transition through the standard "gaps–labyrinth–spots" sequence (Figure 1) with decreasing rainfall. When they reach the patchy "spots" phase, small regions of grass are isolated in the environment and dynamically decoupled, and a catastrophic shift may be imminent ([**24**]).

Here, we employ a topological approach to quantifying population spatial patterns and tracking their changes over time. We focus on two types of spatial features of the populations: connected components and holes, counted by the first and second Betti numbers. Figure 1 illustrates the Betti numbers of example population patterns, showing the gaps, labyrinth, and spots phases, respectively. The images are binary and each two-dimensional patch is categorized as either occupied (black) or unoccupied (white). We compute Betti numbers for the patterns given by the union of the black patches. Since each patch is considered to be a closed square, two edge or corner adjacent patches overlap and their union forms a connected set.

Analyzing a high dimensional system via Betti numbers has several advantages. First, the usage of Betti numbers reduces the dimensionality of the system to two coarse-grain measurements, β_0 and β_1. Although we reduce the dimensionality of the system, we are retaining information about the spatial patterns, which provides different insights than tracking, e.g., total population. Lastly, we can also use this technique for a wide variety of additional applications, such as the analysis of datasets and/or model validation. For example, Chung & Day ([**4**]) obtain topological measurements of three-dimensional firn data (the stage between snow

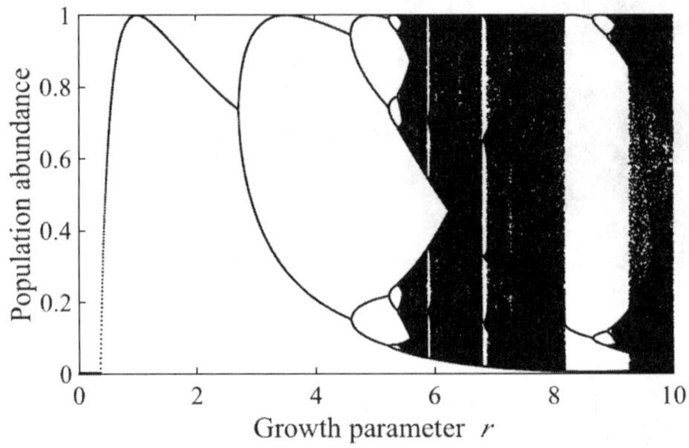

FIGURE 2. Bifurcation diagram for the normalized Ricker map equation.

and ice), which are used to accurately depict the "pores" in the firn (one- and two-dimensional holes) to inform gas transfer models. We also envision using this technique on GIS satellite images to track changes in, e.g., vegetation cover (additionally, see [**8, 18**] for applications of cubical homology and Betti numbers to the study of pattern formation and evolution in convection and phase separation models).

2. Background

2.1. Population model. We use a density-dependent coupled patch population model, where growth and dispersal occur on a 2-dimensional $N \times N$ lattice. Patch-wise population abundances are recorded in an $N \times N$ matrix X. That is, for $1 \leq i, j \leq N$, $X(i, j)$ is the population abundance in patch (i, j). The growth phase occurs first, where the population in each individual patch reproduces, independently of population values in other patches. A set portion of the population in each patch is then dispersed uniformly amongst the four nearest (co-dimension one) neighboring patches.

The growth phase is modeled using a *normalized Ricker map* $f : [0, \infty) \mapsto [0, 1]$ ([**23**]), given by

$$f(x) = rxe^{(1-rx)}$$

where $r > 0$ is the *growth parameter*. This map exhibits similar behavior to the well-known logistic map, including a period-doubling cascade and chaos for certain values of r (see Figure 2).

The normalized Ricker map is applied patch-wise so that for input population abundance matrix X_n, $\bar{X}_n$, where

$$\bar{X}_n(i, j) = f(X_n(i, j))$$

is the abundance matrix following the growth phase and n is the time index.

In the dispersal phase, a set fraction of the population in each patch, dictated by the *dispersal parameter* $0 \leq d \leq 1$, disperses symmetrically into the four nearest neighboring patches on the lattice. This results in $\bar{\bar{X}}_n$, given input population abundance matrix X_n and $\bar{X}_n$ as defined above, where

$$\bar{\bar{X}}_n(i,j) = (1-d)\bar{X}_n(i,j) + \frac{d}{4}\Big[\bar{X}_n(i-1,j) + \bar{X}_n(i+1,j) + \bar{X}_n(i,j+1) + \bar{X}_n(i,j-1)\Big].$$

Here, $\bar{X}_n(l,k) := 0$ if l and/or k is equal to 1 or $N+1$.

After the dispersal phase, we apply a *local extinction threshold* patch-wise, so

$$X_{n+1}(i,j) = \begin{cases} \bar{\bar{X}}_n(i,j) & \text{if } \bar{\bar{X}}_n(i,j) \geq \epsilon \\ 0 & \text{otherwise.} \end{cases}$$

Therefore, if the population abundance on a patch dips below the local extinction value of ϵ, then that patch experiences a local extinction event resulting in an abundance of zero. In this way we introduce an allee effect, as patches with low population abundances will be set to zero and will be unsuccessful at reproduction in the following generation.

As we keep the fitness parameter r and dispersal parameter d to be the same across all patches, the spatial domain is assumed homogeneous and there is no preferred directionality to the dispersal. If part of the population disperses outside of the confines of the $N \times N$ lattice, it is considered lost to the system. This is an absorbing boundary condition. Thus, the 2-dimensional domain represents the habitable range of the modeled population, and outside of the domain the species cannot survive due to environmental conditions, resource availability, etc.

In future work we explore additional mechanisms of dispersal, as well as stochastic growth and dispersal, to make the model more applicable to real-world systems. However, for this proof-of-concept paper, we choose this simple, yet dynamically rich, model.

It is natural to visualize a population distribution of this type as a greyscale image where each pixel corresponds to a patch and the greyscale value on that pixel corresponds to the abundance on that patch. Since we will be focusing on computing topological information on sets, we must first extract a region of interest. In this work, we focus on the collection of patches that have positive abundance values. We will call these patches "occupied" and color them black. Since patches with an abundance of "0" are already white, this corresponds to converting the greyscale image for the population distribution to a binary image (see Figure 3 for example greyscale and binary images). The black set given by the union of the occupied patches is the set or *pattern* we will study.

2.2. Computational homology and Betti numbers. In this work, the key approach we employ for measuring spatial population patterns utilizes cubical homology. We will focus on computing homological information about the collection of occupied (black) patches. By construction, these patches live on a cubical lattice and the collection may be specified as a list of closed, two-dimensional squares (cubes) of the form $[i,i+1] \times [j,j+1]$ for some integers $1 \leq i,j \leq N$. That is, for population abundance matrix X, the *collection of occupied patches*, $\mathcal{X}^+$, is

$$\mathcal{X}^+ := \{[i,i+1] \times [j,j+1] |\, X(i,j) > 0\}.$$

We note here that a subset of patches may be chosen differently, for example by defining the super level set $\mathcal{X}^\tau := \{[i,i+1] \times [j,j+1] | X(i,j) > \tau\}$ for some threshold $\tau > 0$. The corresponding *population pattern* (also known as the *topological realization of* $\mathcal{X}^+$) is

$$|\mathcal{X}^+| := \bigcup_{[i,i+1]\times[j,j+1]\in\mathcal{X}} [i,i+1]\times[j,j+1],$$

that is, the spatial pattern formed by the collection of occupied patches.

Betti numbers, β_k, arise in the field of homology as the ranks of the free parts of homology groups. These are computable for cubical sets; see [**28**] for further explanation. In what follows, we use the CHomP package (Computational Homology Project software, available at chomp.rutgers.edu) to compute all Betti numbers. Beyond being computable, what is important for our purposes is the fact that Betti numbers count holes of various dimensions. For a collection of two-dimensional cubes $\mathcal{X}^+$, $\beta_0 = \beta_0(|\mathcal{X}^+|)$ is the number of connected components (0-dimensional holes) in $|\mathcal{X}^+|$, $\beta_1 = \beta_1(|\mathcal{X}|^+)$ is the number of 1-dimensional holes in $|\mathcal{X}^+|$, and all other Betti numbers are 0. In our context, β_0 gives the number of connected regions of populated patches while β_1 gives the number of enclosed dead (non-populated) regions. See Figure 1 for example population patterns and their corresponding Betti numbers.

3. Results

The following three subsections describe sample results to illustrate this method. In the first subsection we demonstrate how we use Betti numbers as a low-dimensional metric for interpreting spatiotemporal behavior of the model. In the second subsection we view an extinction event through the lens of Betti numbers, and in the third subsection we use Betti numbers to characterize the model parameter space.

3.1. Examining spatiotemporal dynamics via Betti numbers. We explore how model dynamics are interpreted through Betti numbers, focusing on two dispersal examples, keeping all other parameters fixed. The lattice size is set to $N = 21$ for easy visualization, the growth parameter (r) is set to 8 (within the chaotic range of the normalized Ricker map). Recalling that the population in a given patch can fluctuate between zero and one, the local extinction threshold (ϵ) is set to 0.06. The two dispersal parameters used are $d = 0.1$ (low dispersal) and $d = 0.5$ (high dispersal). The initial condition consists of assigning each patch in the lattice a random value between zero and one, drawn from a uniform distribution. The initial condition chosen for both trials is depicted in Figure 3 (top row).

The spatial scale of the population distribution is dependent on the dispersal parameter, which we illustrate in Figure 3. The greyscale images in the middle row are snapshots of one generation of the population model, out of 100 total saved generations (each population matrix is the 100th iterate). The spatial structures of the left image are noticeably smaller than the right image, emulating an irregular checkerboard pattern. The differences in pattern are also reflected in the Betti numbers, where the low dispersal population has more connected components and fewer holes, indicating that the spatial structure is composed of more disconnected regions of populated patches that do not directly communicate via dispersal, at least under one iteration of the map. In contrast, the high dispersal population has one connected component and many holes, indicating that most of the patches in the lattice are occupied and connected via dispersal. This is shown in the bottom row of Figure 3, where the population distributions have been converted to the corresponding (binary) population patterns.

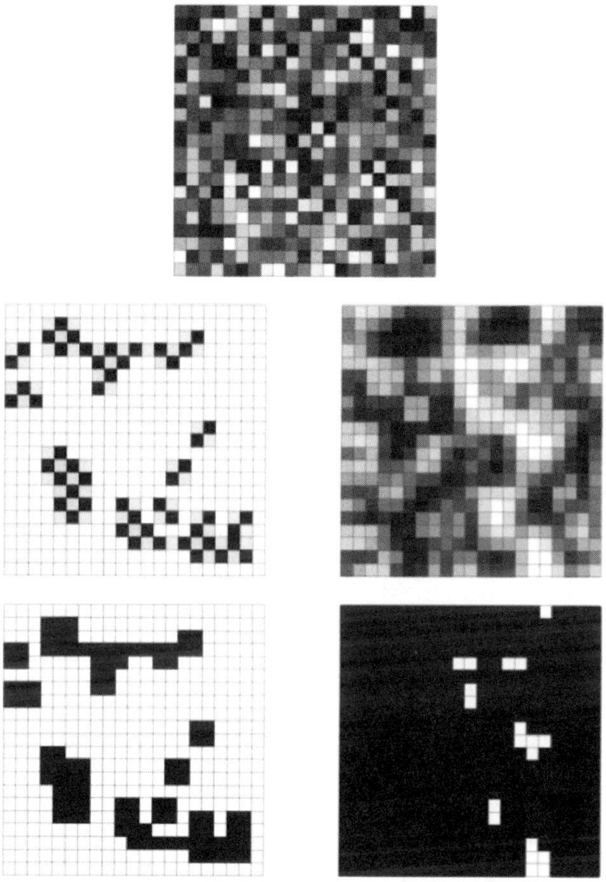

FIGURE 3. Example population distributions for low dispersal (middle row, left image, $d = 0.1$) and high dispersal (middle row, right image, $d = 0.5$). Each population distribution is taken from the 100th iterate of a model run with initial condition shown at top. The bottom row depicts the corresponding (binary) population patterns for the distributions in the middle row. Betti numbers are computed for the patterns given by the union of the closed, black (occupied) patches. For the left image, bottom row, $\beta_0 = 7$ and $\beta_1 = 0$. For the right image, bottom row, $\beta_0 = 1$ and $\beta_1 = 5$. Patch boundaries are displayed for visualization purposes.

Betti numbers provide a mechanism for quantitatively assessing spatial patterns in a population distribution, and Betti number time series provide a mechanism for assessing the long-term dynamics of a population via its changing spatial patterns. It must be noted, however, that Betti numbers are a coarse-grain method for observing dynamical changes and information is inevitably lost. For example, Figure 4 displays the Betti number time series for the parameter combinations illustrated in Figure 3. The low dispersal population appears to achieve steady-state, with

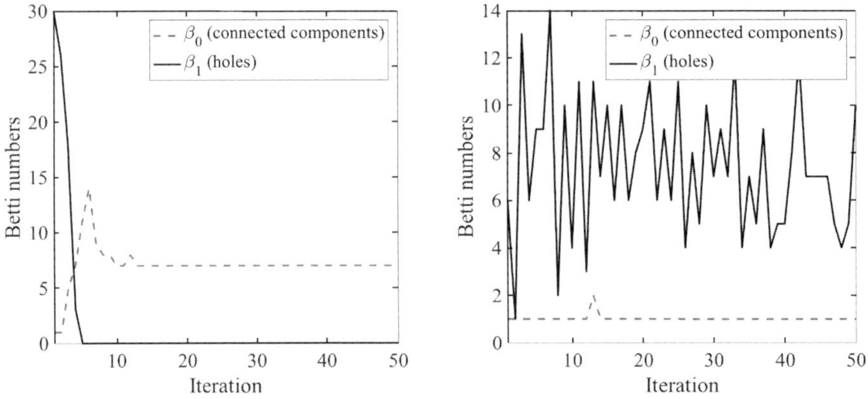

FIGURE 4. Betti number time series for the parameter combinations illustrated in Figure 3, low dispersal on the left ($d = 0.1$) and high dispersal on the right ($d = 0.5$). Via the Betti number time series, we observe that the low dispersal parameter combination appears to achieve steady-state (in terms of Betti numbers), with zero holes and seven connected components. We refer to this as a *topological steady state*. That is, the topology of the pattern of occupied patches is fixed, although the abundance of individual patches, as well as location and number of occupied patches, may continue to fluctuate. In contrast, the high dispersal parameter combination does not achieve steady state, but on average maintains a low number of connected components and many holes. Fluctuating Betti numbers means that the abundance values are necessarily fluctuating as well.

three connected components and zero holes. If, however, we observe the time series of the population distribution itself (non-binary), the individual patches that compose the larger connected components continuously flip between high- and low-density populations (due to the density dependence), thus, the true dynamics of this parameter combination are nearly period-two (there are small fluctuations in the total population, so this parameter combination is not actually period-two and, in fact, does not repeat in 100 iterates).

For the high dispersal parameter combination (Figure 4, right image), the Betti number time series illustrates the continuously changing spatial structure of the population. The number of connected components is consistently low, reflecting the fact that most of the patches in the lattice are occupied, while the number of holes remains high, on average, but fluctuates from generation to generation. Thus, the spatial pattern of this population is continuously changing and does not achieve any type of steady-state.

3.2. Critical transitions and Betti numbers. Observing an extinction event via Betti number time series reveals a characteristic manner in which the population goes extinct. We use a parameter combination which results in a global extinction event ($d = 0.1$, $r = 8$, $\epsilon = 0.07$, lattice size $N = 21$). As before, for the initial condition, each patch is assigned a random value between zero and one, drawn from a uniform distribution. Starting with a random initial condition in which all patches are occupied, there is a rapid increase in the number of connected components and a rapid decrease in the number of holes under iteration of the model. This signifies the breakup of large connected components into smaller decoupled components as the population undergoes fragmentation. The number of one-dimensional holes then drops to zero as the number of connected components continues to increase, representing both an increase in the number of connected regions of occupied patches and a topological simplification of these regions. Thus, regions of occupied patches become increasingly fragmented as they decouple from neighboring components and decrease in size and topological complexity. Finally, the connected components reach a maximum value before dropping, eventually to zero. In this final stage, the remaining occupied patches are isolated and receive no additional population from neighboring patches (as the neighboring patches are empty). The isolated patches cannot sustain their local populations and go extinct.

We illustrate this general trend in Figure 5, which displays the Betti number time series of an extinction event. There are three example population patterns taken from three different iterations of the time series.

3.3. Characterizing the model parameter space. We now measure asymptotic dynamics in the model over a large portion of the parameter space, using "traditional" methods first and then progressing to Betti numbers. We first estimate the number of occupied patches typical for a given parameter combination after many iterations of the map. In order to obtain this information, we run the model using 200 random initial conditions and average the results from the 200 trials. As before, the initial conditions are constructed by assigning each patch in the lattice a uniformly distributed random number between zero and one. Information is obtained from the 1000th iterate of each trial. As illustrated in Figures 4 and 5, model dynamics tend to rapidly settle, so we expect 1000 iterations to be sufficient for estimating long-term dynamical behavior.

Figure 6 displays the average number of occupied patches over a significant portion of the parameter space in d and ϵ ($0 \leq d \leq 0.5$ and $0 \leq \epsilon \leq 0.2$). Growth rate and lattice size are fixed ($r = 8$ and $N = 21$). Lower local extinction (ϵ) values yield a high average number of occupied patches, indicating a healthy population. For local extinction (ϵ) equal to zero, the average number of occupied patches is equal to 441 (all patches occupied), as the Ricker map does not allow the population in a patch to map to zero. Large dispersal (d) values are able to maintain a high average number of occupied patches over a larger range of ϵ values. Regions of grey in the figure indicate an average value of zero, thus, the population experiences global extinction in all trials in these regions.

To better define the "sawtooth" boundary between the global extinction (grey) region and the rest of the parameter space in Figure 6, we determine the *global extinction line*, i.e., the smallest local extinction parameter ϵ for a given dispersal d which leads to global extinction for all 200 trials. As before, each trial is iterated 1000 times. We additionally determine the largest possible ϵ, for a given d, in which all 200 trials are still alive after 1000 iterations (i.e., the *population persistence line*). Figure 7 displays the global extinction and population persistence lines for the tested parameter space.

To obtain spatial information about the population dynamics below the global extinction line, we calculate average β_0 and β_1 values over the parameter space used

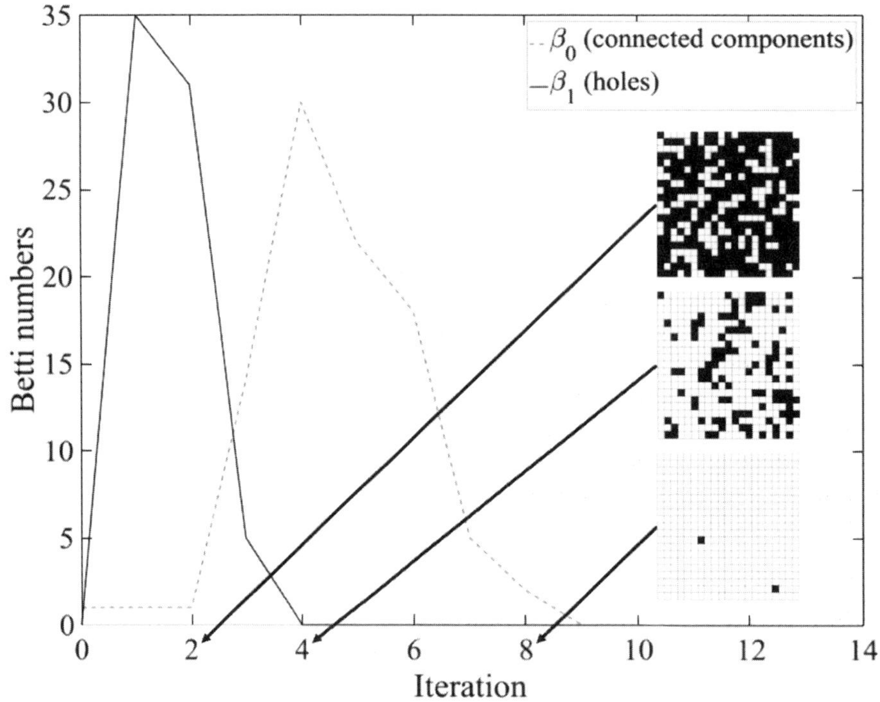

FIGURE 5. Betti number time series for an extinction event. Parameters are: $N = 21$, $r = 8$, $d = 0.1$. Initial condition is random, similar to the initial condition in Figure 3, local extinction threshold $\epsilon = 0.07$. En route to a global extinction event, the Betti number time series displays a characteristic shape, where first the number of holes (β_1) drops to zero while the number of connected components (β_0) increases (signaling a decoupling of the populated regions), and then connected components drop to zero as the decoupled regions cannot maintain population abundance without immigration from neighboring patches. The figure contains three snapshots of the population patterns en route to extinction, at iterations (generations) two, four, and eight.

in Figures 6 and 7. As before, β_0 and β_1 values are obtained from the 1000th iterates of each trial, with 200 total trials. Figure 8 displays the β_0 and β_1 averages (top row), maxima (middle row), and minima (bottom row). The maxima display the highest β_0 and β_1 values observed for each parameter combination over 200 trials, and the minima display the smallest β_0 and β_1 values observed for each parameter combination over 200 trials. As in Figure 7, the solid black line indicates, for a given d, the smallest ϵ for which all 200 trials go extinct, and is included as a reference to easily compare across figures.

We observe areas of higher β_0 averages near the global extinction line, which rapidly drop to zero as they cross the line. For example, using $d = 0.2$ and moving vertically upward in the β_0 averages plot, when ϵ is low, most patches in the lattice

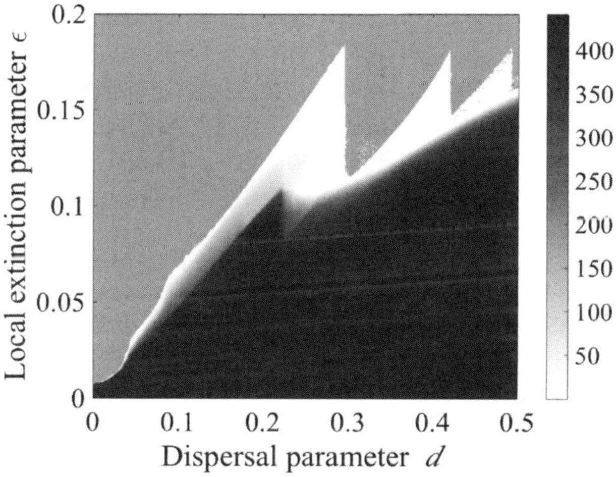

FIGURE 6. Average number of occupied patches over 200 trials with random initial conditions. Lattice size $N = 21$ (so total number of patches equals 441), growth rate $r = 8$. For lower ϵ values, the average number of occupied patches is higher, indicating a healthy population. An average of zero is designated grey (all trials experience global extinction in the grey regions).

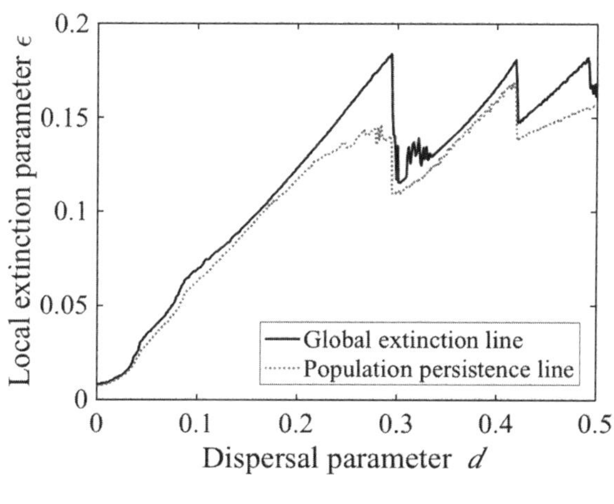

FIGURE 7. Illustrating regions in the parameter space with persistence versus extinction, based on the local extinction parameter ϵ versus the dispersal parameter d. Growth rate is $r = 8$, lattice size $N = 21$. For each parameter combination, 200 trials with random initial conditions were run for 1000 iterations. Above the solid black line, all 200 trials experience global extinction within 1000 iterations. Below the dotted red line, all 200 trials persist over 1000 iterations.

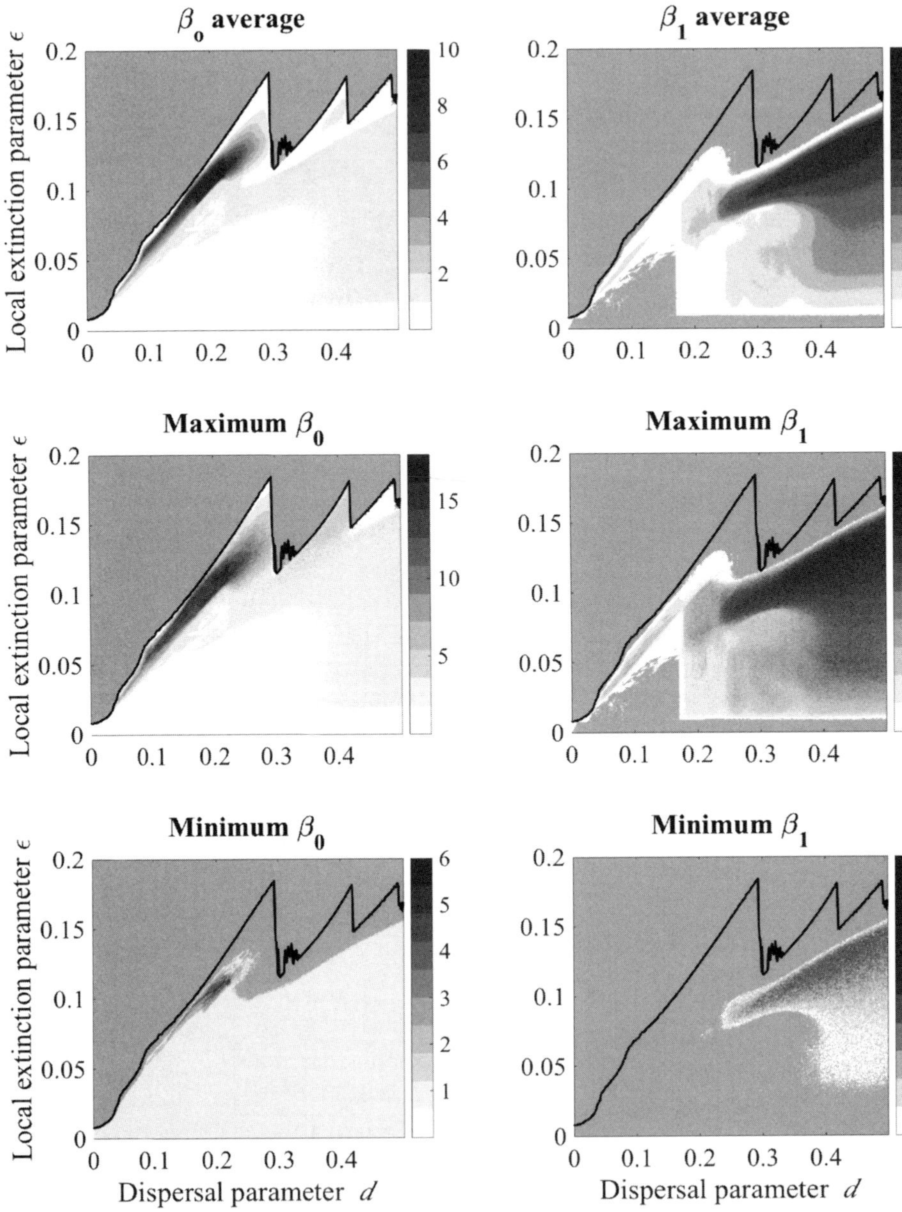

FIGURE 8. Averages, maxima, and minima for β_0 (left) and β_1 (right). For each parameter combination, 200 trials with random initial conditions were run for 1000 iterations each, information obtained from the last iteration. Top row: Betti number averages over the 200 trials. Middle row: maximum β_0 and β_1 values observed over 200 trials. Bottom row: minimum β_0 and β_1 values observed over 200 trials. The growth rate is $r = 8$, lattice size $N = 21$. Grey indicates a value of zero. The black solid line indicates, for a given dispersal d, the smallest ϵ for which all trials go extinct. Please note the different scales of each colorbar.

are occupied, and β_0 is small. As ϵ increases, β_0 also increases as the large regions of occupied patches are broken up into smaller disconnected components. As the limit of global extinction approaches, the number of disconnected components rapidly drops to zero. We observe that the β_1 average is also lower for small ϵ, as most patches are occupied and thus there are few holes. As ϵ increases and more patches experience local extinction, more holes are created (β_1 increases). Finally, as the global extinction limit is reached, the population consists exclusively of disconnected small components, and the number of holes drops to zero. As in the global extinction example above, this is a topological simplification of connected regions of occupied patches.

We can characterize patterns by type based on β_0 and β_1 values in different regions of the parameter space. Figure 9 illustrates several example patterns. The average β_0 and β_1 plots from the top row of Figure 8 have been overlapped to provide a guide for regions of high β_0 versus high β_1. The four example regions are not meant to provide a full picture of the parameter space or provide the exact boundary of each region. The four spatial patterns illustrate, from bottom to top, a continuous population ($\beta_0 = 1$, $\beta_1 = 0$), a "swiss cheese" population (β_0 low, β_1 high), a population comprised of multiple disconnected subpopulations with some holes (β_0 and β_1 high/nonzero), and a population consisting solely of isolated islands (β_0 high, $\beta_1 = 0$). Figure 9 additionally illustrates the shortcomings of relying on β_0 and β_1 alone. A continuous population (bottom spatial distribution) and a population with a single connected component both yield $\beta_0 = 1$ and $\beta_1 = 0$.

4. Discussion

Using a simple deterministic population model, we illustrate how Betti numbers are a useful metric for characterizing spatial patterns in a high-dimensional system. Calculation of β_0 and β_1 allows us to reduce the dimensionality of the system, while still retaining important information about the spatial patterns of the population. The characteristic changes in β_0 and β_1 over time en route to global extinction suggest promising applications of Betti numbers as a tool in the prediction of critical transitions. Betti number time series can also be utilized to assess the coarse-grain dynamics and stability of a population over time, which may be more relevant to ecologists and managers than assessing the population dynamics in a mathematically rigorous/higher dimensional way, e.g., in Figure 4 the left Betti number time series appears stable but the population is quasi-periodic (nearly period-2).

This work also illustrates the shortcomings of relying solely on information about β_0 and β_1 to assess the health or predict the dynamics of a population, particularly when attempting to ascertain such information from a single point in time. For example, a population distribution with $\beta_0 = 1$ and $\beta_1 = 0$ can describe either a single connected component in an otherwise empty lattice, or a contiguous population in which each patch in the lattice is occupied. Either of these population distributions may appear stable over time with respect to their Betti number time series, but one may be more susceptible to extinction than the other. Additional information about the system is required in order to assess the population. We ideally envision Betti number time series being used as an additional tool in the critical transition prediction toolbox, in addition to the techniques mentioned previously (e.g., [**25**]).

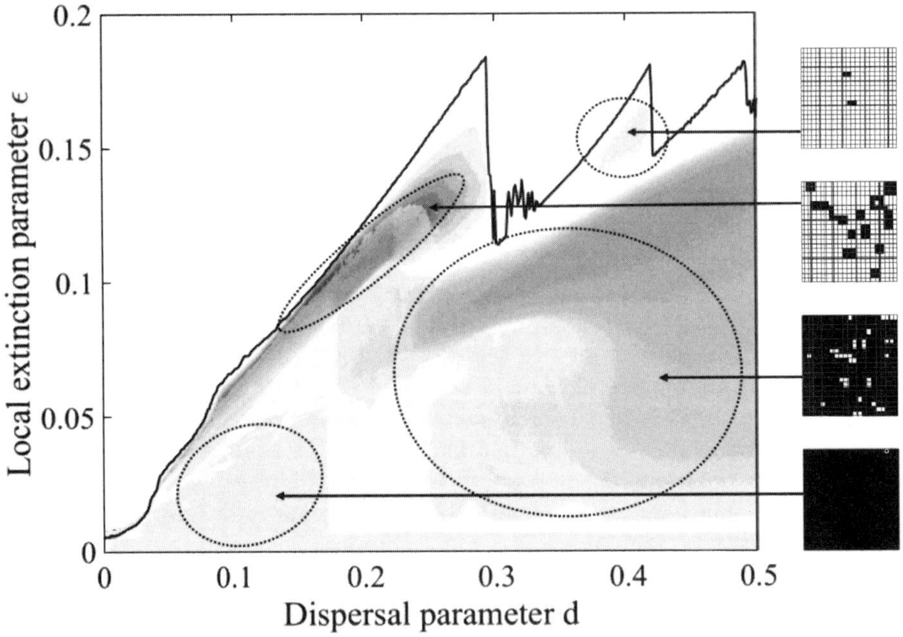

FIGURE 9. Four example spatial patterns from different regions of the parameter space. β_0 (blue) and β_1 (red) values are indicated by the transparent colored regions, overlapping the two plots from the top row of Figure 8. As before, the black solid line indicates the parameter combination for which all 200 trials go extinct. From bottom to top, (1) regions of $\beta_0 = 1$ and $\beta_1 = 0$ imply a continuous population (one large connected component), (2) regions of β_0 low and β_1 high imply a "swiss cheese" population, (3) regions of β_0 and β_1 high/nonzero imply multiple disconnected subpopulations, (4) regions of $\beta_0 > 1$ and $\beta_1 = 0$ imply isolated islands of populations containing no holes.

While the deterministic growth and symmetric dispersal model utilized here is a convenient model for illustrating the dynamics of a spatially distributed population and the interpretation of said dynamics via Betti numbers, the model does not provide sufficient ecological realism for direct application to many real-world populations. However, the coupled patch lattice model offers a high level of flexibility and is adaptable to a variety of local growth and population dispersal scenarios. The tools presented here for measuring resulting population patterns remain applicable, and it would be interesting to use them to study more ecologically realistic models, as well as GIS and other digital images of population patterns.

For each of the figures above, we chose a random initial condition in which each patch in the lattice is assigned a random value between 0 and 1 (recalling that the normalized Ricker model maps the population exclusively to [0,1]). Using this random initial condition, we constructed figures of the parameter space which exhibit a characteristic "sawtooth" shape with complicated structure (Figures 6, 7, 8, and 9). We find that the structure of these figures is echoed in alternate initial conditions. Using the same parameter space as Figures 6 through 9, we ran the model using a bivariate Gaussian distribution as the initial condition, where the Gaussian is centered in the center of the lattice, i.e., the population is more concentrated in the center. The Gaussian initial condition also displays the sawtooth pattern with similar locations of high β_0 and high β_1 regions, as in Figure 8. Dramatically changing the initial condition, however, dramatically alters the figures. Using, e.g., an initial condition that consists of an empty lattice except for one occupied patch in the corner (invasion scenario), the complicated structures are absent. The invading population either successfully spreads across the lattice or dies, and so the structures in Figures 6 through 9 that arise from using an initial condition with an established population (with spatial variability) are absent.

A global thresholding value of zero was used for all of the work presented here. In other words, for binary processing of population distribution images, a patch in the lattice was classified as either unoccupied (value of zero) or occupied (value greater than zero). For tracking a global extinction event this choice is natural, as we are concerned with the transition from populated to globally unpopulated. However, for different types of critical transitions or for data analysis, a global thresholding value of zero may not be the most natural choice. Any choice of global thresholding value can seem somewhat arbitrary, and the value chosen may dramatically impact the topological features of the resulting population pattern. For example, Figure 10 displays the same population distribution processed with two different global thresholding values. The leftmost image shows the population distribution, X, in greyscale, the middle image shows, $|\mathcal{X}^{0.05}|$, the population processed with a global thresholding value of 0.05, and the right image shows, $|\mathcal{X}^{0.5}|$, the same population processed with a global thresholding value of 0.5. The topological features of the two patterns differ dramatically. To avoid this complication, in future work we will employ cubical persistent homology (see e.g., **[21]** and references therein), which measures topological features across different thresholding values for a given population distribution, tracking the appearance and disappearance of features as the threshold is varied.

The potential uses of topological analysis in spatial ecology are numerous and diverse. We have shown how Betti numbers can be used to exhibit characteristic changes occurring in a population during a critical transition, characterize spatial patterns in a model, and reduce the dimensionality of a high-dimensional system. Betti numbers provide important spatial information that is usually absent in other low dimensional system measurements, such as total population abundance. Thus, we see it as a powerful tool for understanding and predicting high dimensional spatially explicit systems. As the dynamics of ecological systems become more volatile due to global climate change, and as we gain the ability to analyze larger and larger spatial data sets with increased computing power, we will require additional methods of analysis. We believe Betti numbers can serve as an important tool to aid in the understanding of dynamically complex, high dimensional spatial systems.

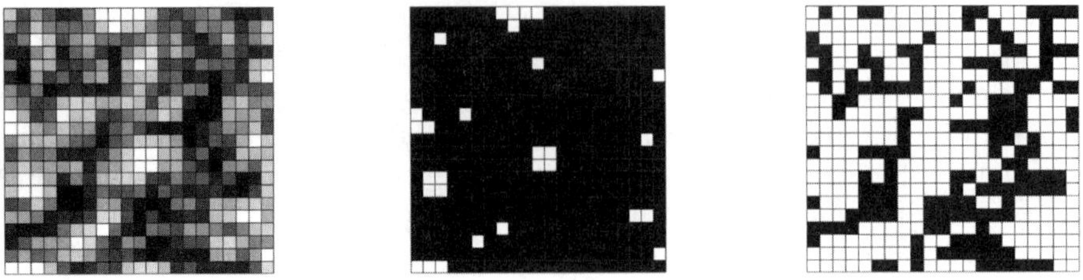

FIGURE 10. The choice of global thresholding value affects the topology of the spatial pattern. Moving from left to right: (1) original greyscale image of the population distribution, (2) the population pattern processed with a global thresholding value of 0.05, (3) the population pattern processed with a global thresholding value of 0.5. For image (2), $\beta_0 = 1$ and $\beta_1 = 9$, for image (3), $\beta_0 = 10$ and $\beta_1 = 7$. Parameters are $d = 0.45$, $\epsilon = 0.1$, $r = 8$, lattice size $N = 21$.

5. Acknowledgements

S. Day and L. Storch would like to acknowledge the work of B. Holman ([**12**]), whose undergraduate thesis provided preliminary results in this direction. S. Day's research was partially sponsored by the Army Research Office and was accomplished under Grant Number W911NF-18-1-0306. The views and conclusions contained in this document are those of the authors and should not be interpreted as representing the official policies, either expressed or implied, of the Army Research Office or the U.S. Government. The U.S. Government is authorized to reproduce and distribute reprints for Government purposes notwithstanding any copyright notation herein.

References

[1] Magnus Bakke Botnan and Michael Lesnick, *Algebraic stability of zigzag persistence modules*, Algebr. Geom. Topol. **18** (2018), no. 6, 3133–3204, DOI 10.2140/agt.2018.18.3133. MR3868218

[2] Kevin Buchin, Maike Buchin, Marc van Kreveld, Bettina Speckmann, and Frank Staals, *Trajectory grouping structure*, Algorithms and data structures, Lecture Notes in Comput. Sci., vol. 8037, Springer, Heidelberg, 2013, pp. 219–230, DOI 10.1007/978-3-642-40104-6_19. MR3126359

[3] Hugues Chaté and Paul Manneville, *Spatio-temporal intermittency in coupled map lattices*, Phys. D **32** (1988), no. 3, 409–422, DOI 10.1016/0167-2789(88)90065-6. MR980197

[4] Yu-Min Chung and Sarah Day, *Topological fidelity and image thresholding: A persistent homology approach*, J. Math. Imaging Vision **60** (2018), no. 7, 1167–1179, DOI 10.1007/s10851-018-0802-4. MR3832139

[5] Gregory S. Cochran, Thomas Wanner, and Paweł Dłotko, *A randomized subdivision algorithm for determining the topology of nodal sets*, SIAM J. Sci. Comput. **35** (2013), no. 5, B1034–B1054, DOI 10.1137/120903154. MR3106488

[6] P. Corcoran and C. B. Jones, *Modelling topological features of swarm behaviour in space and time with persistence landscapes*, IEEE Access **5** (2017), 18534–18544.

[7] Sarah Day, William D. Kalies, and Thomas Wanner, *Verified homology computations for nodal domains*, Multiscale Model. Simul. **7** (2009), no. 4, 1695–1726, DOI 10.1137/080735722. MR2539195

[8] Marcio Gameiro, Konstantin Mischaikow, and Thomas Wanner, *Evolution of pattern complexity in the Cahn-Hilliard theory of phase separation*, Acta Materialia **53** (2005), no. 3, 693–704.

[9] Karna Gowda, Yuxin Chen, Sarah Iams, and Mary Silber, *Assessing the robustness of spatial pattern sequences in a dryland vegetation model*, Proc. A. **472** (2016), no. 2187, 20150893, 25, DOI 10.1098/rspa.2015.0893. MR3488682

[10] Karna Gowda, Hermann Riecke, and Mary Silber, *Transitions between patterned states in vegetation models for semiarid ecosystems*, Phys. Rev. E **89** (2014), 022701.

[11] Alan Hastings, *Complex interactions between dispersal and dynamics: Lessons from coupled logistic equations*, Ecology **74** (1993), no. 5, 1362–1372.

[12] Benjamin R. Holman, *Topological characterization of extinction in a coupled Ricker patch model*, Undergraduate Honors Thesis (2011), `scholarworks.wm.edu/honorstheses/393/`.

[13] Chih-hao Hsieh, Christian Reiss, John R. Hunter, John R. Beddington, Robert May, and George Sugihara, *Fishing elevates variability in the abundance of exploited species*, Nature **443** (2006), 859–62.

[14] Kunihiko Kaneko, *Lyapunov analysis and information flow in coupled map lattices*, Phys. D **23** (1986), no. 1-3, 436–447, DOI 10.1016/0167-2789(86)90149-1. MR876920

[15] Kunihiko Kaneko, *Spatiotemporal chaos in one- and two-dimensional coupled map lattices*, Phys. D **37** (1989), no. 1-3, 60–82, DOI 10.1016/0167-2789(89)90117-6. MR1024382

[16] Kunihiko Kaneko, *Supertransients, spatiotemporal intermittency and stability of fully developed spatiotemporal chaos*, Physics Letters A **149** (1990), 105–112.

[17] Kunihiko Kaneko, *Overview of coupled map lattices*, Chaos **2** (1992), no. 3, 279–282, DOI 10.1063/1.165869. MR1184469

[18] Kapilanjan Krishan, Huseyin Kurtuldu, Michael F. Schatz, Marcio Gameiro, Konstantin Mischaikow, and Santiago Madruga, *Homology and symmetry breaking in Rayleigh-Bénard convection: Experiments and simulations*, Physics of Fluids **19** (2007), no. 11, 117105.

[19] Fabio A. Labra, Nelson A. Lagos, and Pablo A. Marquet, *Dispersal and transient dynamics in metapopulations*, Ecology Letters **6** (2003), no. 3, 197–204.

[20] John F. McLaughlin, Jessica J. Hellmann, Carol L. Boggs, and Paul R. Ehrlich, *Climate change hastens population extinctions*, Proceedings of the National Academy of Sciences **99** (2002), no. 9, 6070–6074.

[21] Konstantin Mischaikow and Vidit Nanda, *Morse theory for filtrations and efficient computation of persistent homology*, Discrete Comput. Geom. **50** (2013), no. 2, 330–353, DOI 10.1007/s00454-013-9529-6. MR3090522

[22] Christian N. K. Anderson, Chih-hao Hsieh, Stuart Sandin, Roger Hewitt, Anne Hollowed, John Beddington, Robert May, and George Sugihara, *Why fishing magnifies fluctuations in fish abundance*, Nature **452** (2008), 835–9.

[23] W. E. Ricker, *Stock and recruitment*, Journal of the Fisheries Research Board of Canada **11** (1954), no. 5, 559–623.

[24] Max Rietkerk, Stefan C. Dekker, Peter C. de Ruiter, and Johan van de Koppel, *Self-organized patchiness and catastrophic shifts in ecosystems*, Science **305** (2004), no. 5692, 1926–1929.

[25] Marten Scheffer, Jordi Basecompte, William A. Brock, Victor Brovkin, Stephen R. Carpenter, Vasilis Dakos, Hermann Held, Egbert H. van Nes, Max Rietkerk, and George Sugihara, *Early-warning signals for critical transitions*, Nature **461** (2009), 53 – 59.

[26] Ricard V. Sole, Jordi Bascompte, and Joaquim Valls, *Nonequilibrium dynamics in lattice ecosystems: Chaotic stability and dissipative structures*, Chaos: An Interdisciplinary Journal of Nonlinear Science **2** (1992), no. 3, 387–395.

[27] David J. T. Sumpter, *Collective animal behavior*, Princeton University Press, 2010.

[28] Kaczynški Tomasz, Misăjkov Konstantin Mihail., and Marian Mrozek, *Computational homology*, Springer, 2010.

[29] Chad M. Topaz, Lori Ziegelmeier, and Tom Halverson, *Topological data analysis of biological aggregation models*, PLOS ONE **10** (2015), no. 5, 1–26.

[30] Steven M. White and K. A. Jane White, *Relating coupled map lattices to integro-difference equations: Dispersal-driven instabilities in coupled map lattices*, J. Theoret. Biol. **235** (2005), no. 4, 463–475, DOI 10.1016/j.jtbi.2005.01.026. MR2158280

[31] Frederick H. Willeboordse, *The spatial logistic map as a simple prototype for spatiotemporal chaos*, Chaos: An Interdisciplinary Journal of Nonlinear Science **13** (2003), no. 2, 533–540.

[32] Derin B. Wysham and Alan Hastings, *Sudden shifts in ecological systems: intermittency and transients in the coupled Ricker population model*, Bull. Math. Biol. **70** (2008), no. 4, 1013–1031, DOI 10.1007/s11538-007-9288-8. MR2391177

Department of Mathematics, William & Mary, Williamsburg, VA
Email address: laura.s.storch@gmail.com

Department of Mathematics, William & Mary, Williamsburg, VA
Email address: sldayx@wm.edu

Contemporary Mathematics
Volume **736**, 2019
https://doi.org/10.1090/conm/736/14834

Relating singularly perturbed rational maps to families of entire maps

Joanna Furno and Lorelei Koss

1. Introduction

In [**19**], Hawkins and the authors of this paper analyzed the dynamics of a sequence of families of rational maps $f_{a,d}(z) = az(1 + 1/(zd))^d$ and the limiting family as $d \to \infty$, $f_a(z) = aze^{1/z}$. In this paper, we study the dynamics of related sequences of families of rational maps and their limiting families.

For $a \in \mathbb{C}^* = \mathbb{C}\setminus\{0\}$ and $0 < m < d$, define the rational function

$$f_{a,m,d}(z) = az^m\left(1 + \frac{1}{dz}\right)^d.$$

This family is conjugate to the family $S_\lambda(z) = \lambda^{m-1}z^m(1+1/z)^d$, studied in [**31, 32**], with the conjugating map $z \mapsto dz$ and $a = (d\lambda)^{m-1}$. As mentioned in [**31, 32**], these families are also related to the family of singular perturbations $R_\lambda(z) = z^m + \lambda/z^{d-m}$, which has been extensively studied (see, for example, [**3, 4, 6, 12, 23, 33**]). To be precise, they are quasiconjugate via an intermediate map $T_\lambda(z) = z^m(1 + \lambda/z)^d$ and the conjugations

$$\lambda S_\lambda(z) = T_\lambda(\lambda z) \quad \text{and} \quad T_\lambda(z^d) = R_\lambda(z)^d. \tag{1.1}$$

For a fixed a and m, the pointwise convergence of exponential maps gives

$$\lim_{d\to\infty} f_{a,m,d}(z) = f_{a,m}(z) = az^m e^{1/z}.$$

This convergence is also uniform on compact subsets of $\mathbb{C}^*$ and extends continuously and holomorphically to the point at ∞. The function $f_{a,m}$ is conjugate to the family $g_{\lambda,m}(z) = \lambda z^m e^z$, studied extensively in [**17, 18, 20**], with conjugating map $z \mapsto -1/z$ and $a = (-1)^{m-1}/\lambda$. In [**17, 18, 20**], major landmarks in the parameter plane (such as some of the capture components defined in Section 2) switch between the right and left sides of the parameter plane, based on the parity of m. Since $a = (-1)^{m-1}/\lambda$, this switching does not occur in our families.

Previous studies have investigated polynomial mappings converging to entire functions [**5, 16, 20, 28**]. With one exception, none of the functions $f_{a,m,d}$ are polynomials, and the functions discussed here exhibit different behavior from that of polynomials. For example, there are parameter values for which the Julia set of $f_{a,m,d}$ is a Sierpiński curve. Further, the functions $f_{a,m,d}$ when $m > 1$ can exhibit

2010 *Mathematics Subject Classification.* 37F10, 37F45, 30D05.

dynamical behavior different from when $m = 1$. In particular, Julia sets of $f_{a,1,d}$ are either Cantor or connected [**19**]. For $f_{a,m,d}$ with $m > 1$, there are parameter values for which the Julia set of $f_{a,m,d}$ consists of uncountable many quasicircles around the origin (the first example of these types of Julia sets dates to McMullen [**26**]). To our knowledge, no studies have been done connecting $f_{a,m,d}$ with $f_{a,m}$ when $m > 1$.

In Section 2, we describe the critical set and give some basic results and definitions. In Section 3, we discuss properties of the parameter plane in the cases when the orbit of the free critical point is bounded and unbounded. In particular, we present how the orbits of the free critical point of $f_{a,m,d}$ and the limiting map $f_{a,m}$ influence properties of the Julia sets. Finally, in Section 4, we discuss properties of the parameter planes of $f_{a,m,d}$ and $f_{a,m}$.

2. Basic properties of the functions

We begin with a brief discussion of the critical points of these families and provide some important definitions. Recall that we defined the functions $f_{a,m,d}$ for m and d with $0 < m < d$. In this case, we have a critical point at $-1/d$ of order $d-1$ and a simple critical point at $c_d = (d-m)/(md)$. Moreover, we have $-1/d \neq (d-m)/(md)$ and $-1/d \mapsto 0 \mapsto \infty \circlearrowleft$. The case when $m = 1$ was studied in [**19**], and so we consider $m \geqslant 2$ in this paper. Unlike the $m = 1$ case, if $m \geqslant 2$, then there is critical point at ∞ of order $m-1$. For the sake of simplicity, we also focus on the case where $d \geqslant m+2$. In this case, 0 is a critical point of order $d-m-1$. Since we are interested in understanding the limiting behavior as $d \to \infty$, this is not a strong restriction. Since $f_{a,m,d}$ has critical orbit $-1/d \mapsto 0 \mapsto \infty \circlearrowleft$, c_d is the only free critical point for $f_{a,m,d}$.

Considering $f_{a,m}$ as a map on $\mathbb{C}^*_\infty$, the sphere with the origin removed, the point at infinity is a fixed critical point and an asymptotic value. Moreover, there is another critical point at $c_0 = 1/m$, with critical value $v = ae^m/m^m$, and c_0 is the only free critical orbit for $f_{a,m}$. Note that $c_d = (d-m)/(md) \to 1/m$ as $d \to \infty$.

THEOREM 2.1. *Each map $f_{a,m,d}$ and $f_{a,m}$ has a superattracting fixed point at ∞. In addition, exactly one of the following can occur.*

(1) *there exists one attracting or superattracting cycle;*
(2) *there exists one cycle of Siegel disks;*
(3) *there exists one parabolic cycle;*
(4) *there are no additional Fatou cycles.*

PROOF. For $f_{a,m,d}$, the critical points $-1/d$ and 0 have finite forward orbits that land on the superattracting fixed point at ∞. Since there is at most one critical point, c_d, with an infinite forward orbit, there cannot be any Herman rings [**34**]. Thus the free critical point c_d can be associated with an attracting, parabolic, or Siegel disk cycle.

The transcendental entire maps $g_{\lambda,m}(z) = \lambda z^m e^z$ do not have Herman rings [**1**], Baker domains [**14**], or wandering domains [**15**], so its conformal conjugate $f_{a,m}$ cannot exhibit these behaviors. Thus the free critical point c_0 could only be associated with an attracting, parabolic, or Siegel disk cycle. □

Most of the results about $f_{a,m,d}$ and $f_{a,m}$ focus on when the free critical point iterates to the superattracting component at infinity. We present some notation for this situation. Let $m \geqslant 2$ be fixed. Let $A_{a,d}$ denote the basin of attraction of the superattracting fixed point at infinity for $f_{a,m,d}$, and let $A^*_{a,d}$ be the connected component of $A_{a,d}$ that contains ∞. We define the following capture components in the parameter plane for $f_{a,m,d}$:

$$\begin{aligned} \mathcal{C}^0_d &= \{a : c_d \in A^*_{a,d}\} \\ \mathcal{C}^n_d &= \{a : f^n_{a,m,d}(c_d) \in A^*_{a,d}, f^k_{a,m,d}(c_d) \notin A^*_{a,d}, 0 \leqslant k < n\} \end{aligned}$$

Similarly, let A_a denote the basin of attraction of the superattracting fixed point at infinity for $f_{a,m}$, and let A^*_a be the connected component of A_a that contains ∞. We again define capture components in the parameter plane, now for $f_{a,m}$:

$$\begin{aligned} \mathcal{C}^0 &= \{a : c_0 \in A^*_a\}, \\ \mathcal{C}^n &= \{a : f^n_{a,m}(c_0) \in A^*_a, f^k_{a,m}(c_0) \notin A^*_a, 0 \leqslant k < n\}. \end{aligned}$$

We point out that the notation for the families of functions, basins of attraction, and capture zones are similar, but that the subscript d appears only in the notation for the rational maps. We have also tried to improve the clarity of results in this paper by including the notation (RAT) for results on the family of rational maps, (EXP) for results on the limiting exponential family, and (LIM) for results relating the behavior of the rational family as $d \to \infty$ to the behavior of the exponential family.

The dynamics of polynomials that converge to exponential maps was first studied in [**5**]. A map is conjugate to a polynomial if and only if there is a critical point that is forward and backwards invariant. When $a = 1/4$, $m = 1$, and $d = 2$, the map $f_{1/4,1,2}$ is conjugate to the Tchebychev polynomial of degree 2 [**22**], and this is the only map in the family when $m = 1$ that is conjugate to a polynomial [**19**]. For $f_{a,m,d}$ when $1 < m < d$, none of the critical points $-1/d, 0$, or ∞ are backwards invariant since $-1/d \mapsto 0 \mapsto \infty \circlearrowleft$. Lemma 3.6 shows the exact parameters for which c_d is fixed. But since c_d is a simple critical point, there are $d-2$ other distinct points that map to c_d, so c_d is not backwards invariant. Thus there are no polynomials in the family $f_{a,m,d}$ when $m > 1$.

In Lemma 2.2, we give an escape radius for each map. In Section 3.2, we use the escape radii to understand the relationship between attracting cycles of $f_{a,m,d}$ and $f_{a,m}$.

LEMMA 2.2. *Fix $m \geqslant 2$ and $a \in \mathbb{C}^*$.*

(RAT) Let $r_d = \max\left\{1, \left(\frac{1}{|a|}\left(\frac{d}{d-1}\right)^d\right)^{\frac{1}{m-1}}\right\}$. *Then $\{z \in \mathbb{C} : |z| > r_d\}$ is contained in $A^*_{a,d}$.*

(EXP) Let $r_\infty = \max\left\{1, (e/|a|)^{\frac{1}{m-1}}\right\}$. *Then $\{z \in \mathbb{C} : |z| > r_\infty\}$ is contained in A^*_a.* [**17**]

(LIM) The sequence $\left\{\left(\frac{1}{|a|}\left(\frac{d}{d-1}\right)^d\right)^{\frac{1}{m-1}}\right\}_{d=m+2}^{\infty}$ *is decreasing to the limit* $(e/|a|)^{\frac{1}{m-1}}$.

PROOF. (RAT) First, note that, if $|z| \geqslant 1/d$, then

$$\begin{aligned}|f_{a,m,d}(z)| &= |a||z|^m \left|1 + \frac{1}{dz}\right|^d \\ &\geqslant |z|^{m-1}\left(1 - \frac{1}{d|z|}\right)^d \cdot |a||z|.\end{aligned}$$

Since $|z|$ is a nonnegative real number, we want to understand the real valued function

$$h(x) = x^{m-1}\left(1 - \frac{1}{dx}\right)^d.$$

Note that $h(1/d) = 0$. Using $d > m > 0$ and

$$h'(x) = \frac{\left(1 - \frac{1}{dx}\right)^d x^{m-2}(1 + d - m + dx(m-1))}{dx - 1},$$

if $x > 1/d$ then $h'(x) > 0$, so $h(x)$ is positive and strictly increasing for $x > 1/d$.

We break the proof into two cases, based on the size of $|a|$.

Suppose $|a| \leqslant \left(\dfrac{d}{d-1}\right)^d$. Then $r_d = \left(\dfrac{1}{|a|}\left(\dfrac{d}{d-1}\right)^d\right)^{\frac{1}{m-1}} \geqslant 1$. If $|z| > r_d > 1/d$, then

$$\begin{aligned}|f_{a,m,d}(z)| &\geqslant h(|z|)|a||z| \\ &> h\left(\left(\frac{1}{|a|}\left(\frac{d}{d-1}\right)^d\right)^{\frac{1}{m-1}}\right)|a||z| \\ &= \left(\frac{d}{d-1}\right)^d\left(1 - \frac{1}{d\,r_d}\right)^d |z|.\end{aligned}$$

Since $r_d > 1$, we also have $\left(\frac{d}{d-1}\right)^d\left(1 - \frac{1}{d\,r_d}\right)^d \geqslant 1$, so $h(|z|)|a| > 1$. Hence $|f^n_{a,m,d}(z)| \to \infty$ as $n \to \infty$, so $z \in A_{a,d}$.

Suppose $|a| > \left(\dfrac{d}{d-1}\right)^d$. Then $r_d = 1 > \left(\dfrac{1}{|a|}\left(\dfrac{d}{d-1}\right)^d\right)^{\frac{1}{m-1}}$. If $|z| > r_d > 1/d$, then

$$\begin{aligned}|f_{a,m,d}(z)| &\geqslant h(1)|a||z| \\ &= \left(\frac{d-1}{d}\right)^d |a||z|.\end{aligned}$$

Since $\left(\frac{d-1}{d}\right)^d |a| > 1$, we have $|f^n_{a,m,d}(z)| \to \infty$ as $n \to \infty$, so $z \in A_{a,d}$.

In either case, $\{z \in \mathbb{C} : |z| > r_d\} \subset A_{a,d}$. Since $\{z \in \mathbb{C}_\infty : |z| > r_d\}$ is a neighborhood of ∞, the punctured neighborhood $\{z \in \mathbb{C} : |z| > r_d\}$ is contained in the connected component $A^*_{a,d}$.

(LIM) Calculation of derivatives shows that

(1) $x \mapsto x/(x-1)$ is greater than 1 and decreasing for $x > 1$,
(2) $x \mapsto x^d$ is positive and increasing for $x > 0$ and $d \geqslant 1$,
(3) $x \mapsto x/|a|$ is positive and increasing for $x > 0$ because $|a| > 0$, and
(4) $x \mapsto x^{1/(m-1)}$ is defined and increasing for $x > 0$ because $m \geqslant 2$.

We use induction to show that $s_d = (d/(d-1))^d$ is decreasing for $d \geqslant 2$. For the base case, we observe that $s_2 = 4 > 3.375 = s_3$.

For the induction step, suppose $s_d > s_{d+1}$ for some $d \geqslant 2$. Since $x \mapsto x/(x-1)$ is greater than 1 and decreasing for $x > 1$, we have $1 < (d+1)/d \leqslant d/(d-1)$. Thus,

$$\begin{aligned}\left(\frac{d+1}{d}\right)^{d+1} &\leqslant \left(\frac{d}{d-1}\right)^{d+1} \\ &< \left(\frac{d}{d-1}\right)^{d}.\end{aligned}$$

Thus, it follows from induction that s_d is decreasing for $d \geqslant 2$. By composition, the sequence $\left\{\left(\frac{1}{|a|}\left(\frac{d}{d-1}\right)^d\right)^{\frac{1}{m-1}}\right\}_{d=m+2}^{\infty}$ is decreasing.

It is well known that $1/(d/(d-1))^d = (1-1/d)^d$ converges to e^{-1} as $d \to \infty$. Thus,

$$\lim_{d\to\infty}\left(\frac{1}{|a|}\left(\frac{d}{d-1}\right)^d\right)^{1/(m-1)} = \left(\frac{1}{|a|e^{-1}}\right)^{1/(m-1)} = \left(\frac{e}{|a|}\right)^{1/(m-1)}.$$

□

In particular, Lemma 2.2 (LIM) says that $r_\infty \leqslant r_d \leqslant r_{m+2}$ for all $d \geqslant m+2$, which implies that the immediate basins $A^*_{a,d}$ and A^*_a all contain a common neighborhood of ∞.

3. Dynamical Plane

Fix $m \geqslant 2$. Suppose $d \geqslant m+2$, which ensures that 0 is a critical point. We split our discussion of the dynamical plane into two cases based on whether the orbit of the free critical point is bounded or unbounded.

3.1. Unbounded free critical orbit. In this section, we describe the dynamical plane in the case when the orbit of c_d or c_0 is unbounded. We begin with some important definitions.

A subset of the Riemann sphere is said to be a *Cantor set* if it is non-empty, closed, perfect, and totally disconnected. Cantor Julia sets are common, even in the family of quadratic polynomial maps, and we also see them in the family $f_{a,m,d}$.

Julia sets of entire transcendental maps can exhibit a complicated behavior called a Cantor bouquet, as first shown in [**11**], and we present the formal definition here. For a fixed $N \in \mathbb{N}$, let

$$\Sigma_N = \{s = (s_0 s_1 s_2 \ldots) | s_j \in \{0, 1, 2, \ldots, N-1\} \text{ for each } j\}$$

denote the space of one-sided infinite sequences of N symbols. The map $\sigma : \Sigma_N \to \Sigma_N$ denotes the *right-shift map* defined by $\sigma(s_0 s_1 s_2 \ldots) = (s_1 s_2 \ldots)$.

An invariant set C_N of $J(f)$ is an *N-Cantor Bouquet* if there is a homeomorphism $h : \Sigma_N \times [0, \infty) \to C_N$ that satisfies the following conditions:

(1) For the projection $\pi : \Sigma_N \times [0, \infty) \to \Sigma_N$,
$$\pi \circ h^{-1} \circ f \circ h(s, t) = \sigma(s).$$

(2) For each $s \in \Sigma_N$,
$$\lim_{t \to \infty} h(s, t) = +\infty.$$

(3) If $t > 0$, then for each $s \in \Sigma_N$,
$$\lim_{j \to \infty} f^j(h(s, t)) = +\infty.$$

Once $s \in \Sigma_N$ has been fixed, the curve $\{h(s,t)|t > 0\}$ is called a *tail* and $h(s, 0) = z_s$ is called its *endpoint*. The union of a tail with its endpoint is known as a *hair* associated to s, and we say that the hair *lands at* z_s. The set
$$C = \overline{\bigcup_{N \geqslant 0} C_N}$$
is called a *Cantor bouquet*.

We also define what it means for a point in the Julia set of a rational map or an entire transcendental function f to be accessible from the Fatou set. A point $z_0 \in J(f)$ is *accessible from the Fatou set* if there is a continuous curve $\gamma : [0, 1) \to \mathbb{C}$ for which γ_t lies in the Fatou set for all t and $\lim_{t \to 1^-} \gamma(t) = z_0$.

For the families, $f_{a,m,d}$ and $f_{a,m}$, we see Cantor or Cantor bouquet behavior when the free critical point lies in the immediate basin of the superattracting fixed point at infinity.

THEOREM 3.1. *(RAT) If $a \in \mathcal{C}_d^0$, then the Julia set of $f_{a,m,d}$ is a Cantor set, and the action of $f_{a,m,d}$ on the Julia set is isomorphic to the Bernoulli shift on the space of d symbols. Moreover, $A_{a,d} = A_{a,d}^*$.* [**12, 26, 31, 32**]

(EXP) If $a \in \mathcal{C}^0$, then the Julia set of $f_{a,m}$ is a Cantor bouquet, and so the Julia set is disconnected and nonlocally connected. Moreover, $A_a = A_a^*$. [**17**] *The hairs in the Julia set are pairwise disjoint. The accessible points in the Julia set are the endpoints of the hairs.* [**20**]

We display some Julia sets illustrating the results stated in Theorem 3.1 in Figure 1 when $m = 2$ and $a = 0.6$. Points in the plane are colored based on how many iterates it takes for the orbit to get large. Figures 1a, 1b, and 1c show the Cantor Julia sets of $f_{a,m,d}$ for $d = 2, 4$, and $d = 10$ respectively, and Figure 1d shows the Cantor bouquet Julia set (contained in the black region) for the limiting map $f_{a,m}$.

Next, we move to a discussion of $\mathcal{C}_d^n$ and $\mathcal{C}^n$ when $n \geqslant 2$. We will see in Theorem 4.2 that $\mathcal{C}_d^1 = \varnothing$ and $\mathcal{C}^1 = \varnothing$.

A subset of the Riemann sphere is called a *Sierpiński curve* if it is compact, connected, locally connected, has empty interior, and has the property that complementary domains are bounded by pairwise disjoint simple curves.

THEOREM 3.2. *(RAT) If $a \in \mathcal{C}_d^2$, then the Julia set of $f_{a,m,d}$ consists of uncountably many quasicircles around the origin. If $a \in \mathcal{C}_d^n$, $n \geqslant 3$, then the Julia set of $f_{a,m,d}$ is a Sierpiński curve.* [**12, 26, 31, 32**]

(EXP) If $a \in \mathcal{C}^n$, $n \geqslant 2$, then A_a has infinitely many components. Connected components different from A_a^* are not bounded away from 0. [**17**] *This implies that*

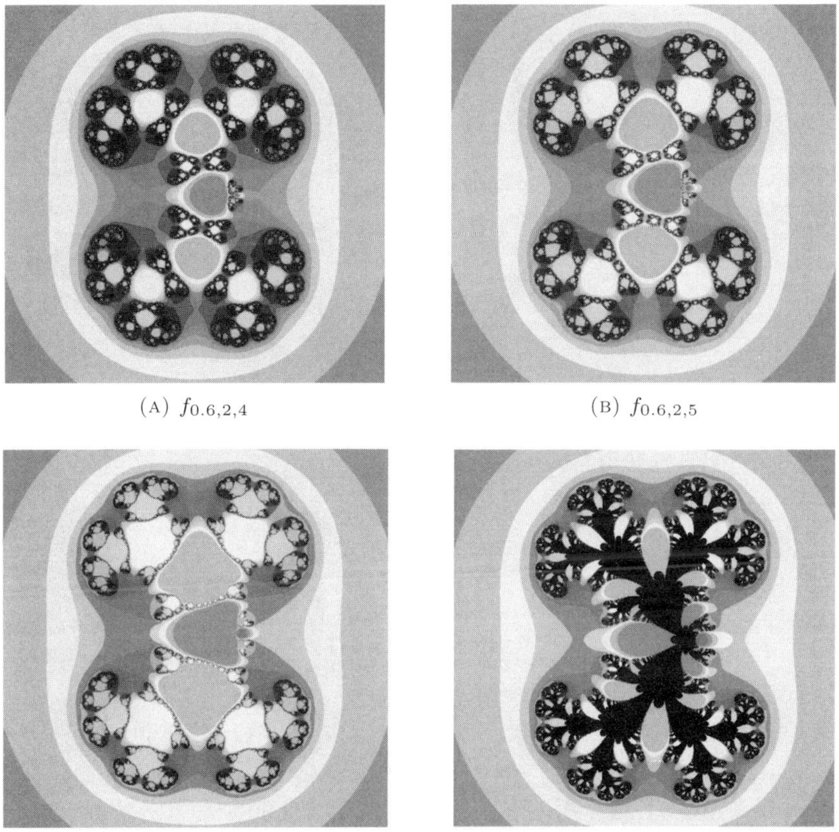

(A) $f_{0.6,2,4}$ (B) $f_{0.6,2,5}$

(C) $f_{0.6,2,10}$ (D) $f_{0.6,2}$

FIGURE 1. Cantor and Cantor bouquet Julia sets when $m = 2$ and $a = 0.6$

the Julia set is not locally connected at any point [**2, 24**]. *If $a \in \mathcal{C}^n$, $n \geqslant 2$, then the Julia set of $f_{a,m}$ is the union of hairs, each one of them homeomorphic to $[0, \infty)$. Some of the hairs may share the same (non-escaping) endpoint. The accessible points in the Julia set are the set of endpoints lying in ∂A_a^* and all its preimages. On the other hand, for any $N \in \mathbb{N}$, there exists an N-Cantor bouquet that contains only non-accessible points.* [**20**]

The region $\mathcal{C}_d^2$, where the Julia set consists of uncountably many quasicircles around the origin, is often called the McMullen domain. McMullen first described Julia sets exhibiting this type of behavior in 1988 [**26**].

We illustrate some Julia sets illustrating the results stated in Theorem 3.2 in Figures 2 and 3. First, Figure 2 shows examples when when $m = 2$ and $a = -0.2$. Points in the plane are colored based on how many iterates it takes for the orbit to get large. Figures 2a, 2b, and 2c show the dynamical plane for $f_{a,m,d}$ for $d = 4, 5$, and $d = 10$ respectively. Figure 2d shows the Julia set (shown in black) for the limiting map $f_{a,m}$.

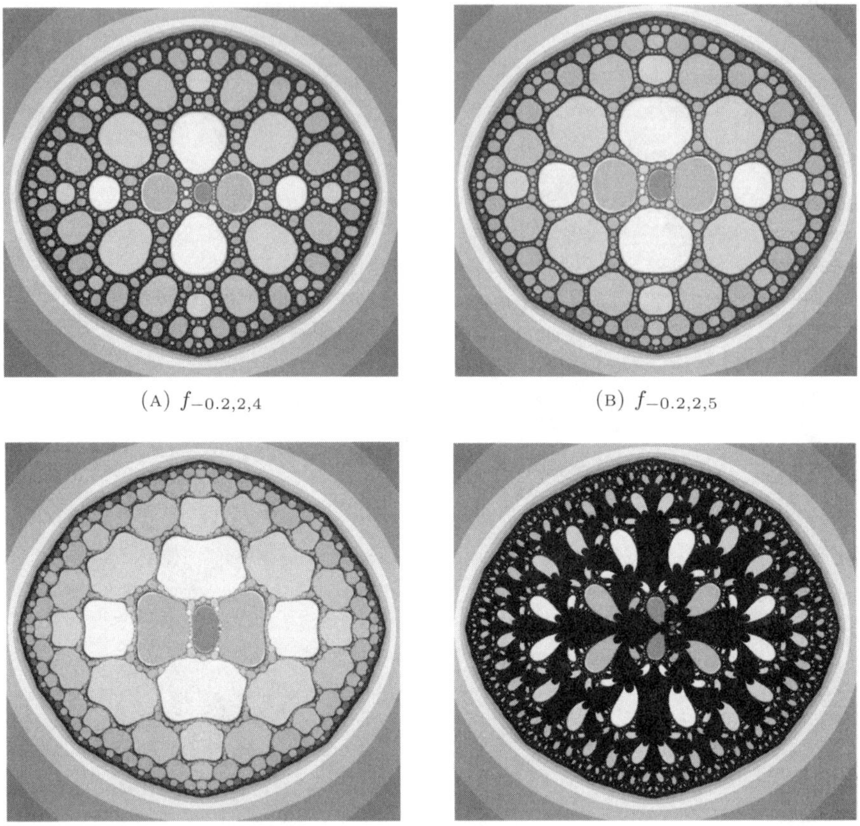

(A) $f_{-0.2,2,4}$ (B) $f_{-0.2,2,5}$

(C) $f_{-0.2,2,10}$ (D) $f_{-0.2,2}$

FIGURE 2. Sierpinski Julia sets and their limit when $m = 2$ and $a = -0.2$

Figure 3 shows examples when when $m = 2$ and $a = .0001$. Points in the plane are colored based on how many iterates it takes for the orbit to get large. Figures 3a, 3b, and 3c show the Julia sets of $f_{a,m,d}$ for $d = 5, 6$ and $d = 10$ respectively. In these cases, the Julia sets consist of uncountably many quasicircles around the origin. Figure 3d shows the Julia set (shown in black) for the limiting map $f_{a,m}$.

Next, we have a summary on the connectivity of Fatou components when the orbit of the free critical point is unbounded. In particular, some maps $f_{a,m,d}$ have a Fatou component that is not simply connected, but as we take the limit as d approaches infinity, the Fatou components of the limiting map $f_{a,m}$ are simply connected.

THEOREM 3.3. *(RAT) If $a \in \mathcal{C}_d^0$ (the Cantor set region), then there are connected components of the Fatou set that are infinitely connected. If $a \in \mathcal{C}_d^2$ (the McMullen domain), then there are connected components of the Fatou set that are doubly connected. If $a \in \mathcal{C}_d^n$, $n \geqslant 3$, all components of the Fatou set are simply connected.*

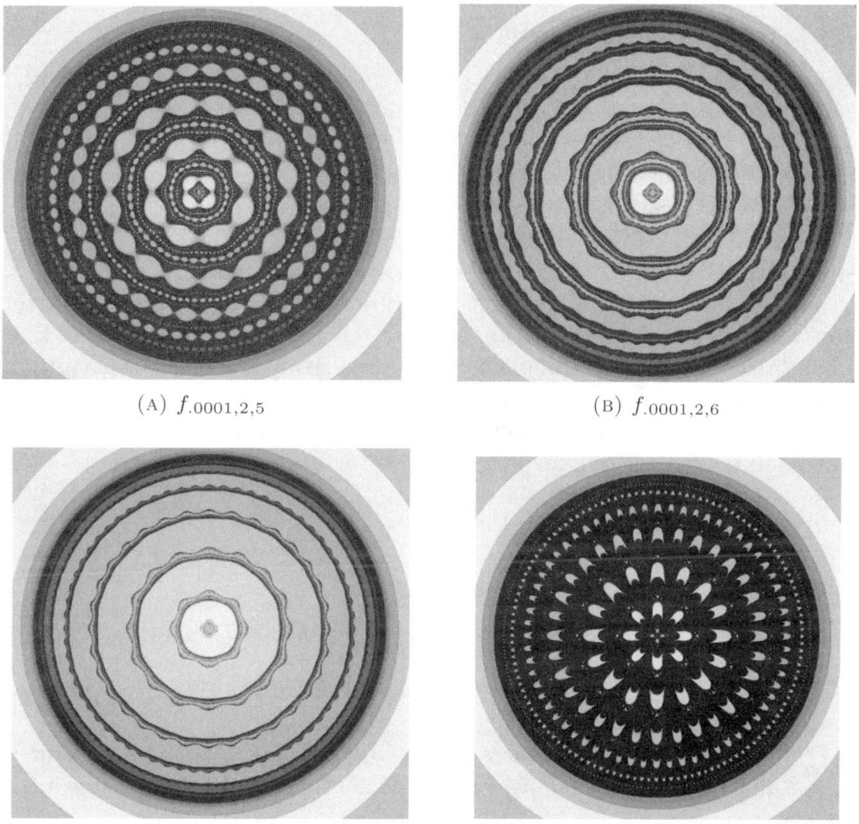

(A) $f_{.0001,2,5}$ (B) $f_{.0001,2,6}$

(C) $f_{.0001,2,10}$ (D) $f_{.0001,2}$

FIGURE 3. Julia sets when $m = 2$ and $a = .0001$

(EXP) All connected components of the Fatou set of $f_{a,m}$ are simply connected [**18**].

We conclude with a theorem that gives conditions on a under which the boundary of the component at infinity is a quasicircle.

THEOREM 3.4. *(RAT) For $|a| < a^* = m^{m-1}(d-m)^{d-m+1}/d^{d-m+1}$, the boundary of $A^*_{a,d}$ is a quasicircle* [**6, 31**].

(EXP) If $|a| < (m-1)^{m-1}/e^{m-1}$, then the boundary of A^*_a is a quasi-circle [**17**]. *Let $\mathcal{U}$ be the connected component of $\mathbb{C}\backslash\overline{\mathcal{C}^0}$. If $a \in \mathcal{U}$, then the boundary of A^*_a is a quasicircle. In particular, if $a \in \mathcal{C}^n$ for any $n \geqslant 2$, then the boundary of A^*_a is a quasicircle.* [**20**]

3.2. Bounded free critical orbit. In this section, we describe the dynamical plane in the case when the orbit of c_d or c_0 is bounded. We begin with a theorem about the connectivity of the Julia set.

THEOREM 3.5. *Suppose the free critical point has bounded orbit.*

(RAT) All connected components of the Fatou set of $f_{a,m,d}$ are simply connected and $J(f_{a,m,d})$ is connected.

(EXP) All connected components of the Fatou set of $f_{a,m}$ are simply connected and $J(f_{a,m}) \cup \{0\}$ is connected [**18**].

PROOF. (RAT) In [**23**], Hawkins proved that every Fatou component of $R_\lambda(z) = z^m + \lambda/z^{d-m}$ contained at most one critical point and $J(R_\lambda)$ is connected. Suppose that $J(f_{a,m,d})$ is not connected. Since $f_{a,m,d}$ has no Herman rings, to have a non-simply connected component of the Fatou set, there must be a component of the Fatou set that contains two different critical points (see [**29**]). Since the orbit of c_d is bounded, c_d cannot lie in a Fatou component with $0, -1/d$, or ∞. So at least two of $0, -1/d$, or ∞ lie in the same Fatou component U, say 0 and ∞ (the other two possibilities follow by the same argument). Then there exists a curve in U from 0 to ∞. Since the conjugation between $f_{a,m,d}$ and R_λ described in Equation 1.1 is conformal away from 0 and ∞, this implies that there is a curve from 0 to ∞ in the corresponding Fatou component for R_λ. However, this would imply that the Julia set for R_λ was disconnected, contradicting [**23**]. Hence, 0 and ∞ are in separate components for our maps as well, so the Julia set is connected. □

Next, we discuss parameter values for which $f_{a,m,d}$ and $f_{a,m}$ have a finite superattracting fixed point.

LEMMA 3.6. *Fix $m \geqslant 2$.*

(RAT) For $d > m$, if $a = m^{m-1}(1-\frac{m}{d})^{d-(m-1)}$, then $f_{a,m,d}$ has a superattracting fixed point at $c_d = (d-m)/(md)$.

(EXP) If $a = m^{m-1}e^{-m}$, then $f_{a,m}$ has a superattracting fixed point at $c_0 = 1/m$.

(LIM) As $d \to \infty$, the parameter values at which $f_{a,m,d}$ has a superattracting fixed point at c_d converge to the parameter value at which $f_{a,m}$ has a superattracting fixed point at c_0.

PROOF. (RAT) Fix m. First, we show that the critical point c_d is fixed when $a = m^{m-1}(1-\frac{m}{d})^{d-(m-1)}$.

$$\begin{aligned} f_{a,m,d}(c_d) &= f_{a,m,d}\left(\frac{d-m}{md}\right) \\ &= m^{m-1}\left(1-\frac{m}{d}\right)^{d-(m-1)}\left(\frac{d-m}{md}\right)^m\left(1+\frac{1}{d(\frac{d-m}{md})}\right)^d \\ &= \frac{d-m}{md}. \end{aligned}$$

(EXP) Next, we verify that the critical point $c_0 = 1/m$ is fixed for $f_{a,m}$ when $a = m^{m-1}e^{-m}$.

$$f_{a,m}(c_0) = f_{a,m}\left(\frac{1}{m}\right) = m^{m-1}e^{-m}\left(\frac{1}{m}\right)^m e^{\frac{1}{\frac{1}{m}}} = \frac{1}{m}.$$

(LIM) Taking the limit of parameter values gives

$$\begin{aligned} \lim_{d\to\infty} m^{m-1}\left(1-\frac{m}{d}\right)^{d-(m-1)} &= \lim_{d\to\infty} m^{m-1}\left(1-\frac{m}{d}\right)^d\left(\frac{d}{d-m}\right)^{m-1} \\ &= m^{m-1}e^{-m}. \end{aligned}$$

□

We illustrate some Julia sets with superattracting fixed points in Figure 4, where $m = 2$ and the parameter a is calculated as in Lemma 3.6. Figures 4a, 4b, and 4c show the Julia sets of $f_{a,m,d}$ for $d = 2, 4$, and $d = 10$ respectively, and Figure 4d shows the Julia set for the limiting map $f_{a,m}$. Black regions indicate points which iterate to the superattracting fixed point. Other colors indicate information about how many iterates it takes for the orbit to get large. Any point on the boundary of where a black region meets a colored/grey region is in the Julia set. In each figure, the fixed finite critical point lies in the largest black region on the right.

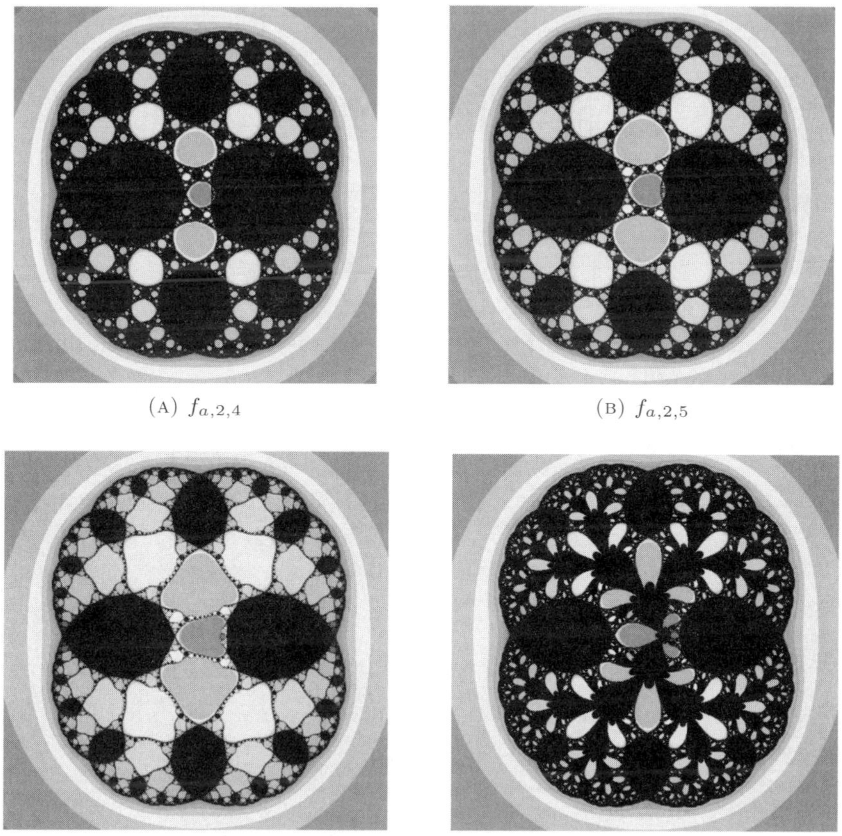

(A) $f_{a,2,4}$ (B) $f_{a,2,5}$

(C) $f_{a,2,10}$ (D) $f_{a,2}$

FIGURE 4. Julia sets when $f_{a,2,d}$ or $f_{a,2}$ has a finite superattracting fixed point, where a is calculated from Lemma 3.6.

Next, we show that every attracting cycle must contain a point that is bounded away from the origin. This result will be used to relate attracting cycles of $f_{a,m,d}$ and $f_{a,m}$.

LEMMA 3.7. *For $d \geqslant m + 2$, if $z_1, z_2, \ldots, z_n$ is an attracting periodic orbit for $f_{a,m,d}$, then there exists an i such that*

$$|z_i| > \frac{d - (m+1)}{(m+1)d} \geqslant \frac{1}{(m+1)(m+2)}.$$

PROOF. We begin by expressing the derivative of $f_{a,m,d}$ in terms of $f_{a,m,d}$.

$$f'_{a,m,d}(z) = a\left[mz^{m-1}\left(1+\frac{1}{dz}\right)^{d} - z^{m-2}\left(1+\frac{1}{dz}\right)^{d-1}\right] \tag{3.1}$$

$$= f_{a,m,d}(z)\frac{(mdz+m-d)}{z(dz+1)}.$$

Thus for any point z_i in the cycle,

$$\left|\prod_{i=1}^{n} f'_{a,m,d}(z_i)\right| = \left|\prod_{i=1}^{n} f_{a,m,d}(z_i)\frac{\prod_{i=1}^{n}(mdz_i+m-d)}{\prod_{i=1}^{n} z_i \prod_{i=1}^{n}(dz_i+1)}\right|$$

$$= \left|\prod_{i=1}^{n} z_{i+1}\frac{\prod_{i=1}^{n}(mdz_i+m-d)}{\prod_{i=1}^{n} z_i \prod_{i=1}^{n}(dz_i+1)}\right|$$

Since $z_1 = z_{n+1}$, we have $\prod_{i=1}^{n} z_{i+1}/z_i = 1$. Thus

$$\left|\prod_{i=1}^{n} f'_{a,m,d}(z_i)\right| = \prod_{i=1}^{n}\left|m - \frac{d}{dz_i+1}\right|.$$

The cycle is attracting, so the product must be less than one. This implies that there must be a term that is less than 1, so for some i,

$$\left|m - \frac{d}{dz_i+1}\right| < 1.$$

Note that if k is a point for which $|m-k| < 1$, then $|k| < m+1$. This implies that

$$\left|\frac{d}{dz_i+1}\right| < m+1,$$

so

$$d < (m+1)|dz_i+1| < (m+1)d|z_i| + (m+1).$$

This implies

$$|z_i| > \frac{d-(m+1)}{d(m+1)} \geqslant \frac{1-\frac{m+1}{m+2}}{m+1} = \frac{1}{(m+1)(m+2)}$$

since $d \geqslant m+2$. □

Recall that ∞ is a superattracting fixed point for $f_{a,m}$ and for $f_{a,m,d}$ for all $d \geqslant m+2$. Lemma 3.7 and Lemma 2.2 allow us to relate finite attracting cycles of $f_{a,m,d}$ to finite attracting cycles of $f_{a,m}$.

THEOREM 3.8. *Fix $m \geqslant 2$ and $a \in \mathbb{C}^*$.*

(1) *If $f_{a,m}$ has a finite attracting cycle of period k on $\mathbb{C}^*$, then there exists $d_0 \in \mathbb{N}$ such that, for all $d \geqslant d_0$, $f_{a,m,d}$ has a finite attracting cycle of period k.*
(2) *If $f_{a,m,d}$ has a finite attracting cycle of period k on $\mathbb{C}_\infty$ for infinitely many $d \in \mathbb{N}$, then $f_{a,m}$ has a finite cycle of period dividing k that is either attracting or neutral.*

PROOF. (1): This follows from uniform convergence on compact sets and Hurwitz's Theorem.

(2): For $a \in \mathbb{C}^*$, suppose that $f_{a,m,d}$ has an attracting cycle of period k for infinitely many $d \in \mathbb{N}$. By Lemma 3.7, for each such $d \geqslant 3$, there exists a periodic point z_d of period k such that $|z_d| > \frac{1}{(m+1)(m+2)}$. Since z_d is a point in a finite attracting cycle, it cannot be in $A^*_{a,d}$. Thus, Lemma 2.2 implies that $|z_d| < r_{m+2}$. Since the infinite sequence z_d with $d \in \mathbb{N}$ is contained in a compact subset of the Riemann sphere (namely $\left\{z \in \mathbb{C} : \frac{1}{(m+1)(m+2)} \leqslant |z| \leqslant r_{m+2}\}\right\}$), it has a finite, nonzero accumulation point z^*. By the convergence of $f_{a,m,d}$ to $f_{a,m}$, we have $f^k_{a,m}(z^*) = z^*$ and $|(f^k_{a,m})'(z^*)| \leqslant 1$. □

LEMMA 3.9. *Let $A(z^*)$ denote the attracting basin of a point $z^* \in \mathbb{C}^*_\infty$ that is an attracting periodic point under $f_{a,m}$. Let z^*_d be ∞ if z^* is ∞ or an attracting periodic point for $f_{a,m,d}$ from Theorem 3.8 (1) if z^* is finite. Let $A(z^*_d)$ be the attracting basin of z^*_d under $f_{a,m,d}$. Then, for every $K \subset A(z^*)$ that is compact in the spherical metric, there exists $d_1 \in \mathbb{N}$ such that $K \subset A(z^*_d)$ for all $d \geqslant d_1$.*

PROOF. This follows immediately from Lemma 3(b) of [**25**], after conjugating the families by $z \mapsto -1/z$. □

Similarly, after conjugating our families by $z \mapsto -1/z$, we see that ∞ is in the Julia set of $g_{\lambda,m}(z) = \lambda z^m e^z$ because it is a transcendental entire function. Since the conjugation does not change the Hausdorff metric (when the Hausdorff metric is defined in terms of the spherical metric), the following Theorem 3.10 from [**25**] implies that, if $a \in \mathbb{C}^*_\infty$ is in a capture component $\mathcal{C}^n$ or $f_{a,m}$ has a finite attracting cycle, then $J(f_{a,m,d})$ converges to $J(f_{a,m})$ in the Hausdorff metric as $d \to \infty$.

THEOREM 3.10 (Krauskopf, Kriete [**25**]). *Suppose g_d and g are meromorphic functions from $\mathbb{C}$ to $\mathbb{C}_\infty$ such that g_d converges uniformly on compact sets of $\mathbb{C}$ to g as $d \to \infty$. If $F(g)$ is the union of basins of attracting periodic orbits and $\infty \in J(g)$, then $J(g_d)$ converges to $J(g)$ in the Hausdorff metric as $d \to \infty$.*

4. Parameter Plane

In this section, we describe results on parameter spaces of $f_{a,m,d}$ and $f_{a,m}$. We begin with an illustration of some of the relevant parameter planes. Figure 5 shows four parameter planes when $m = 2$. Figures 5a, 5b, and 5c show the parameter planes of $f_{a,2,d}$ for $d = 2, 4$, and $d = 10$ respectively, and Figure 5d shows the parameter plane for the limiting map $f_{a,2}$. Black regions indicate parameters for which the free critical point has bounded orbit. Other colors/shades indicate information about how many iterates it takes for the free critical point to get large. Figure 6 illustrates four different parameter planes for $f_{a,m}$ when $m = 2, 3, 4$, and 10.

The theory of polynomial-like mappings was developed by Douady and Hubbard [**13**]. In 2000, McMullen showed that, for generic families of rational maps, small copies of the Mandelbrot set are dense in the bifurcation locus [**27**]. The ideas developed in [**13, 26**] have previously been applied to S_λ in [**31, 32**] and $g_{\lambda,m}$ for $m = 1, 2$ [**16, 21**]. Since the proofs are local arguments, they extend to the maps discussed here.

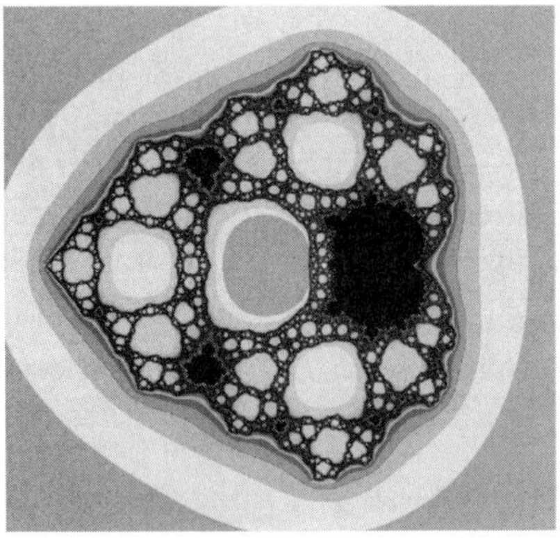

(A) $f_{a,2,4}$

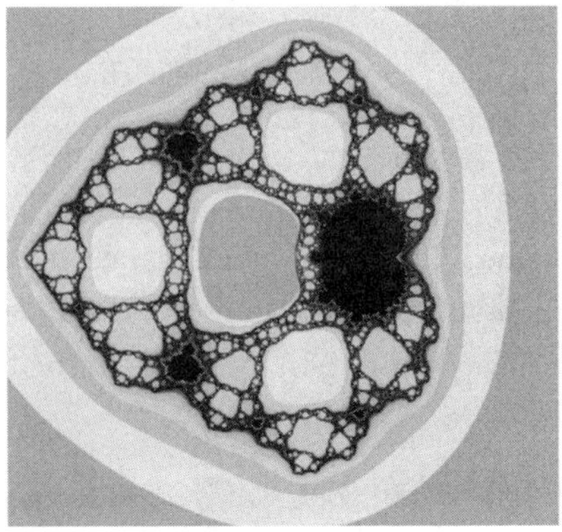

(B) $f_{a,2,5}$

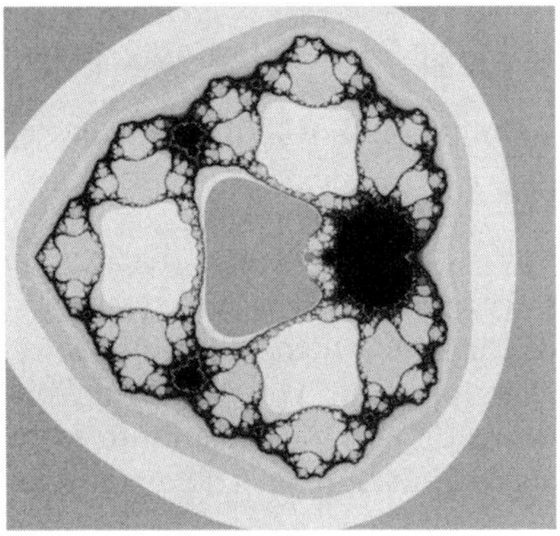

(C) $f_{a,2,10}$

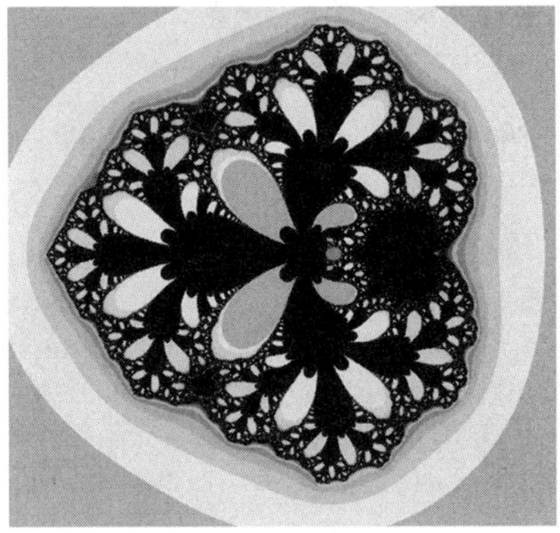

(D) $f_{a,2}$

FIGURE 5. Parameter planes when $m = 2$.

THEOREM 4.1. *(RAT)* $\mathbb{C}^* \setminus (\cup_{n=0}^{\infty} \mathcal{C}_d^n)$ *contains infinitely many copies of the Mandelbrot set.* [**31**, **32**]

(EXP) $\mathbb{C}^* \setminus (\cup_{n=0}^{\infty} \mathcal{C}^n)$ *contains infinitely many copies of the Mandelbrot set.* [**16**, **21**]

To understand the capture components themselves, we first note that there are no parameters a for which $f_{a,m,d}(c_d)$ (resp. $f_{a,m}(c_0)$) lies in the superattracting fixed component at infinity and c_d (resp. c_0) does not.

THEOREM 4.2. *(RAT)* $\mathcal{C}_d^1 = \varnothing$. [**31**]
(EXP) $\mathcal{C}^1 = \varnothing$. [**17**]

Also, the McMullen domain does not appear in every parameter space. Specifically, $\mathcal{C}_d^2$ is empty when $m = 2, d = 4$ [**4**], so Figure 5a does not have a McMullen domain. The McMullen domain is easiest to see in Figure 5c as the very small dark orange/grey region intersecting the real axis.

THEOREM 4.3. *(RAT) If* $1/(d-m) + 1/m < 1$, *then* $\mathcal{C}_d^2 \cup \{0\}$ *is a simply connected domain. For* $m = 2, d = 4$, $\mathcal{C}_d^2 = \varnothing$. [**4**, **6**, **12**, **31**]

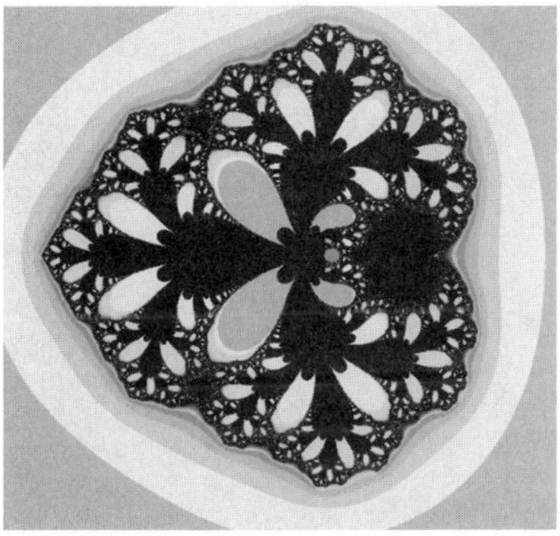

(A) $f_{a,2}$

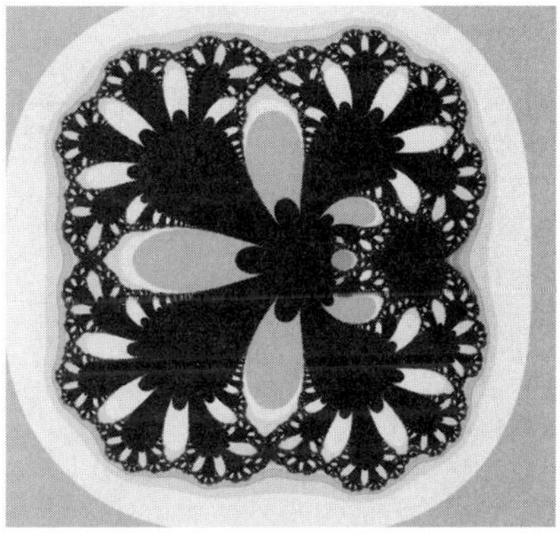

(B) $f_{a,3}$

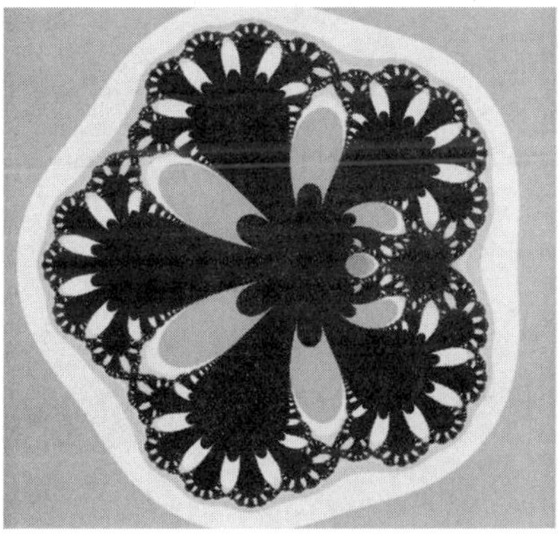

(C) $f_{a,4}$

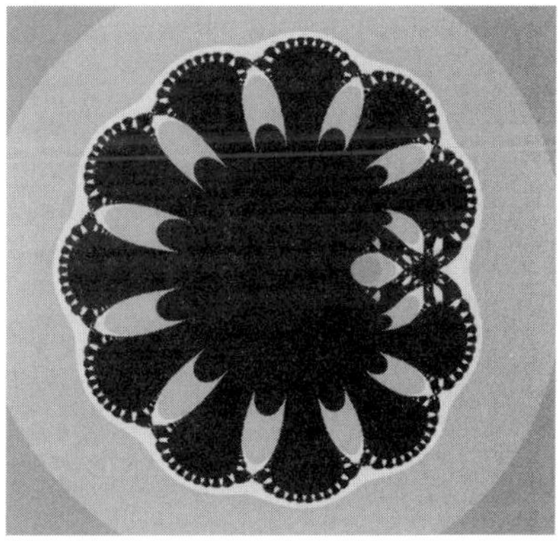

(D) $f_{a,10}$

FIGURE 6. Parameter planes for $f_{a,m}$. Figure (A) shows the window $[-1,1]\times[-1,1]$, Figure (B) shows the window $[-1.6,1.1]\times[-1.3,1.3]$, Figure (C) shows the window $[-4.7,3]\times[-3.6,3.6]$, and Figure (D) shows $[200,000,120,000]\times[-150,000,150,000]$

(EXP) The set $\mathcal{C}^2$ contains a set approaching the asymptotic value at 0 from the right. All connected components of $\mathcal{C}^2$ are simply connected. [**17, 18**]

The colored/grey region extending to infinity is the set $\mathcal{C}_d^0$ in Figures 5a, 5b, and 5c and $\mathcal{C}^0$ in Figure 5d.

THEOREM 4.4. *(RAT) $\mathcal{C}_d^0 \cup \{\infty\}$ is a simply connected domain. It contains the neighborhood* $\left\{a \in \mathbb{C}_\infty : |a| > \max\left\{\left(\frac{d}{d-1}\right)^d, m^m\left(\frac{d-m}{d}\right)^{d-m}\right\}\right\}$. *In addition, $\mathcal{C}_d^0$ is bounded away from 0.*

(EXP) $\mathcal{C}^0 \cup \{\infty\}$ is a simply connected domain. [**18**] *It contains the neighborhood $\{a \in \mathbb{C}_\infty : |a| > \max\{e, (m/e)^m\}\}$. It is bounded away from the asymptotic value at 0.* [**17**] *(See* [**17**] *also for explicit bounds.)*

(LIM) $\lim_{d\to\infty}\left(\frac{d}{d-1}\right)^d = e$ *and* $\lim_{d\to\infty} m^m\left(\frac{d-m}{d}\right)^{d-m} = \left(\frac{m}{e}\right)^m$

PROOF. (RAT) Suppose $a \in \mathbb{C}$ such that

$$|a| > \max\left\{\left(\frac{d}{d-1}\right)^d, m^m\left(\frac{d-m}{d}\right)^{d-m}\right\}.$$

By Theorem 4.2, $f_{a,m,d}(c_d) \in A^*_{a,d}$ implies that $c_d \in A^*_{a,d}$ and $a \in \mathcal{C}^0_d$. Since $|a| > m^m((d-m)/d)^{d-m}$, we have

$$\begin{aligned}
|f_{a,m,d}(c_d)| &= |a|\left(\frac{d-m}{md}\right)^m\left(1+\frac{1}{(d-m)/m}\right)^d \\
&= |a|\left(\frac{d-m}{md}\right)^m\left(\frac{d}{d-m}\right)^d \\
&= |a|\left(\frac{1}{m}\right)^m\left(\frac{d}{d-m}\right)^{d-m} > 1.
\end{aligned}$$

Since $|a| > \left(\frac{d}{d-1}\right)^d$, we have $r_d = 1$ and $f_{a,m,d}(c_d) \in A^*_{a,d}$ by Lemma 2.2.

When $m = 2$ and $d = 4$, the connectedness locus of the Julia set is contained in a disk around the origin [**3**]. Using Theorem 4.3, we know that when $1/(d-m)+1/m < 1$, then $\mathcal{C}^2_d \cup \{0\}$ is a simply connected domain, so $\mathcal{C}^0_d$ is bounded away from 0.

(LIM) We have

$$\lim_{d\to\infty}\left(\frac{d}{d-1}\right)^d = \lim_{d\to\infty}\left(\frac{d-1}{d}\right)^{-d} = e,$$

and

$$\begin{aligned}
\lim_{d\to\infty} m^m\left(\frac{d-m}{d}\right)^{d-m} &= \lim_{d\to\infty} m^m\left(\frac{d-m}{d}\right)^{d-m} \\
&= \lim_{d\to\infty} m^m\left(1+\frac{-m}{d}\right)^d\left(1+\frac{-m}{d}\right)^{-m} \\
&= \left(\frac{m}{e}\right)^m
\end{aligned}$$

□

Colored/grey regions in Figures 5a, 5b, and 5c that are bounded and don't contain the origin represent capture components $\mathcal{C}^n_d$ for $n \geqslant 3$.

THEOREM 4.5. *(RAT) For $n \geqslant 3$, the capture component $\mathcal{C}^n_d$ is nonempty and consists of finitely many connected components* [**7, 30, 32**].

(EXP) For $n \geqslant 3$, $\mathcal{C}^n$ has infinitely many connected components extending to 0. All the connected components of $\mathcal{C}^n$ are simply connected [**17, 18, 20**].

We note that there are fairly explicit descriptions of $\mathcal{C}^0$, $\mathcal{C}^2$, and $\mathcal{C}^3$ for the family $g_{\lambda,m}(z) = \lambda z^m e^z$ in [**17**], but we do not add them here.

THEOREM 4.6. *Fix $m \geqslant 2$ and $a \in \mathbb{C}^*$. If $a \in \mathcal{C}^n$ for some $n \in \mathbb{N}$, then there exists $d_0 \in \mathbb{N}$ such that, for all $d \geqslant d_0$, $a \in \bigcup_{k=0}^{n} \mathcal{C}^k_d$.*

PROOF. Fix $m \geqslant 2$ and $a \in \mathbb{C}^*$. Suppose $a \in \mathcal{C}^n$ for some $n \in \mathbb{N}$. Then there exists a path γ from $f_{a,m}^n(c_0)$ to ∞ contained in A_a^*. Moreover, there exists a compact neighborhood K of $f_{a,m}^n(c_0)$ contained in A_a^*. Then $K \cup \gamma$ is a subset of A_a that is compact in the spherical metric. By Lemma 3.9, there exists $d_1 \in \mathbb{N}$ such that $K \cup \gamma \subset A_{a,d}$ for all $d \geqslant d_1$. Since $K \cup \gamma$ is connected and contains ∞, we can conclude that $K \cup \gamma \subset A_{a,d}^*$. By the convergence of $f_{a,m,d}$ to $f_{a,m}$, there exists $d_2 \in \mathbb{N}$ such that $f_{a,m,d}^n(c_d) \in K$ for all $d \geqslant d_2$. Taking $d_0 \geqslant \max\{d_1, d_2\}$, we find that $f_{a,m,d}^n(c_d) \in A_{a,d}^*$ for all $d \geqslant d_0$. Hence, for each $d \geqslant d_0$, $a \in \mathcal{C}_d^k$ for some $0 \leqslant k \leqslant n$. □

It would be easy to strengthen Theorem 4.6 to show that $a \in \mathcal{C}_d^n$ for all sufficiently large $d \in \mathbb{N}$ (and to prove a converse statement) if Lemma 3.9 had a second part similar to Theorem 3.8, stating that compact sets that are eventually contained in the attracting basins for $f_{a,m,d}$ are then also contained in a corresponding basin of $f_{a,m}$. One difficulty in proving this kind of statement is that a point might be in $A_{a,d}$ for all sufficiently large $d \in \mathbb{N}$, but still require larger and larger iterates to reach the immediate basin $A_{a,d}^*$ as d increases. We conclude by giving some evidence for this type of behavior in our families.

We note that for $f_{a,m,d}$, it is straightforward to solve explicitly for parameter values for the center of the capture component $\mathcal{C}_d^3$. For any $2 \leqslant m \leqslant d-2$, if $a = -\frac{m^m}{d}\left(\frac{d-m}{d}\right)^{d-m}$, then $c_d \to -1/d$. For a fixed $m \geqslant 2$, note that these centers converge to 0 as $d \to \infty$. In Figure 5, the capture components $\mathcal{C}_2^3$, $\mathcal{C}_3^3$, and $\mathcal{C}_4^3$ are visible as the first large component to the left of the origin. In Figure 5 (C), we can see the part of the capture component on the negative real axis starting to pinch toward the origin. In Figure 5 (D), this same region of the negative real axis no longer appears to intersect capture components. Theorem 4.7 rigorously describes this observation for even $m \geqslant 2$ and an interval in the negative real axis.

THEOREM 4.7. *For each even $m \geqslant 2$ and for all $a \in \mathbb{R}$ such that $-(m/e)^{m-1} < a < 0$, the parameter a is not in the capture component $\mathcal{C}^n$ for all $n \in \mathbb{N}$.*

PROOF. Fix an even $m \geqslant 2$. Fix $a \in \mathbb{R}$ such that $-(m/e)^{m-1} < a < 0$. We use induction to show that $-(e^{-m/e})^{n-1}(e/m) < f_{a,m}^n(1/m)$ for all $n \in \mathbb{N}$. Moreover, since m is even and a is a negative real number, $f_{a,m}(x) < 0$ for all nonzero real numbers x. In particular, we have $f_{a,m}^n(1/m) < 0$ for all $n \in \mathbb{N}$. Thus, the orbit $f_{a,m}^n(1/m)$ converges to 0, not ∞, so $a \notin \mathcal{C}^n$ for any $n \in \mathbb{N}$.

For the base case, the assumptions that m is even and $a > -(m/e)^{m-1}$ imply that

$$f_{a,m}(1/m) > -\left(\frac{m}{e}\right)^{m-1}\left(\frac{1}{m}\right)^m e^m = -\frac{e}{m}.$$

For the induction step, suppose $-(e^{-m/e})^{n-1}(e/m) < f_{a,m}^n(1/m) < 0$ for some integer $n \geqslant 1$. If m is even, then $g(x) = x^{m-1}e^{1/x}$ is increasing for $x < 0$. Thus, if $-e/m < x < 0$, then

$$\begin{aligned} f_{a,m}(x) &= axg(x) \\ &> -\left(\frac{m}{e}\right)^{m-1} xg(-e/m) \\ &= e^{-m/e}x. \end{aligned}$$

In particular, we have $f_{a,m}^{n+1}(1/m) > e^{-m/e} f_{a,m}^{n}(1/m) > -(e^{-m/e})^n (e/m)$. □

In [**8**–**10**], Devaney studies patterns of Mandelbrot sets and capture components of $R_\lambda(z) = z^m + \lambda/z^{d-m}$ for even $m \geqslant 2$ and odd $d \geqslant m+3$. In particular, in [**10**], he constructs infinitely many Mandelbrot sets and capture components along the negative real axis. Then he observes that, for parameters not in either of these types of sets, the free critical orbits have an infinite itinerary that is neither approaching an attracting periodic cycle nor escaping.

CONJECTURE 4.8 (Devaney [**10**]). *Let $m \geqslant 2$ be even and $d \geqslant m+3$ be odd. The set of parameters λ on the negative real axis such that the orbits of the free critical points under R_λ are neither approaching an attracting periodic cycle nor escaping is a set of singletons.*

Note that when $m \geqslant 2$ is even, parameters on the negative real axis for R_λ correspond through the quasiconjugacy to parameters on the negative real axis for $f_{a,m,d}$. Hence, it follows that there is a similar pattern of Mandelbrot sets and capture components for $f_{a,m,d}$, for even $m \geqslant 2$ and odd $d \geqslant m+3$. We note that Theorem 4.7 implies that the analog of Conjecture 4.8 for the family $f_{a,m,d}$ does not pass to the limit. When $m \geqslant 2$ is even, the negative real axis of the parameter space for $f_{a,m}$ contains an entire interval of parameters for which the orbit of the free critical point is neither approaching an attracting periodic cycle nor escaping.

References

[1] I. N. Baker, *The domains of normality of an entire function*, Ann. Acad. Sci. Fenn. Ser. A I Math. **1** (1975), no. 2, 277–283. MR0402044

[2] I. N. Baker and P. Domínguez, *Some connectedness properties of Julia sets*, Complex Variables Theory Appl. **41** (2000), no. 4, 371–389, DOI 10.1080/17476930008815263. MR1785150

[3] Paul Blanchard, Robert L. Devaney, Daniel M. Look, Monica Moreno Rocha, Pradipta Seal, Stefan Siegmund, and David Uminsky, *Sierpinski carpets and gaskets as Julia sets of rational maps*, Dynamics on the Riemann sphere, Eur. Math. Soc., Zürich, 2006, pp. 97–119, DOI 10.4171/011-1/5. MR2348957

[4] Paul Blanchard, Robert L. Devaney, Daniel M. Look, Pradipta Seal, and Yakov Shapiro, *Sierpinski-curve Julia sets and singular perturbations of complex polynomials*, Ergodic Theory Dynam. Systems **25** (2005), no. 4, 1047–1055, DOI 10.1017/S0143385704000380. MR2158396

[5] Clara Bodelón, Robert L. Devaney, Michael Hayes, Gareth Roberts, Lisa R. Goldberg, and John H. Hubbard, *Dynamical convergence of polynomials to the exponential*, J. Differ. Equations Appl. **6** (2000), no. 3, 275–307, DOI 10.1080/10236190008808229. MR1785056

[6] Robert L. Devaney, *Structure of the McMullen domain in the parameter planes for rational maps*, Fund. Math. **185** (2005), no. 3, 267–285, DOI 10.4064/fm185-3-5. MR2161407

[7] Robert L. Devaney, *The McMullen domain: satellite Mandelbrot sets and Sierpinski holes*, Conform. Geom. Dyn. **11** (2007), 164–190, DOI 10.1090/S1088-4173-07-00166-X. MR2346215

[8] Robert L. Devaney, *A Mandelpinski maze for rational maps of the form $z^n + \lambda/z^d$*, Indag. Math. (N.S.) **27** (2016), no. 5, 1042–1058, DOI 10.1016/j.indag.2015.10.004. MR3573746

[9] Robert L. Devaney, *Mandelpinski spokes in the parameter planes of rational maps*, J. Difference Equ. Appl. **22** (2016), no. 2, 330–342, DOI 10.1080/10236198.2015.1092525. MR3474986

[10] Robert L. Devaney, *Mandelpinski structures in the parameter planes of rational maps*, Ergodic theory, dynamical systems, and the continuing influence of John C. Oxtoby, Contemp. Math., vol. 678, Amer. Math. Soc., Providence, RI, 2016, pp. 133–150. MR3589819

[11] Robert L. Devaney and Michał Krych, *Dynamics of* $\exp(z)$, Ergodic Theory Dynam. Systems **4** (1984), no. 1, 35–52, DOI 10.1017/S014338570000225X. MR758892

[12] Robert L. Devaney, Daniel M. Look, and David Uminsky, *The escape trichotomy for singularly perturbed rational maps*, Indiana Univ. Math. J. **54** (2005), no. 6, 1621–1634, DOI 10.1512/iumj.2005.54.2615. MR2189680

[13] Adrien Douady and John Hamal Hubbard, *On the dynamics of polynomial-like mappings*, Ann. Sci. École Norm. Sup. (4) **18** (1985), no. 2, 287–343. MR816367

[14] A. È. Erëmenko and M. Yu. Lyubich, *The dynamics of analytic transformations* (Russian), Algebra i Analiz **1** (1989), no. 3, 1–70; English transl., Leningrad Math. J. **1** (1990), no. 3, 563–634. MR1015124

[15] A. È. Erëmenko and M. Yu Lyubich, *Dynamical properties of some classes of entire functions*, Ann. Inst. Fourier (Grenoble) **42** (1992), no. 4, 989–1020. MR1196102

[16] Núria Fagella, *Limiting dynamics for the complex standard family*, Internat. J. Bifur. Chaos Appl. Sci. Engrg. **5** (1995), no. 3, 673–699, DOI 10.1142/S0218127495000521. MR1345989

[17] Núria Fagella and Antonio Garijo, *Capture zones of the family of functions* $\lambda z^m \exp(z)$, Internat. J. Bifur. Chaos Appl. Sci. Engrg. **13** (2003), no. 9, 2623–2640, DOI 10.1142/S0218127403008120. MR2013117

[18] Núria Fagella and Antonio Garijo, *The parameter planes of* $\lambda z^m \exp(z)$ *for* $m \geqslant 2$, Comm. Math. Phys. **273** (2007), no. 3, 755–783, DOI 10.1007/s00220-007-0265-8. MR2318864

[19] Joanna Furno, Jane Hawkins, and Lorelei Koss, *Rational families converging to a family of exponential maps*, J. Fractal Geom. **6** (2019), no. 1, 89–108, DOI 10.4171/JFG/70. MR3910543

[20] Antonio Garijo, Xavier Jarque, and Mónica Moreno Rocha, *Joining polynomial and exponential combinatorics for some entire maps*, Publ. Mat. **54** (2010), no. 1, 113–136, DOI 10.5565/PUBLMAT_54110_06. MR2603591

[21] Antonio Garijo, Xavier Jarque, and Jordi Villadelprat, *An effective algorithm to compute Mandelbrot sets in parameter planes*, Numer. Algorithms **76** (2017), no. 2, 555–571, DOI 10.1007/s11075-017-0270-8. MR3704881

[22] Jane Hawkins, *Lebesgue ergodic rational maps in parameter space*, Internat. J. Bifur. Chaos Appl. Sci. Engrg. **13** (2003), no. 6, 1423–1447, DOI 10.1142/S021812740300731X. MR1992056

[23] Jane M. Hawkins, *Proof of a folklore Julia set connectedness theorem and connections with elliptic functions*, Conform. Geom. Dyn. **17** (2013), 26–38, DOI 10.1090/S1088-4173-2013-00252-X. MR3019711

[24] Masashi Kisaka, *On the local connectivity of the boundary of unbounded periodic Fatou components of transcendental functions*, Sūrikaisekikenkyūsho Kōkyūroku **988** (1997), 113–119. Complex dynamical systems and related areas (Japanese) (Kyoto, 1996). MR1605767

[25] Bernd Krauskopf and Hartje Kriete, *Hausdorff convergence of Julia sets*, Bull. Belg. Math. Soc. Simon Stevin **6** (1999), no. 1, 69–76. MR1674702

[26] Curt McMullen, *Automorphisms of rational maps*, Holomorphic functions and moduli, Vol. I (Berkeley, CA, 1986), Math. Sci. Res. Inst. Publ., vol. 10, Springer, New York, 1988, pp. 31–60, DOI 10.1007/978-1-4613-9602-4_3. MR955807

[27] Curtis T. McMullen, *The Mandelbrot set is universal*, The Mandelbrot set, theme and variations, London Math. Soc. Lecture Note Ser., vol. 274, Cambridge Univ. Press, Cambridge, 2000, pp. 1–17. MR1765082

[28] Shunsuke Morosawa, *Dynamical convergence of a certain polynomial family to* $f_a(z) = z + e^z + a$, Ann. Acad. Sci. Fenn. Math. **40** (2015), no. 1, 449–463, DOI 10.5186/aasfm.2015.4028. MR3329154

[29] Franz Peherstorfer and Christoph Stroh, *Connectedness of Julia sets of rational functions*, Comput. Methods Funct. Theory **1** (2001), no. 1 [On table of contents: 2002], 61–79, DOI 10.1007/BF03320977. MR1931603

[30] P. Roesch, *On capture zones for the family* $f_\lambda(z) = z^2 + \lambda/z^2$, Dynamics on the Riemann sphere, Eur. Math. Soc., Zürich, 2006, pp. 121–129, DOI 10.4171/011-1/6. MR2348958

[31] Norbert Steinmetz, *On the dynamics of the McMullen family* $R(z) = z^m + \lambda/z^l$, Conform. Geom. Dyn. **10** (2006), 159–183, DOI 10.1090/S1088-4173-06-00149-4. MR2261046

[32] Norbert Steinmetz, *Sierpiński curve Julia sets of rational maps*, Comput. Methods Funct. Theory **6** (2006), no. 2, 317–327, DOI 10.1007/BF03321617. MR2291139

[33] Yingqing Xiao and Weiyuan Qiu, *The rational maps* $F_\lambda(z) = z^m + \lambda/z^d$ *have no Herman rings*, Proc. Indian Acad. Sci. Math. Sci. **120** (2010), no. 4, 403–407, DOI 10.1007/s12044-010-0044-x. MR2761768

[34] Fei Yang, *Rational maps without Herman rings*, Proc. Amer. Math. Soc. **145** (2017), no. 4, 1649–1659, DOI 10.1090/proc/13336. MR3601556

Department of Mathematics, University of Houston, 3551 Cullen Blvd., Room 641, Philip Guthrie Hoffman Hall, Houston, TX 77204, USA
Email address: jfurno@math.uh.edu

Department of Mathematics and Computer Science, Dickinson College, P. O. Box 1773, Carlisle, Pennsylvania 17013, USA
Email address: koss@dickinson.edu

Contemporary Mathematics
Volume **736**, 2019
https://doi.org/10.1090/conm/736/14837

Stability of Cantor Julia sets in the space of iterated elliptic functions

Jane Hawkins

ABSTRACT. Starting with a real lattice $\Lambda \subset \mathbb{C}$ that is of one of several shapes, and the Weierstrass elliptic $\wp$ function with poles at lattices points, which we denote by $\wp_\Lambda$, we show that there are many maps of the form $a\wp_\Lambda + b$ that are stable under perturbation of all parameters (Λ, a, b) and have Cantor Julia sets. We use this and other stability results to describe moduli space of order 2 elliptic functions with poles at the lattice points.

1. Introduction

There is a well established theory of the dynamics of iterated meromorphic functions [**1**, **2**, **6**], and elliptic functions provide an interesting class of examples. An elliptic function is meromorphic and periodic with respect to a lattice $\Lambda \subset \mathbb{C}$; in particular the author and others have shown that the dynamics depend on the lattice Λ, both its shape and size, and also on the function, (see [**3**],[**8**]–[**13**],[**15**] – [**18**] for example). In this paper we discuss stability properties and moduli space of a large family of elliptic functions.

The building blocks of all elliptic functions are the Weierstrass elliptic $\wp$ function and its derivative. In this paper we explore the moduli space of:

$$\mathfrak{D} = \{\text{order 2 elliptic functions with double poles at lattice points}\}.$$

We let $\wp$ denote the classical Weierstrass $\wp$ function, which will be defined and discussed in some detail below. The starting point of our study is that for any given lattice Λ, $\wp \in \mathfrak{D}$, and so are the maps $\wp + b$, and $a\wp + b$, for complex constants $a \neq 0$ and b. It is a classical result, as shown for example in [**7**], and proved in Proposition 2.3 below, that this is all that can occur.

A lattice Λ is a group of complex numbers generated by two linearly independent non-zero vectors in $\mathbb{C}$; we write $\Lambda = [\lambda_1, \lambda_2]$, and $\Lambda = \{m\lambda_1 + n\lambda_2 \colon m, n \in \mathbb{Z}\} \subset \mathbb{C}$. Let $\widehat{\mathbb{C}} = \mathbb{C} \cup \{\infty\}$ denote the Riemann sphere. The quotient space $\mathbb{C}/\Lambda$ determines a torus. An *elliptic function* $f\colon \mathbb{C} \to \mathbb{C}_\infty$ is a meromorphic function in $\mathbb{C}$ which is periodic with respect to a lattice Λ. If f is elliptic, $f(z + \lambda) = f(z)$ for all $\lambda \in \Lambda, z \in \mathbb{C}$, and the set $z + \Lambda = \{z + \lambda,\ \lambda \in \Lambda\}$ is called *the residue class of* z.

2010 *Mathematics Subject Classification.* Primary 37F45, 30D05.

Key words and phrases. Complex dynamics, elliptic functions, stability.

The author was supported in part by NSF Grant DMS 1600746 .

This means that $\mathfrak{D}$ is locally homeomorphic to a ball in $\mathbb{C}^4$; there are identifications up to complex conjugacy that reduce the space, and one (complex) dimensional slices can be visualized. We discuss some of the structure of $\mathfrak{M} = \mathfrak{D}/\sim$, where $f \sim g$, $f, g, \in \mathfrak{D}$ if and only if f is conformally conjugate to g; this is what we refer to as *moduli space* or reduced parameter space for $\mathfrak{D}$. For example, the author and Moreno Rocha showed in [**13**] that b can be parametrized by $\mathbb{C}/\Lambda$, which is topologically a torus, and further reductions are possible when the lattice is triangular.

Throughout this paper, we write $\wp_\Lambda$ when we want to denote the Weierstrass elliptic $\wp$ function with period lattice Λ, except when the statement in which it appears holds independently of the lattice. Despite the complicated structure of moduli space, we prove that there are many regions in moduli space, i.e., open sets in $\mathfrak{M}$, with the property that for every $(\Lambda, a, b) \in U \subset \mathfrak{M}$, the Julia set $J(a\wp_\Lambda + b)$ is a Cantor set. In the last section we discuss hyperbolic components in $\mathfrak{M}$ with connected Julia sets, typically with b near 0, extending existing results on $\wp_\Lambda$.

2. Preliminary background and definitions

We say two lattices $\Lambda_1 = [\lambda_1, \lambda_2]$ and $\Lambda_2 = [\omega_1, \omega_2]$ are *similar* if the ratio $\tau_1 = \lambda_2/\lambda_1$ is equal to $\tau_2 = \omega_2/\omega_1$; call the common value τ. Then $\Lambda_1 = \lambda_1[1, \tau]$ and $\Lambda_2 = \omega_1[1, \tau]$; since there are many choices of generators, we choose a generator so that $\mathrm{Im}(\tau) > 0$. Clearly $\Lambda_1 = k\Lambda_2$ for some nonzero $k \in \mathbb{C}$ exactly when they are similar. Similarity is an equivalence relation between lattices, and an equivalence class of lattices is called a *shape*. In Figure 1 we show a region S of the upper half plane with the property that every lattice shape is represented exactly once, by $[1, \tau], \tau \in S$ (see [**11**] or [**7**] for more details).

For any Λ, the *Weierstrass elliptic function* $\wp$, is defined by

$$\wp(z) = \frac{1}{z^2} + \sum_{\lambda \in \Lambda \setminus \{0\}} \left(\frac{1}{(z-\lambda)^2} - \frac{1}{\lambda^2} \right), \tag{2.1}$$

$z \in \mathbb{C}$; it is an even elliptic function with poles of order 2. Its derivative is an odd elliptic function which is also periodic with respect to Λ. Analogous to the case for sine and cosine functions, the elliptic functions $\wp$ and its derivative $\wp'$, are related to each other via a differential equation, namely:

$$\wp'(z)^2 = 4\wp(z)^3 - g_2\wp(z) - g_3, \tag{2.2}$$

where

$$g_2(\Lambda) = 60 \sum_{\lambda \in \Lambda \setminus \{0\}} \lambda^{-4} \text{ and } g_3(\Lambda) = 140 \sum_{\lambda \in \Lambda \setminus \{0\}} \lambda^{-6}. \tag{2.3}$$

That is, $y = \wp(z)$ is a solution to the differential equation:

$$\left(\frac{dy}{dz} \right)^2 = 4y^3 - g_2 y - g_3. \tag{2.4}$$

By differentiating both sides of Eqn (2.2) and solving for $\wp''$, we have the identity:

$$\wp''(z) = 6\wp^2(z) - \frac{1}{2}g_2. \tag{2.5}$$

The numbers $g_2(\Lambda)$ and $g_3(\Lambda)$ are invariants of the lattice Λ in the following sense: if $g_2(\Lambda_1) = g_2(\Lambda_2)$ and $g_3(\Lambda_1) = g_3(\Lambda_2)$, then $\Lambda_1 = \Lambda_2$. Furthermore given

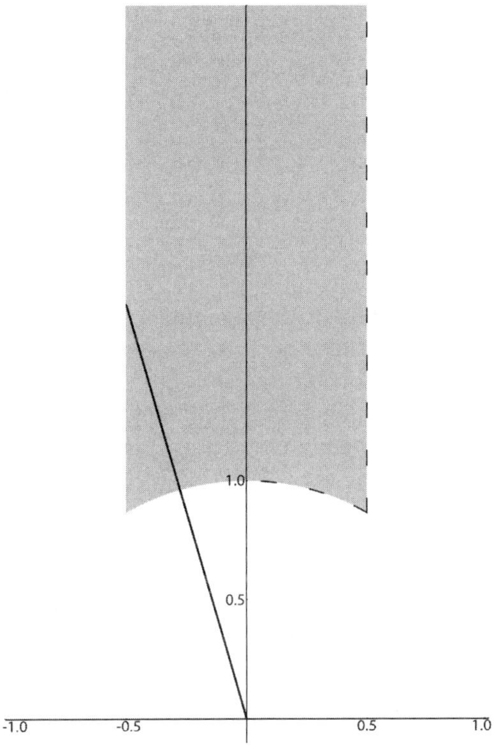

FIGURE 1. Every lattice Λ is similar to one of the form $\Omega = [1, \tau]$ where τ is in the shaded region S. The dotted boundary curves are not in S.

any g_2 and g_3 such that $g_2^3 - 27g_3^2 \neq 0$ there exists a lattice Λ having $g_2 = g_2(\Lambda)$ and $g_3 = g_3(\Lambda)$ as its invariants [**7**]. For $\Lambda_\tau = [1, \tau]$, the functions $g_i(\tau) = g_i(\Lambda_\tau), i = 2, 3$, are analytic functions of τ in the open upper half plane $\text{Im}(\tau) > 0$ ([**7**], Theorem 3.2).

From Eqn (2.3) we have the following homogeneity in the invariants g_2 and g_3.

LEMMA 2.1. *For lattices* Λ_1 *and* Λ_2, $\Lambda_2 = k\Lambda_1 \Leftrightarrow$

$$g_2(\Lambda_2) = k^{-4} g_2(\Lambda_1) \quad \textit{and} \quad g_3(\Lambda_2) = k^{-6} g_3(\Lambda_1).$$

A lattice Λ is said to be *real* if $\Lambda = \overline{\Lambda} := \{\overline{\lambda} : \lambda \in \Lambda\}$, where $\overline{z}$ denotes the complex conjugate of $z \in \mathbb{C}$, and the next result is standard.

PROPOSITION 2.2. The following are equivalent:

(1) Λ is a real lattice;
(2) $\wp_\Lambda(\overline{z}) = \overline{\wp_\Lambda(z)}$;
(3) $g_2, g_3 \in \mathbb{R}$.

Given any Λ, for $k \in \mathbb{C}\backslash\{0\}$, using Eqn (2.1), the following homogeneity properties hold:

$$\begin{aligned} \wp_{k\Lambda}(ku) &= \frac{1}{k^2}\wp_\Lambda(u) \\ \wp'_{k\Lambda}(ku) &= \frac{1}{k^3}\wp'_\Lambda(u) \end{aligned} \tag{2.6}$$

Each lattice shape determines an entire parameter space of lattices, obtained by varying either $k \in \mathbb{C}$, or g_2 and g_3 (staying within the shape class), and this has been studied by many authors, including [**8**]–[**12**], [**15, 16, 23**]. Those studies focus on the diversity and bifurcations of the dynamics of iterating the function $\wp_\Lambda$, as Λ varies. In this paper we change the focus to look at stable regions in a larger space of elliptic functions; in other words, are there open sets in $\mathfrak{D}$ for which we see the same dynamics? The following shows that family $\mathfrak{D}$ provides a natural class to consider; details of the proof can be found in ([**7**], Chapter 1) for example.

Proposition 2.3. Every elliptic function of order 2 with double poles at lattice points is of the form: $f(z) = a\wp_\Lambda(z) + b$, $a, b \in \mathbb{C}$, $a \neq 0$.

Proof. We assume by hypothesis that f is elliptic with double poles at the lattice points, so that each pole is in the residue class $0 + \Lambda$. Additionally, $f(z)$ must be even, so vanishes in a pair of residue classes $\pm\kappa + \Lambda$ for some κ. Then the function $g(z) = \wp_\Lambda(z) - \wp_\Lambda(\kappa)$ is elliptic with the same zeros and (double) poles as f. Every nonconstant elliptic function must have poles by Liouville's Theorem, so since $f/g(z)$ is an elliptic function with no poles, it is therefore a nonzero constant. This means that $f(z) = a(\wp_\Lambda(z) - \wp_\Lambda(\kappa)) = a\wp_\Lambda(z) + b$ as claimed. $\square$

The space of elliptic functions of order 2 with double poles has complex dimension 4; for the lattice Λ we need two coordinates, a point in the region shown in Figure 1 and a complex multiple of that. We can denote this pair by (τ, k) to correspond to the lattice $\Lambda = k[1, \tau]$. In addition we need a pair $(a, b) \in \mathbb{C}^2$, with $a \neq 0$; there is also a further reduction since it is enough to choose b from one fundamental region for Λ. There are other identifications when we consider a reduced space (no two maps conformally conjugate), but the maps move holomorphically as we vary each parameter. For example, one can parametrize the lattices Λ by the invariants $(g_2, g_3) \in \mathbb{C}^2$, with singularities at the locus of points $g_2^3 - 27g_3^2 = 0$, or we could use the critical values e_1, e_2, e_3, with the identifications: e_j's are all distinct and $\sum e_j = 0$. Therefore we obtain a complex manifold M of complex dimension 4. Our focus will be in a neighborhood of $a = 1$ with Λ a real lattice (however, see also Theorem 7.3), since by varying only Λ and b, we already obtain a parameter space with two complex dimensions. Up to now, if we consider Julia sets associated to iterating the meromorphic function $\wp_\Lambda$, for Λ any lattice, all connectivity results show that $J(\wp_\Lambda)$ is connected. However the connectivity of $J(\wp_\Lambda)$ has not been established for all lattices yet.

The main results of this paper are summarized by the following. We give the needed definitions immediately below.

Theorem 2.4 (Main Result). *If $\Lambda = [1, \tau]$ is a lattice that is square, real rectangular, or real triangular, then there exists some $k \in \mathbb{R}$ and $b \in \mathbb{C}$ such that in a neighborhood of $(\tau, k, a, b) \subset \mathbb{C}^4$ all Julia sets are Cantor and $a\wp_{k\Lambda} + b$ is J-stable. That is, the map lies in a hyperbolic component of moduli space.*

THEOREM 2.5. *There is no square lattice* Λ *for which* $\wp_\Lambda$ *is J-stable.*

We now turn to definitions and proofs of the main results.

2.1. Real period lattices for $\wp_\Lambda$**.** For most of this paper we assume that $\Lambda = [\lambda_1, \lambda_2]$, with $\lambda_1 > 0$ and λ_2 lying in the upper half plane. A closed, connected subset Q of $\mathbb{C}$ is a *fundamental region* for Λ if for each $z \in \mathbb{C}$, Q contains at least one point in the same Λ-orbit as z, and no two points in the interior of Q are in the same Λ-orbit of $z \mapsto z + \lambda$, with $\lambda \in \Lambda$. If Q is a fundamental region for Λ, then for any $s \in \mathbb{C}$, the set

$$Q + s = \{z + s : z \in Q\}$$

is also a fundamental region. If Q is a parallelogram, in we call Q a *period parallelogram* for Λ.

If Λ is a real lattice, a fundamental region Q can be chosen to be a rectangle with two sides parallel to the real axis and two sides parallel to the imaginary axis, or a rhombus with sides of the form $(0, \lambda), (0, \overline{\lambda})$ for some nonzero $\lambda \in \mathbb{C}$ [**7**]. While the property of being a real lattice is not invariant under the similarity relation, multiplication by i, or any real or purely imaginary number, preserves this property. We find the following result useful in what follows.

LEMMA 2.6. *If* Λ *is a real lattice with invariants* (g_2, g_3), *then* Λ *is similar to a lattice* Ω *with* $g_3 = g_3(\Omega) \geq 0$ *and* $g_2(\Omega) = g_2(\Lambda)$.

PROOF. Assume that $(g_2, g_3) \in \mathbb{R}^2 \backslash \mathcal{E}$, where $\mathcal{E}$ is the curve given by $x^3 = 27y^2$. If $g_3 < 0$, then we set $\Omega = i\Lambda$. Then by Lemma 2.1, $g_3(\Omega) = i^{-6} g_3(\Lambda) = -g_3(\Lambda) > 0$, and g_2 is left unchanged by multiplication by i^{-4}. $\square$

REMARK 2.7. We make the following observations which come directly from classical identities. These have dynamical significance, these are also discussed in more detail in the sources mentioned above.

Critical points: For any lattice Λ, $\wp_\Lambda$ has infinitely many simple critical points, one at each half lattice point, and we denote them by $c_1 + \Lambda, c_2 + \Lambda$, and $c_3 + \Lambda$, where

$$c_1 = \frac{\lambda_1}{2}, c_2 = \frac{\lambda_2}{2}, c_3 = \frac{\lambda_1 + \lambda_2}{2}.$$

Since the generators are not unique we adopt the convention that when Λ is real rectangular, λ_1 is real and λ_2 is purely imaginary. When Λ is real rhombic we choose $\lambda_2 = \overline{\lambda_1}$, and λ_1 in the first quadrant. We denote the set of all critical points by $\mathrm{Crit}(\wp_\Lambda)$.

Critical values: $\wp_\Lambda$ has three critical values $e_j = \wp_\Lambda(c_j)$ satisfying, for Λ real, at least one $e_j \in \mathbb{R}$. Also, one of these hold: $e_2 < e_3 < 0$ (if $g_3 > 0$), $e_2 < 0 < e_3 < e_1$ (if $g_3 < 0$), or $e_3 = 0$ (if $g_3 = 0$). In the third case, $e_2 = -e_1 = \sqrt{g_2}/2$ and $e_3 = 0$, with e_2 and e_1 both real in the *rectangular square* case and complex conjugates in the *rhombic square* case.

Critical value relations: Since for any lattice Λ, e_1, e_2, e_3 are the distinct zeros of Equation (2.2), we have these critical value relations:

$$\wp'_\Lambda(z)^2 = 4(\wp_\Lambda(z) - e_1)(\wp_\Lambda(z) - e_2)(\wp_\Lambda(z) - e_3). \tag{2.7}$$

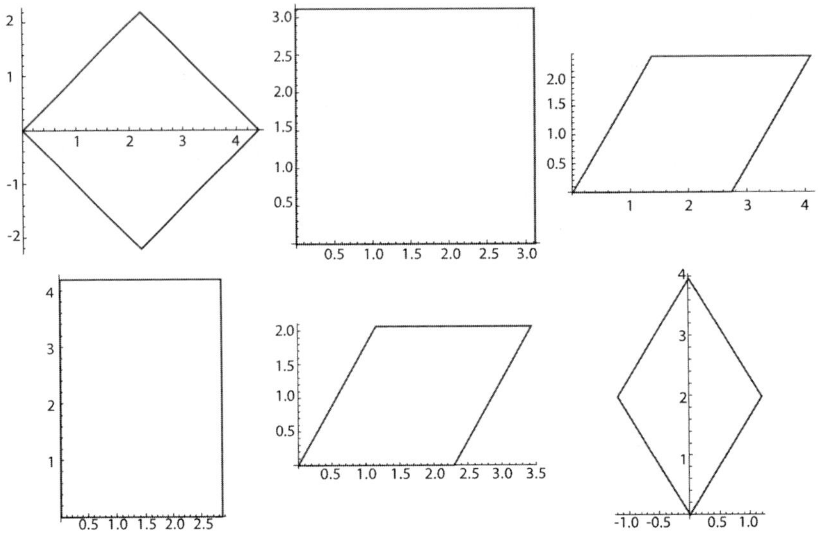

FIGURE 2. A fundamental region for each of the possible real lattice shapes, going clockwise from upper left: rhombic square ($g_2 < 0, g_3 = 0$), rectangular square ($g_2 > 0, g_3 > 0$), triangular ($g_2 = 0, g_3 > 0$), vertical rhombic ($g_2 < 0, g_3 > 0$), horizontal rhombic ($g_2 > 0, g_3 > 0$), and real rectangular ($g_2 > 0, g_3 > 0$),

Equating like terms in Equations (2.2) and (2.7), we obtain

$$
\begin{aligned}
e_1 + e_2 + e_3 &= 0, \\
e_1e_3 + e_2e_3 + e_1e_2 &= \frac{-g_2}{4}, \\
e_1e_2e_3 &= \frac{g_3}{4}.
\end{aligned} \tag{2.8}
$$

If we consider the polynomial coming from Equation (2.2),

$$q(u) = 4u^3 - g_2u - g_3, \tag{2.9}$$

a cubic polynomial of the form (2.9) has discriminant:

$$\triangle(g_2, g_3) = g_2^3 - 27g_3^2. \tag{2.10}$$

Real rectangular: The lattice Λ is real rectangular if and only if $\triangle(g_2, g_3) > 0$ (which forces $g_2 > 0$).

Rectangular square: Λ is real rectangular square if and only if the roots of q are $0, \pm\sqrt{g_2}/2$, and then we have: $e_3 = 0$ and $e_1 = \sqrt{g_2}/2 = -e_2 > 0$.

Non-square Rectangular: For real rectangular with $g_3 > 0$, we have $e_1 > 0$, $e_2 < 0$, and $e_2 < e_3 < 0$.

Real rhombic: The lattice Λ is real rhombic if and only if $\triangle(g_2, g_3) < 0$.

Vertical real rhombic: Setting c_1 to be a real critical point, and e_1 the corresponding critical value, $e_1 > 0$, $e_2 = e_1/2 + \zeta i$, for some $\zeta \in \mathbb{R}$ (non-zero), and $e_3 = \overline{e_2}$.

Rhombic square: In this case, $e_1 = 0$ and $e_2 = ib$, with $e_3 = \overline{e_2}$ and $b > 0$.

Our results center around Λ being a real lattice, and g_3 is non-negative by Lemma 2.6. The possible lattice shapes under consideration are shown in Figure 2, though the results we obtain imply these shapes can be perturbed so that the symmetry with respect to the real axis is lost. Despite the fact that a nonconstant analytic function f in a region D does not have any interior maximum points, it is useful for some examples to find the critical values of $\wp'$. The function $\wp'$ is also elliptic, having the same poles as $\wp$, so the moduli of the critical values only provide a maximum value of the derivative in some local settings; these can give useful extreme values of $\wp'$ on one real dimensional compact sets.

PROPOSITION 2.8. The critical values of $\wp'$ are all the values of

$$\left[-g_3 \pm \left(\frac{g_2}{3}\right)^{3/2}\right]^{\frac{1}{2}}.$$

PROOF. The critical points of $\wp'$ are the points $u \in \mathbb{C}$ such that $\wp''(u) = 0$, or equivalently by (2.5), where $\wp^2(u) = g_2/12$.

Setting $y = \wp(u)$; we have, using (2.2)

$$(\wp'(u))^2 = 4y^3 - g_2 y - g_3 \tag{2.11}$$

and $12y^2 = g_2$. Therefore we can write $y = \sqrt{\frac{g_2}{12}} = \pm\frac{1}{2}\left(\frac{g_2}{3}\right)^{1/2}$, which we plug into the right side of (2.11) to see that

$$\wp'(u) = \left[-g_3 \pm \left(\frac{g_2}{3}\right)^{3/2}\right]^{\frac{1}{2}}.$$

□

In [**13**] Proposition 2.8 was used to determine the maximum modulus of $\wp_\Lambda{}'$ along some one real dimensional lines, but it is not a tool whose usefulness is apparent in general, due to the Maximum Modulus Principle.

3. Julia sets

3.1. Fatou and Julia sets for elliptic functions. Definitions and properties of Julia sets for meromorphic functions are discussed in [**1, 2, 5**] and [**6**]. Let $f\colon \mathbb{C} \to \mathbb{C}_\infty$ be a meromorphic function where $\mathbb{C}_\infty = \mathbb{C} \cup \{\infty\}$ is the Riemann sphere. The *Fatou set* $F(f)$ is the set of points $z \in \mathbb{C}_\infty$ such that $\{f^n\colon n \in \mathbb{N}\}$ is defined and normal in some neighborhood of z. The *Julia set* is the complement of the Fatou set on the sphere, $J(f) = \mathbb{C}_\infty \backslash F(f)$. Since $\mathbb{C}_\infty \backslash \overline{\bigcup_{n\geq 0} f^{-n}(\infty)}$ is the largest open set where all iterates are defined and $f(\mathbb{C}_\infty \backslash \overline{\bigcup_{n\geq 0} f^{-n}(\infty)}) \subset \mathbb{C}_\infty \backslash \overline{\bigcup_{n\geq 0} f^{-n}(\infty)}$, then Montel's theorem implies that

$$J(f) = \overline{\bigcup_{n\geq 0} f^{-n}(\infty)}.$$

If f is any elliptic function with period lattice Λ, the *singular set* $Sing(f)$ of f is the set of critical values of f and their limit points. A function is called *Class S* if f has only finitely many critical (and asymptotic) values; for each lattice Λ, every elliptic function with period lattice Λ is of Class S. If f is elliptic with critical values $\{v_1, v_2, \ldots, v_m\}$, then the *postcritical set* of f is:

$$P(f) = \overline{\bigcup_{n \geq 0} f^n(v_1 \cup v_2 \cdots \cup v_m)}.$$

DEFINITION 3.1. If any elliptic function f with period lattice Λ has a component $W \subset F(f)$ which contains a simple closed loop which forms the boundary of a fundamental region for Λ, then W is a *double toral band.*

The definition of a *hyperbolic elliptic function* is the same as that for a rational map, namely, $J(f) \cap P(f) = \emptyset$, and this is equivalent to uniform expansion on $J(f)$. [**11**].

THEOREM 3.2 ([**11**], Sec. 3). *For an elliptic function f, if f is hyperbolic and all the critical values of f are contained in one Fatou component, then all of the following hold:*

- *$J(f_\Lambda)$ is a Cantor set;*
- *W is a double toral band;*
- *there is exactly one attracting fixed point for f_Λ.*

3.2. Cantor Julia Sets and J-stability for $a\wp_\Lambda + b$. Given a lattice Λ, we are interested in the dynamics and Julia sets of order two elliptic functions with poles at every $\lambda \in \Lambda$. Therefore, by Proposition 2.3 we consider the family of maps, for any triple $m = (\Lambda, a, b)$, with Λ a lattice:

$$f_m(z) = a\wp_\Lambda(z) + b,\ a, b \in \mathbb{C},\ a \neq 0. \tag{3.1}$$

By [**13**] it is enough to consider b coming from one fundamental region of Λ, as $a\wp_\Lambda(z) + b + \lambda$ is conformally conjugate to f_m for any $\lambda \in \Lambda$.

The theory of stable families of holomorphic families, started by Mañé, Sad, and Sullivan [**19**] and analyzed in [**20**] was generalized to the setting of meromorphic maps with finite singular set by Keen and Kotus [**14**]. It is discussed in the elliptic setting in [**11**]. We summarize it in terms of our current setting.

THEOREM 3.3. *Let f_m be a holomorphic family of elliptic functions, parametrized by $m \in \mathfrak{M}$, and let m_0 be a point in $\mathfrak{M}$. Then the following are equivalent:*

(1) *The number of attracting cycles of f_m is locally constant at m_0.*
(2) *The maximum period of an attracting cycle of f_m is locally bounded at m_0.*
(3) *The Julia set moves homomorphically at m_0*
(4) *For all m sufficiently close to m_0, every periodic point of f_m is attracting or repelling (or persistently indifferent).*
(5) *In the Hausdorff topology, the Julia set $J(f_m)$ depends continuously on m in a neighborhood of m_0.*
(6) *For $i = 1, 2, 3$, letting $c_i(m)$ denote the residue class of the critical point c_i for f_m (each c_i depends continuously on Λ), the maps*

$$m \mapsto f_m^k(c_i(m)), k = 0, 1, \ldots$$

form a normal family at m_0.

(7) *There is a neighborhood $U \subset \mathfrak{M}$ of m_0 such that for all $m \in U$, $c_i(m) \in J(f_m)$ if and only if $c_i(m_0) \in J(f_m)$.*

TABLE 1. The relationships among values and parameters of of $\wp_\Lambda$ for rhombic square lattices, using $c > 0$; the last column shows the real quarter lattice point values. $\gamma \approx 2.62206$ is a lemniscate constant (see e.g., [**9**]).

Fixed	$\{\mathbf{e_1}, \mathbf{e_2}, \mathbf{e_3}\}$	$\{\mathbf{g_2}, \mathbf{g_3}\}$	**Side length**	$\wp_\Lambda(\mathbf{c_3}/\mathbf{2})$
Standard	$\{-i, i, 0\}$	$\{-4, 0\}$	γ	1
e_1	$\{-ci, ci, 0\}$	$\{-4c^2, 0\}$	$\dfrac{\gamma}{\sqrt{c}}$	c
g_2	$\left\{\dfrac{-\sqrt{-c}}{2}, \dfrac{\sqrt{-c}}{2}, 0\right\}$	$\{-c, 0\}$	$\left(\dfrac{4}{c}\right)^{1/4}\gamma$	$\dfrac{\sqrt{c}}{2}$
side length	$\left\{\dfrac{-\gamma^2}{c^2}i, \dfrac{\gamma^2}{c^2}i, 0\right\}$	$\left\{\dfrac{-4\gamma^4}{c^4}, 0\right\}$	c	$\dfrac{\gamma^2}{c^2}$

The set $\mathfrak{M}^{\text{stab}} \subset \mathfrak{M}$ denotes the *J-stable set* of parameters in the sense that any one of the above conditions are satisfied. From now on we use the term *stable* for J-stable parameters or maps.

THEOREM 3.4 ([**14**]). *For any holomorphic family of elliptic functions defined over the complex manifold $\mathfrak{M}$, $\mathfrak{M}^{\text{stab}}$ is open and dense in $\mathfrak{M}$.*

We give some key corollaries to Theorem 3.3, the first of which uses Theorem 3.2.

COROLLARY 3.5. For $m = (\Lambda, a, b)$ and f_m as in (3.1), if f_{m_0} is hyperbolic and all the critical values of f_{m_0} are contained in one Fatou component, then $f_{m_0} \in \mathfrak{M}^{\text{stab}}$ and $J(f_m)$ is a Cantor set for m near m_0.

COROLLARY 3.6. For $m = (\Lambda, a, b)$ and f_m as in (3.1), if there are 3 distinct attracting cycles for f_{m_0}, then $f_{m_0} \in \mathfrak{M}^{\text{stab}}$ and $J(f_m)$ is connected for m near m_0.

PROOF. This follows from Theorem 3.2, Theorem 3.3 and the fact that a perturbation will not destroy the property of a periodic point being attracting. □

More generally, the following holds.

COROLLARY 3.7. For $m = (\Lambda, a, b)$ and f_m as in (3.1), if f_{m_0} is hyperbolic, then $f_{m_0} \in \mathfrak{M}^{\text{stab}}$.

4. Rhombic square lattices

We first look at the setting where Λ is rhombic square. In this case $g_2 < 0$ and $g_3 = 0$. We note that the map $\wp_\Lambda$ is quite unstable; in [**8**] it was shown that the only real critical value is the prepole 0, and this implies that $J(\wp_\Lambda) = \widehat{\mathbb{C}}$, the whole sphere. The next result illustrates the instability of this setting; Table 1 shows the interrelations among side length and critical values.

THEOREM 4.1 (Instability Theorem). *Given $g_2 < 0$, and the associated lattice $\Lambda = \Lambda(g_2, 0) = [\lambda_1, \overline{\lambda_1}]$, and any $\epsilon > 0$ we can find a constant b, $|b| < \epsilon$, such that*

(1) *$J(\wp_\Lambda) = \widehat{\mathbb{C}}$, and*
(2) *$J(\wp_\Lambda + b)$ has a super-attracting cycle.*
(3) *In (2), we can choose the value $b \in (0, \epsilon)$.*

PROOF. The fact that $J(\wp_\Lambda) = \widehat{\mathbb{C}}$ is just Theorem 1.2 of [**8**]. Using the labelling of the critical points from Remark 2.7, given any small $\epsilon > 0$, since $e_3 = 0$ is a pole of $\wp_\Lambda$, $\wp_\Lambda$ maps $B_\epsilon(0)$ to a set which contains a ball at ∞; i.e., there exists $R > 0$ such that $B_R(\infty) \equiv \{z : |z| > R\} \subset \wp_\Lambda(B_\epsilon(0))$.

Similarly, if we define the meromorphic map $\widetilde{\wp_\Lambda}(t) = \wp_\Lambda(t) + t$, then there exists $R > 0$ such that $B_R(\infty) \subset \widetilde{\wp_\Lambda}(B_\epsilon(0))$ as well; moreover for ϵ small enough, $\widetilde{\wp_\Lambda}|_\mathbb{R} : (-\epsilon, \epsilon) \to (A, \infty]$ for some $A > 0$, and the mapping is surjective. We now consider any $t_N = N\lambda + c_3 \in B_R(\infty)$, (a very large critical point in the residue class of c_3), $\lambda \in \Lambda$. We can then find a point $\varepsilon_0 \in B_\epsilon(0)$ such that $\widetilde{\wp_\Lambda}(\varepsilon_0) = t_N$. If $\lambda \in \mathbb{R}$, then t_N, $\varepsilon_0 \in \mathbb{R}$ as well.

For the map $\wp_\Lambda + \varepsilon_0$, we have the following critcal orbit:

$$c_3 \mapsto 0 + \varepsilon_0 \mapsto \wp_\Lambda(\varepsilon_0) + \varepsilon_0 = \widetilde{\wp_\Lambda}(\varepsilon_0) = t_N = N\lambda + c_3 \mapsto \varepsilon_0,$$

since $\wp_\Lambda(N\lambda + c_3) = \wp_\Lambda(c_3) = 0$.

Therefore $t_N = N\lambda + c_3$ is a superattracting periodic point for $\wp_\Lambda + \varepsilon_0$, of period 2, and choosing $b = \varepsilon_0$ gives the result. □

The same proof gives the following result.

COROLLARY 4.2. *Assume we have any nonzero $g_2 \in \mathbb{C}$, and the associated lattice $\Lambda = \Lambda(g_2, 0) = [\lambda_1, i\lambda_1]$; equivalently suppose Λ is any square lattice. For any $\epsilon > 0$ we can find a b, $|b| < \epsilon$ such that*

(1) *c_3 maps under $\wp_\Lambda$ to a pole;*
(2) *the orbit of c_3 terminates in a superattracting cycle for $\wp_\Lambda + b$.*

PROOF. Since Λ is square, $e_3 = 0$. Setting $c_1 = \lambda_1/2$, $c_2 = i\lambda_1/2$, then $c_3 = c_1 + c_2$. Since 0 is a lattice point, $\wp_\Lambda(c_3)$ is a pole. The proof of (2) is identical to that of Theorem 4.1, except that $\widetilde{\wp_\Lambda}(t) = \wp_\Lambda(t) + t$ is a map on $\mathbb{C}$, and does not leave $\mathbb{R}$ invariant.

The rest of the proof is the same and the superattracting cycle is of the form $\{\varepsilon_0, t_N\}$ for a large critical point $t_N \in B_R(\infty)$ and a small $\varepsilon_0 \in B_\epsilon(0)$.

□

EXAMPLE 4.3. Following the constructive algorithm coming from the proof of Theorem 4.1, if we choose Λ to be the standard rhombic square lattice with $(g_2, g_3) = (-4, 0)$ and $\epsilon = .1$, we can find $\varepsilon_0 \approx .08044 + .0328856i$ so that we have the following critical orbit:

$$c_3 \mapsto \varepsilon_0 \mapsto c_3 + 50\lambda_1 \mapsto \varepsilon_0.$$

We then have a superattracting period 2 orbit containing the critical point $c_3 + 50\lambda_1$; since the basin of attraction for this cycle is extremely small, it is quite difficult to write a computer program accurately showing the Julia set.

Computer experimentation shows that there are many more types of bifurcations occurring for parameters b near 0, so we look for stability away from the poles in this setting. In particular, we look specifically at the constants that will give us superattracting fixed points, keeping in mind that we need g_2 small enough that all the critical values will lie in the immediate attracting basin of that point (see Table 1).

In [**9**], the following result was proved using the basic homogeneity equations and helps determine the entries in Table 1.

PROPOSITION 4.4 ([**9**], Prop. 5.6). If Λ_1 is rectangular square and Λ_2 is rhombic square, both of side length γ, then:

$$\wp_{\Lambda_2}(e^{\pi i/4}z) = -i\wp_{\Lambda_1}(z) \text{ and } \wp'_{\Lambda_2}(e^{\pi i/4}z) = e^{-3\pi i/4}\wp'_{\Lambda_1}(z),$$

so $|\wp'_{\Lambda_2}(e^{\pi i/4}z)| = |\wp'_{\Lambda_1}(z)|$.

For $\Lambda_2 = [\alpha+\alpha i, \alpha - \alpha i]$ rhombic square, we label the critical points $c_1 = \dfrac{\alpha+\alpha i}{2}$, $c_2 = \dfrac{\alpha-\alpha i}{2}$, and $c_3 = \alpha$, $\alpha \in \mathbb{R}$. Then $e_1, e_2 = \pm\sqrt{g_2}/2 = \pm\imath\sqrt{|g_2|}/2$ are purely imaginary and $e_3 = 0$.

We note that if $\Lambda = [\alpha+\alpha i, \alpha-\alpha i]$, $\alpha > 0$, then real quarter lattice points occur at $\alpha/2 + 2m\alpha$, $m \in \mathbb{Z}$.

LEMMA 4.5. *Let $\Lambda = [\alpha+\alpha i, \alpha-\alpha i]$, $\alpha > 0$. Then the following hold:*

(1) *If $g_2 = -4$, then we have the standard lattice $\Lambda(-4,0) = \Gamma = [\beta+\beta i, \beta - \beta i]$ with square side length γ. Therefore $\beta = \gamma/\sqrt{2}$, $\wp_\Lambda(\beta/2) = 1$ and $\wp_\Lambda{}'(\beta/2) = -2\sqrt{2}$.*

(2) *For an arbitrary negative real g_2 the real quarter lattice value of the corresponding lattice is $\wp_\Lambda(\alpha/2) = \sqrt{-g_2}/2 = |e_1| = |e_2|$; moreover if the lattice is $k > 0$ times the standard lattice, $\wp_\Lambda(\alpha/2) = 1/k^2$, and $\wp_\Lambda{}'(\alpha/2) = (-1)(-g_2)^{3/4} = -2\sqrt{2}k^{-3}$.*

PROOF. These statements follow from the classical identities given by Equations (2.6) and (2.8). □

REMARK 4.6. From the classical identities, we have the following for rhombic square lattices $\Lambda = [\alpha+\alpha i, \alpha-\alpha i]$; the dotted lines in Figure 3 illustrate where the map $\wp_\Lambda$ is purely imaginary and the solid lines show where it is real-valued.

- With $c_3 = \alpha \in \mathbb{R}$, and $c_1, c_2 = \frac{1}{2}(-1 \pm i)\alpha$, $\wp_\Lambda$ is real valued on lines through the diagonals that pass through lattice points.
- $\wp_\Lambda$ is purely imaginary on the sides of the squares whose vertices are lattice points.
- Let $u \in \mathbb{C}$. Then

$$\wp_\Lambda(u \pm c_3) = \frac{e_1 \cdot e_2}{\wp_\Lambda(u)} = -\frac{g_2}{4\wp_\Lambda(u)}. \tag{4.1}$$

PROPOSITION 4.7. For any $g_2 \in (-3, 0)$, let ζ denote the closed square boundary of a fundamental region of $\Lambda = \Lambda(g_2, 0)$ formed by line segments from c_3 to ic_3, ic_3 to $-c_3$, $-c_3$ to $-ic_3$, and $-ic_3$ to c_3,. Then $|\wp_\Lambda{}'(z)|_{z\in\zeta} < 1$.

PROOF. Denote by Λ_1 the lattice $e^{\pi i/4}\Lambda$, and by δ the square fundamental region with sides parallel to the axes, with vertices at half lattice points. The curve ζ can be written as $e^{\pi i/4}\delta$, and it was shown in [**13**], and follows from Proposition 2.8, that $|\wp'_{\Lambda_1}(z)|_{z\in\delta} < 1$, so the result follows from Proposition 4.4. □

Figure 4 illustrates Proposition 4.7.

COROLLARY 4.8. Given any $g_2 < 0$, and the corresponding curve ζ given in Proposition 4.7, set $b = c_3$. Then the image of ζ under $\wp_\Lambda + b$ is the vertical line segment from $c_3 - i\dfrac{\sqrt{|g_2|}}{2}$ to $c_3 + i\dfrac{\sqrt{|g_2|}}{2}$.

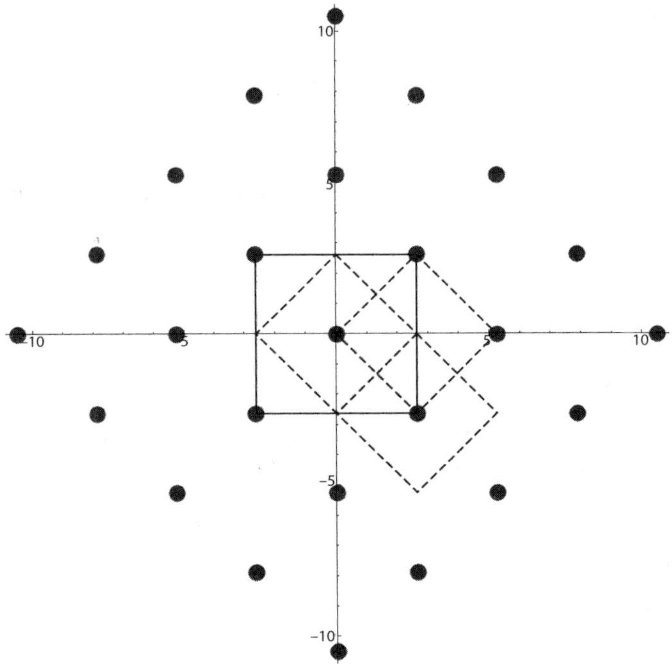

FIGURE 3. We show the lattice points for a real rhombic lattice; the dotted lines show where the map $\wp_\Lambda$ is purely imaginary and the solid lines, including the axes, show where it is real-valued.

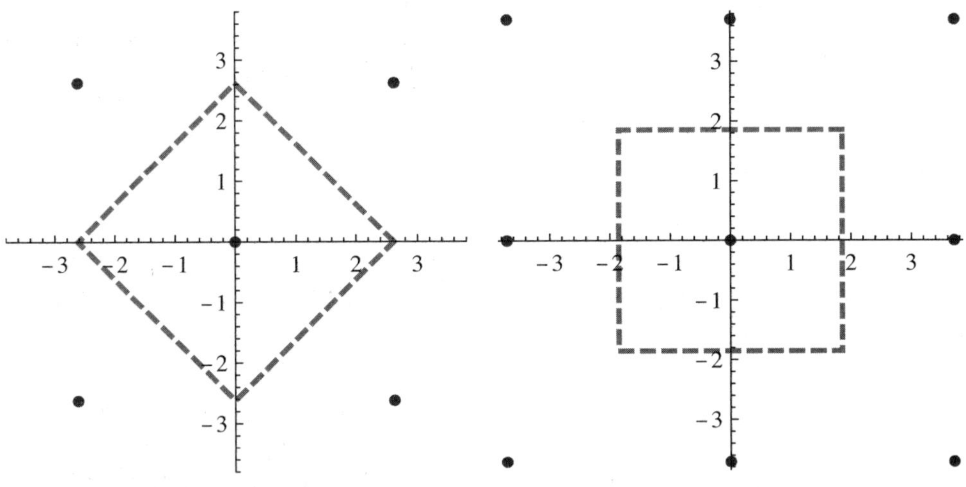

FIGURE 4. On the left, one period square for $\wp_\Lambda$ for Λ rhombic bounded by ζ; ζ gets mapped to the imaginary axis by $\wp_\Lambda$. On the right, δ bounds a fundamental region of $\Lambda_1 = e^{\pi i/4}\Lambda$.

PROOF. We denote the map $\wp_\Lambda + b = f_m$ with $m = (\Lambda, 1, b)$. We can write ζ as $\zeta_1 \cup \zeta_2 \cup \zeta_3 \cup \zeta_4$ where each ζ_i is a line segment joining the points in the order in which we list them: $\zeta_1 = [ic_3, c_3], \zeta_2 = [c_3, -ic_3], \zeta_3 = [-ic_3, -c_3], \zeta_4 = [-c_3, ic_3]$. Since $\zeta_3 = -\zeta_1$, their images are the same under f_m; similarly we have $f_m(\zeta_2) = f_m(\zeta_4)$.

The midpoint of each ζ_j is a critical point in the residue class of c_1 or c_2, and the endpoints of each ζ_j map under f_m to $b = c_3$. From Proposition 4.7 (and its proof) we have that the image of ζ_1 is purely imaginary under $\wp_\Lambda$, so its image under f_m lies along a vertical line $c_3 + iy$, $y \in \mathbb{R}$. Each half segment of ζ_1 maps injectively onto $[c_3 + e_1, c_3]$ and each half segment of ζ_2 maps injectively onto $[c_3, c_3 + e_2]$, which, from Table 1 gives the result.

□

Using Lemma 4.5 and Table 1 we have the following.

LEMMA 4.9. *If* $(g_2, g_3) = (-1, 0)$ *then* $|\wp_\Lambda'(z)| < 1$ *on the line segment* $(c_3/2, 3c_3/2) \subset \mathbb{R}$, $\wp_\Lambda'(c_3/2) = -1$, $\wp_\Lambda'(c_3) = 0$, *and* $\wp_\Lambda(3c_3/2) = 1$. *Moreover,* $c_3 = \gamma$, *so* $c_3/2 > 1$.

PROOF. We apply Lemma 4.5 (2) with $k = \sqrt{2}$. Since the precise value of γ is known: $\gamma = \Gamma(1/4)^2/(2\sqrt{2\pi})$, standard approximations give the last statement (cf. [**21**]). □

We use these results to obtain a concrete map with a stable Cantor Julia set.

PROPOSITION 4.10. If $(g_2, g_3) = (-1, 0)$ using $a = 1$ and $b = c_3 \in \mathbb{R}$, then for the map $f(z) = a\wp(z) + b$, the following hold:

(1) The critical point c_3 is fixed.
(2) The diagonal line segment from ic_3 to c_3 (of length $\sqrt{2}\,\gamma$) is mapped 2-to-1 onto $V = [c_3 - .5i, c_3]$, the vertical line segment. There is one branch point at the midpoint, which is the critical point c_1. Note that one endpoint is $f(c_1) = c_3 - .5i$ and the other is $f(c_3) = c_3$.
(3) The set V is mapped two-to-one by f onto $J = [c_3 - \alpha, c_3] \subset \mathbb{R}$, where $\alpha \in (0, 1/12)$;
(4) The entire interval J converges to c_3 under iteration of f.

PROOF. Since $\Lambda = \Lambda(g_2, g_3)$ is rhombic square, we have that $g_3 = e_3 = 0$; (1) follows from our choice of b, and (2) follows from Corollary 4.8 above.

To prove (3), since $\wp_\Lambda(bi) < 0$ for any nonzero $b \in \mathbb{R}$, and $c_3 \in \mathbb{R}$, f maps V onto the interval in $\mathbb{R}$ of the form $(c_3 - \dfrac{g_2}{4\wp_\Lambda(.5i)}, c_3)$ by applying (4.1) from Remark 4.6. Using the Laurent series expansion about $u = 0$, which simplifies since $g_3 = 0$, we have that:

$$\wp_\Lambda(u) = \frac{1}{u^2} + \frac{g_2}{20}u^2 + \frac{g_2^2}{1200}u^6 + O(z^{10}). \tag{4.2}$$

Under our hypotheses, we have the following:

$$\wp_\Lambda(.5i) = -4 + 0.0125 - \frac{1}{1200}.015625 + O(z^{10}), \tag{4.3}$$

and since it is an alternating series, $-4 < \wp_\Lambda(.5i) < -3.9875 < -3$; and hence (multiplying by 4 and inverting, using $-g_2 = 1$),

$$-1/16 > -g_2/4\wp_\Lambda(.5i) > -1/4(3.9875) > -1/12,$$

so

$$c_3 - \frac{g_2}{4\wp_\Lambda(.5i)} > c_3 - 1/12.$$

This proves (3).

To prove (4), we note that f maps $B = (c_3/2, 3c_3/2)$ onto (c_3, η) for some $\eta > c_3$, and by Lemma 4.9 and (3), $J \subset B$ and $|f'(z)| < 1$ for all $z \in B$. The result now follows since by (2.5), $\wp_\Lambda''$ is always positive on $\mathbb{R}$, so $\wp_\Lambda'$ is strictly increasing on $\mathbb{R}$, and is increasing from a negative value with modulus < 1 to 0 on J, using Lemma 4.9 . Then the Mean Value Theorem on $\mathbb{R}$ gives that for all $z \in J$, $\lim_{n\to\infty} f^n(z) = c_3$. □

We now give the first main theorem of this section.

THEOREM 4.11 (Stability theorem 1). *Let $\Lambda = \Lambda(g_2, 0)$ be a (real) square rhombic lattice (so $g_2 < 0$). Then there exist $k > 0$ and $b \in \mathbb{R}$ so that the map $f_{m_0} = \wp_{k\Lambda} + b$ is hyperbolic, stable in $\mathfrak{M}$, and $J(f_{m_0})$ is a Cantor set in a neighborhood of $m_0 = (k\Lambda, 1, b)$.*

PROOF. Since $g_2 \neq 0$, we set $k = 1/|g_2| > 0$. Then $\Lambda(kg_2, 0)$ is still real rhombic square and satisfies all the hypotheses of Proposition 4.10. Then choosing $b = c_3$, c_3 is a superattracting fixed point and there is a double toral band containing all the critical values. By Theorem 3.2, $J(f)$ is a Cantor set.

Using $m = (\ell, \tau, a, b)$ near m_0 (keeping the lattice $\ell[1, \tau]$ near $k\Lambda$) will not affect the presence of an attracting fixed point, under small perturbations of $m_0 \in \mathfrak{M}$, the map f_m remains hyperbolic and $J(f_m)$ remains a Cantor set. □

4.1. Arbitrary square lattices. We can generalize Theorem 4.11 to any square lattice using the same method, the homogeneity equation (2.6), and applying (4.1).

THEOREM 4.12 (Stability theorem 2). *Let $\Lambda = \Lambda(g_2)$ be any square lattice (so $g_2 \in \mathbb{C}$, $g_2 \neq 0$, and $g_3 = 0$). Then we can find $k > 0$ and $b \in \mathbb{C}$ so that for $m_0 = (k\Lambda, 1, b)$, the map $f_{m_0} = \wp_{k\Lambda} + b$ is hyperbolic, stable in $\mathfrak{M}$, and $J(f_m)$ is a Cantor set for all m in a neighborhood of m_0.*

PROOF. We write $g_2 = r \cdot \exp(2\pi i\theta)$, $\theta \in [0, 1)$ and we show the theorem holds for $k = 1/r$ and writing $k\Lambda = \Lambda_1 = [\lambda_1, i\lambda_1]$. We then choose $b = c_3$, where c_3 is the critical point corresponding to the given value of $g_2(\Lambda_1) = 1/r^4 g_2(\Lambda)$, using the technique and notation above. By our choice, $|g_2(\Lambda_1)| = 1$, and the homogeneity identities give us that $|e_1| = 1/2, e_2 = -e_1$, and $e_3 = 0$. For simplicity, writing f for f_{m_0}, the same steps as in Proposition 4.10 hold, subject to minor modifications.

- The critical point c_3 is fixed by f; this is by our choice of b.
- The boundary of a fundamental region formed by the diagonal line segment S from ic_3 to c_3 (of length $\sqrt{2}\gamma$), T from c_3 to $-ic_3$, and $-S$,$-T$, is mapped onto $V = [c_3 + e_1, c_3] \cup [c_3 + e_2]$, 2 line segments of length .5 meeting at c_3 as follows. Each of S and T gets mapped 2-to-one onto half of V. There is one branch point at the midpoint of S, which is the critical point c_1 (or c_2, depending on the labels). One endpoint is $f(c_1) = c_3 + e_1$ and the other is $f(c_3) = c_3$.
- The set V is mapped two-to-one onto on interval $J = [c_3 - \alpha, c_3] \subset \mathbb{C}$, with $|J| < 1/12$.
- The entire interval J converges to c_3 under iteration of f.

Using this, we obtain the result since all 3 critical points are in the immediate attracting basin of c_3. □

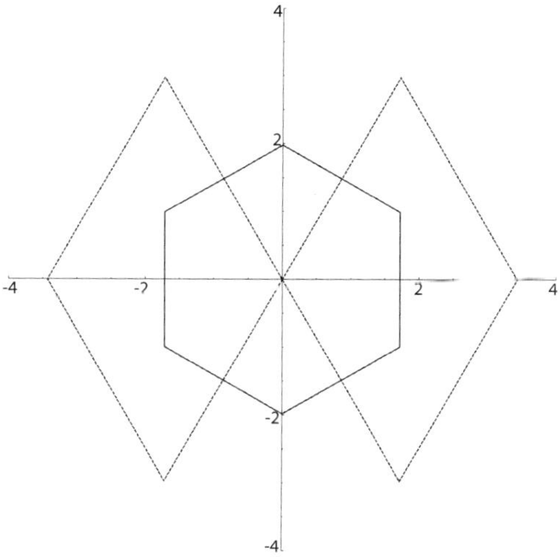

FIGURE 5. Three fundamental regions are shown for real triangular lattice: two are pairs of equilateral triangles and one is a hexagonal period parallelogram.

5. Triangular lattices and stability

We now turn to the case of triangular lattices, where some new techniques are required. Let Λ be a real triangular lattice, so we assume $\Lambda = [\lambda e^{2\pi i/3}, \lambda e^{4\pi i/3}]$, $\lambda > 0$. Because every triangular lattice is similar to a lattice of this form, we assume that $g_3 \in \mathbb{R} \setminus \{0\}$ and $g_2 = 0$. There are various convenient ways to generate the lattice besides the one given above; we summarize a few properties of real triangular lattices here. We denote the cube roots of unity by $\omega = \exp(2\pi i/3)$, so $\omega^2 = \omega^{-1}$, and 1. Then $\Lambda = \omega\Lambda = \omega^2\Lambda$. Then a key identity is, for any $u \in \mathbb{C}$,

$$\wp_\Lambda(\omega u) = \omega \wp_\Lambda(u). \tag{5.1}$$

1. **Critical points:** There are 3 residue classes of critical points for $\wp_\Lambda$. By convention $c_3 > 0$, then $c_3 + \Lambda$ all lie along lines parallel to the real axis. Then $c_2 = (1/2)\omega\lambda$, and $c_1 = \overline{c_2}$ are a conjugate pair of points, each generating a residue class.
2. **Critical values:** The critical values for $\wp_\Lambda$ are $e_j = \left(\frac{g_3}{4}\right)^{1/3}$, where e_3 is the real root, and the other two are chosen to correspond with the labelling of c_j above. Under our hypotheses, $e_3 \in \mathbb{R}$. We also have that $\mathrm{Re}(e_2) = -e_3/2$, which means that all the critical values have explicit formulas when g_3 is known. In our main example we use small values of g_3; e.g., choosing $g_3 = 1/2$, yields $e_3 = 1/2$, and $e_1, e_2 = -1/4 \pm i\sqrt{3}/2$.
3. **Scaling the lattice:** Starting with the *standard lattice* which corresponds to $g_3 = 4$ and $e_3 = 1$, there is an explicit formula for the side length, but we label it as $\lambda^* \approx 2.42865$. Then the scaling works as follows: for a lattice with generators of length $k\lambda^* > 0$ (usually we are interested in $k > 1$), we have $(g_2, g_3) = (0, 4k^{-6})$, which results in critical points of modulus $k\lambda^*/2$, and the new critical values $\tilde{e_j}$ will be $\tilde{e_j} = (1/k^2)e_j$.

4. **Derviatives:** Along vertical lines of the form $(2n+1)c_3+ti$, $t \in \mathbb{R}, n \in \mathbb{Z}$, we have that $\wp_\Lambda$ is real, and $\wp_\Lambda{}'$ is purely imaginary. Also, the maximum value of $\wp_\Lambda$ along these lines is e_3, occurring at c_3, and $\wp_\Lambda$ decreases from e_3 to $-\infty$, periodically, along the lines. The map $\wp_\Lambda{}'$ is 0 at the residue class $[c_3]$, and has critical points, or stationary values, where $\wp_\Lambda{}''(z) = 0$ along these lines. This is discussed below.
5. **Current goal:** In ([**10**], Corollary 3.3) the authors showed that for any triangular lattice, $J(\wp_\Lambda)$ is connected. We want to reduce g_3 to shrink the derivative, which is equivalent to stretching the side length of the corresponding lattice; we then choose a value of b so that the symmetry of the triangular lattice forces all three critical points to be in the basin of attraction of a single attracting fixed point.

When $g_3 > 0$, a fundamental region for Λ is made up of two equilateral triangles, with a common side along the real axis, the line segment $[0, 2c_3]$ and there are other natural choices (see Figure 5). Setting $c_4 = c_1 - c_2$, (purely imaginary and in the residue class of c_3), centers of the equilateral triangles occur at $(\pm 2/3)c_4$.

LEMMA 5.1. *For Λ triangular with $g_3 > 0$, we have that $\wp_\Lambda(u) = 0$ at $z_1, z_2 = \pm\frac{2}{3}c_4$, $z_3, z_4 = \pm\omega\frac{2}{3}c_4$, and $z_5, z_6 = \pm\omega^2\frac{2}{3}c_4$. At each of the zeros of $\wp_\Lambda$, we have $\wp_\Lambda{}'(z_j) = -\sqrt{g_3}i$, and z_j is a stationary point for $\wp_\Lambda{}'$ (i.e., $\wp_\Lambda{}''(z_j) = 0$); moreover $\wp_\Lambda{}^{(3)}(z_j) = 0$ as well.*

PROOF. These statements follow from ([**7**], Section 21) and (2.6). By (2.5), $\wp_\Lambda{}''(z_j) = 6\wp_\Lambda(z_j)^2 = 0$. Differentiating (2.5) gives that $\wp_\Lambda{}^{(3)}(z_j) = 12\wp_\Lambda(z_j)\wp_\Lambda{}'(z_j) = 0$ for $j = 1, 2, \ldots, 6$. □

There is an easy variation of Lemma 5.1 when $g_3 < 0$, replacing c_4 by c_3. We now construct a stable map with Cantor Julia set. We give proofs only of the steps that have not already been proved.

THEOREM 5.2. *Let $(g_2, g_3) = (0, 1/2)$ so Λ is a real triangular lattice, and write $\Lambda = [a + \sqrt{3}ai, a - \sqrt{3}ai], a > 0$. Let $p_0 = z_1 = (2\sqrt{3}a/3)i$ as in Lemma 5.1. Then for the map $f(z) = \wp_\Lambda + p_0$, the following hold.*

(1) *$\wp_\Lambda(p_0) = 0$, so $f(p_0) = p_0$;*
(2) *$\wp_\Lambda{}'(p_0) = -(\sqrt{2}/2)i$, so p_0 is an attracting fixed point.*
(3) *$\wp_\Lambda$ maps the hexagon whose endpoints are $z_1, \ldots, z_6$ onto three line segments of length $1/2$ sharing a common endpoint at p_0. The other endpoints are $e_1 + p_0$, $e_2 + p_0$, and $e_3 + p_0$.*
(4) *All three critical values of f lie in the immediate attracting basin of p_0.*

PROOF. (1) and (2) follow from Lemma 5.1 since $p_0 = z_1$. Using the notation from the lemma, the boundary of a fundamental region formed by the hexagon H with 3 distinguished segments: H_1 from z_1 to z_3, H_3, from z_3 vertically down to z_5, and H_2 from z_5 to $z_2 = -z_1$. They are labelled so that $c_j \in H_j$. Each H_j is mapped by f two-to-one onto a line segment L_j, a line segment of length .5 with a common endpoint of $b \in \mathbb{R}$. There is one branch point at the midpoint of H_j, which is the critical point c_j. One endpoint of each image segment L_j is $f(z_k) = b$, and there are 3 spokes coming out with endpoints $p_0 + e_3$, $p_0 + \omega e_3$, and $b + \omega^2 e_3$.

We apply Eqn (5.1) with $u = z - p_0$; then

$$\wp_\Lambda(\omega(z - p_0)) = \omega\wp_\Lambda(z - p_0). \tag{5.2}$$

By induction on n, we show that $|f^n(c_j)-p_0| = \rho_n$ for $j = 1,2,3$. For $n = 1$, this is just the statement that $|e_j| = 1/2$ for each j. Assume that $|f^{n-1}(c_j) - p_0| = \rho_{n-1}$ for $j = 1,2,3$; then

$$\begin{aligned} f^n(c_j) &= f(f^{n-1}(c_j) \\ &= \wp_\Lambda(f^{n-1}(c_j)) + p_0; \end{aligned}$$

therefore

$$\begin{aligned} |f^n(c_1) - p_0| &= |\wp_\Lambda(f^{n-1}(c_1))| \\ &= |\omega| \cdot |\wp_\Lambda(f^{n-1}(c_2))| = |\omega^2| \cdot |\wp_\Lambda(f^{n-1}(c_3))| \\ &= \rho_n, \end{aligned}$$

which is independent of j. Since the attracting fixed point p_0 must contain one of the critical values e_{j_0}, $\lim_{n\to\infty} \rho_n = 0$, and by the symmetry of the critical orbits about p_0, they all lie in the same attracting basin of p_0. □

We obtain the next stability result as a corollary.

THEOREM 5.3 (Stability Theorem 3). *Let $\Lambda = \Lambda(0, g_3)$ be a real triangular lattice. Then there exists $k \in \mathbb{R}$ and $b \in \mathbb{C}$ so that the map $\wp_{k\Lambda} + b$ is hyperbolic, stable in $\mathfrak{M}$, and $J(\wp_{k\Lambda} + b)$ is a Cantor set in a neighborhood of $m_0 = (k\Lambda, 1, b)$.*

PROOF. Since $g_3 \neq 0$, we set $k = (2g_3)^{1/6}$, taking the real root. Then $k\Lambda = k\Lambda(0, g3) = \Lambda(0, 1/2)$, by Lemma 2.1. We apply Theorem 5.2, choosing $b = p_0$, the corresponding zero for $\wp_{k\Lambda}$. We have a double toral band for the map $f(z) = \wp_{k\Lambda}(z) + p_0$ containing all the critical values, so by Theorem 3.2, $J(f)$ is a Cantor set.

Under small perturbations around the point: $m_0 = (k\Lambda, 1, p_0)$ the map f_m remains hyperbolic and $J(f_m)$ remains a Cantor set. □

Since minor modifications give Theorem 5.2 for $(g_2, g_3) = (0, -1/2)$, using the same techniques and identities, we can always choose the scaling constant to be a positive real number; this is illustrated in Figure 7. We note that when $g_3 = 1/2$, the fixed point p_0 is purely imaginary, and when $g_3 = -1/2$, p_0 is real.

THEOREM 5.4 (Stability Theorem 4). *Let $\Lambda = \Lambda(0, g_3)$ be any triangular lattice. Then there exists $k > 0$ and $b \in \mathbb{C}$ so that the map $\wp_{k\Lambda} + b$ is hyperbolic, stable in $\mathfrak{M}$, and $J(\wp_{k\Lambda} + b)$ is a Cantor set in a neighborhood of $m_0 = (k\Lambda, 1, b)$.*

PROOF. Since $g_3 \neq 0$, we set $k = (2|g_3|)^{1/6}$, choosing the positive real root. The rest of the proof is as in Theorem 5.3. □

EXAMPLE 5.5 (Stable example and conjecture). In this example, for a real triangular lattice, $\Lambda = [0, g_3]$, $g_3 \neq 0$, we describe a method to obtain a triangular lattice such that the map $f = \wp_{k\Lambda} + b$, with $b = c_3 - e_3$ is hyperbolic, has a superattracting fixed point at $c_3 \in \mathbb{R}$, and $J(f)$ is a Cantor set. Assuming this example exists, and we show one in Figure 8, the map f lies in a hyperbolic component of $\mathfrak{M}$. In particular, the value of b is quite different from the earlier examples.

Idea of the proof: We scale Λ and use the lattice $k\Lambda$ given by $k = (2g_3)^{1/6}$, so that we can assume that $g_3 = 1/2$. Therefore without loss of generality, it suffices to prove the result for $g_3 = 1/2$. We follow the steps as in Proposition 4.10, with modifications for the lattice shape:

- The critical point c_3 is fixed; this is by our choice of b.

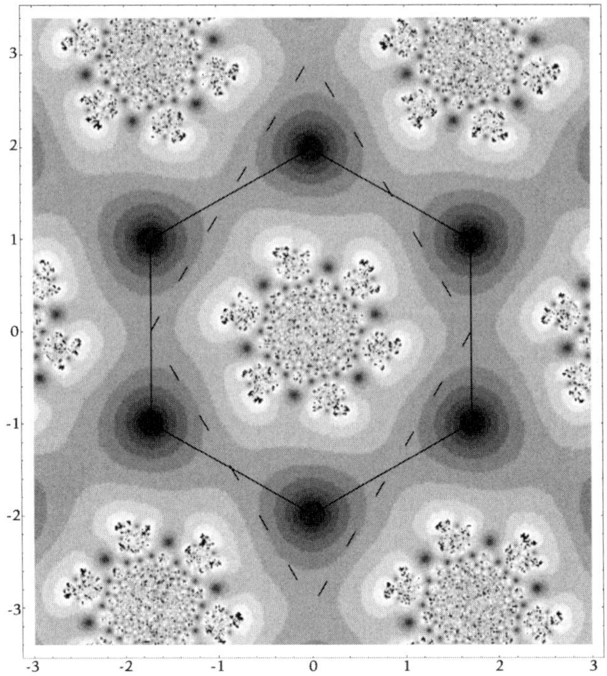

FIGURE 6. A hexagonal fundamental region for a real triangular lattice and a quadrilateral region (dashed). The 6 points of intersection between the two fundamental regions are points where $\wp_\Lambda{}'(c_j) = f'(c_j) = 0$, with f as in Thm 5.2; $J(f)$ is the Cantor set surrounded by white halos.

- The boundary of a fundamental region formed by the hexagon H with 3 distinguished segments: H_1 from z_1 to z_3, H_3, from z_3 vertically down to z_5, and H_2 from z_5 to $z_2 = -z_1$. They are labelled so that $c_j \in H_j$. Each H_j is mapped by f two-to-one onto a line segment L_j, a line segment of length .5 with a common endpoint of $b \in \mathbb{R}$. There is one branch point at the midpoint of H_j, which is the critical point c_j. One endpoint of each image segment L_j is $f(z_k) = b$, and there are 3 spokes coming out with endpoints c_3, $b + \omega c_3$, and $b + \omega^2 c_3$.
- The set $L_1 \cup L_2 \cup L_3$ converges to c_3 under iteration of f.

The difficulty: It remains to justify the last statement above. We have $e_2 = -\frac{1}{4} + i\frac{\sqrt{3}}{4}$ and $e_1 = -\frac{1}{4} - i\frac{\sqrt{3}}{4}$. The map $f = \wp_\Lambda + b$ is holomorphic in a neighborhood of c_3 and its series expansion can be calculated term by term. However, it is difficult to prove explicitly that a disk of radius r, with $r > .5$ lies in the immediate basin of attraction.

Assuming the numerics are correct, the map f is stable, and $J(f)$ with the Julia set of a perturbation are shown in Figure 8.

EXAMPLE 5.6. More generally it seems that this technique works for every real lattice. If we consider $(g_2, g_3) = (1.74063, -0.442781)$, we have a real rhombic

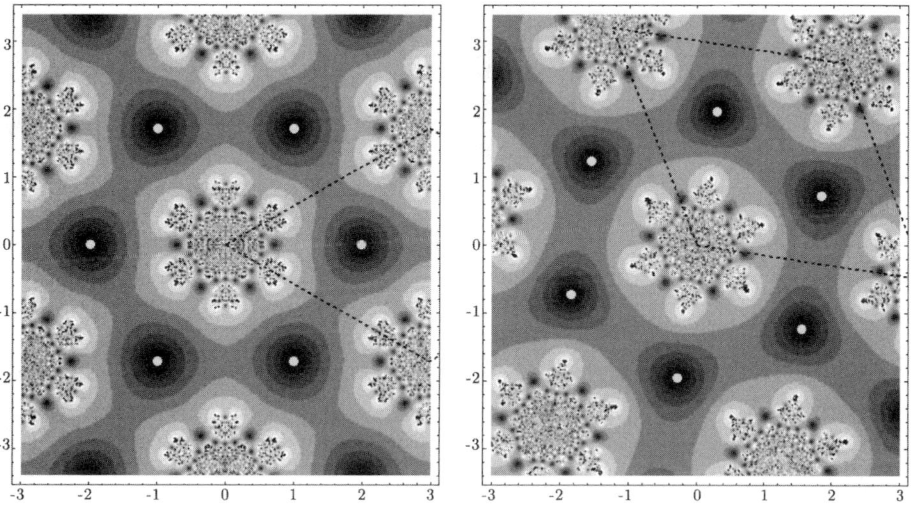

FIGURE 7. Illustration of Thm 5.3 showing Cantor Julia sets for triangular lattices. On the left, $g_3 = -.5$ and on the right, $g_3 = .5\exp(2\pi i/7)$. The centers of the triangles are marked in yellow (white) and the Julia set is the Cantor set of darker points with white halos.

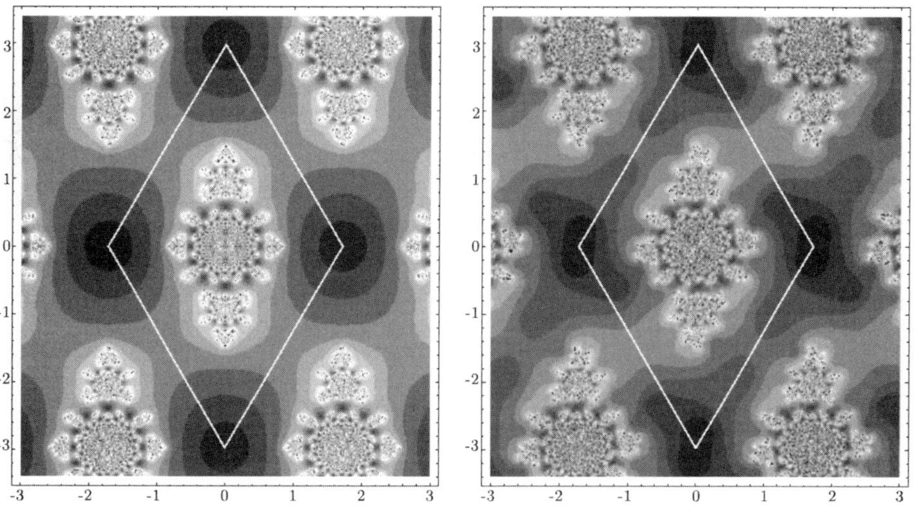

FIGURE 8. $J(\wp_\Lambda + b)$ from Example 5.5, with $b = 1.21732$ (left), and $J(a\wp_\Lambda + \tilde{b})$ (right), both with the real triangular lattice from Fig 6; $a = 1.05 + .15i$ and $\tilde{b} \approx 1.2782 + 0.1826i$.

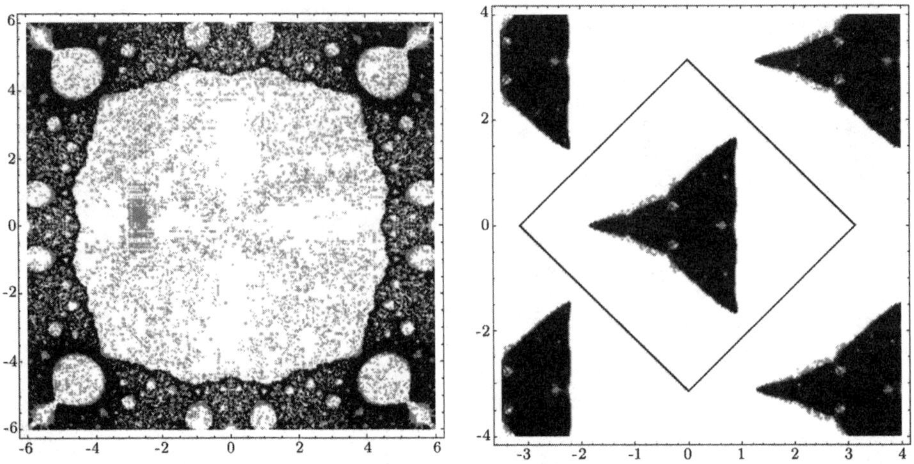

FIGURE 9. On the left we have a-space for the function $a\wp_\Lambda + b$ with $b = 3.11817$ and on the right, b-space, with $a = 1$. We use $(g_2, g_3) = (-1/2, 0)$ so we expect stability to show up around the value 1 on the left and the value $b = 3.11817$ on the right. Black is associated to unstable parameters, and white is associated to stable ones, with Cantor Julia set.

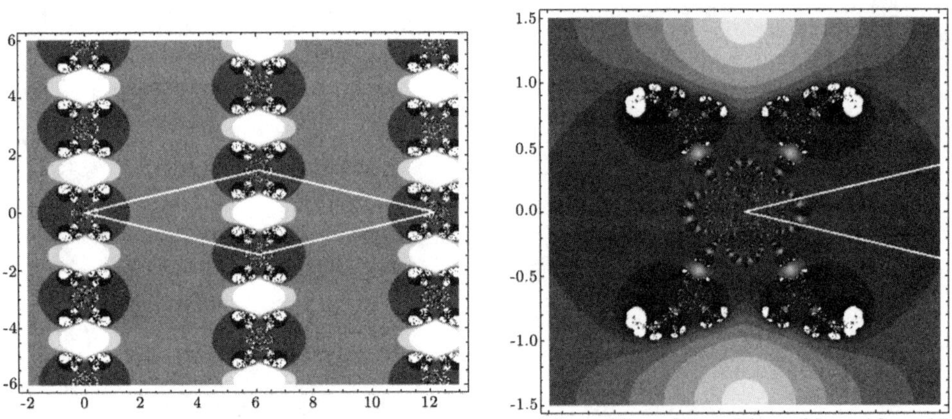

FIGURE 10. $J(\wp_{(2\Lambda_0)} + b)$ from Example 5.6, a Cantor set, for a real rhombic lattice Λ_0. Before scaling, $J(\wp_{\Lambda_0})$ is connected and $F(\wp_{\Lambda_0})$ has a single toral band, as shown in Figure 16.

lattice. We can repeat the technique of previous proofs to find Cantor Julia sets for $f = \wp_\Lambda + b$, with $b = c_3 - e_3$. The Julia set $J(f)$ is shown in Figure 10. The lattice Λ is exactly $2\Lambda_0$, one where $J(\wp_{\Lambda_0})$ is connected; it appears in Section 7.2, Figure 16.

6. Real rectangular lattices

In [**13**], the authors give sufficient conditions for Cantor Julia sets to occur for maps of the form $f(z) = \wp_\Lambda + b$, with Λ a real rectangular lattice, and b a constant lying on the horizontal half lattice line. They also present a number of examples showing that Cantor Julia sets often occur in this setting. The techniques used are very particular to real rectangular lattices and do not easily extend to other lattices. We summarize those results here.

Assume $\Lambda = [\lambda_1, i\lambda_2]$, with $\lambda_1, \lambda_2 > 0$; as usual $c_1 = \lambda_1/2, c_2 = i\lambda_2/2$, and $c_3 = c_1 + c_2$. Then $\wp([c_j]) \in \mathbb{R}$, $e_1 > 0$ and $e_2 < 0$.

LEMMA 6.1. *With the labeling as above, using $b = c_3 - e_3$, the function $f = \wp + b$ has a superattracting fixed point at c_3.*

This is an obvious statement, so it remains to give conditions for which all the critical values are in the attracting basin of the fixed point c_3 of $\wp_\Lambda + b$. We note that $b = \alpha + i\lambda_2/2$ for α real; the key observation is that b lies on the horizontal half lattice line, denoted:

$$L = \{z \in \mathbb{C} : z = t + i\lambda_2/2,\ t \in \mathbb{R}\}. \tag{6.1}$$

The line L contains all critical values of $\wp_\Lambda + b$ (since $e_j \in \mathbb{R}$). With b fixed at $c_3 - e_3$, and defining $f(z) = \wp_\Lambda + b$, we have the following results from [**13**].

LEMMA 6.2. *For any real rectangular lattice, the function $\wp_\Lambda$ maps L into $\mathbb{R}$ and f maps L into L. Moreover, $\wp_\Lambda'(z) \in \mathbb{R}$ if $z \in L$ and reaches a real maximum and minimum on every periodic interval of L.*

PROOF. On L, we have $\wp_\Lambda : L \to [e_2, e_3]$ ([**7**], Chap 1.19). For any $t \in \mathbb{R}$, $\wp_\Lambda : \mathbb{R} \to [e_1, \infty)$, so $f(t + c_2) = \wp_\Lambda(t + c_2) + \alpha + c_2$, which is then of the form $s + i\lambda_2/2 \in L$, since $s \in \mathbb{R}$. A simple argument that the stationary points of $\wp_\Lambda'$, which are the zeros of $\wp_\Lambda''$ occur on L as well (see e.g. [**7**]); that is, using (2.5), there is always a real root of $12\wp(z)^2 - g_2 = 0$ between e_2 and e_3. $\square$

LEMMA 6.3. *The function f given above maps the line $V = \{\lambda_1 + iy \ :\ y \in \mathbb{R}\}$ and $-V$ into L.*

PROOF. $\wp_\Lambda$ takes $\mathbb{R}$ and L to $\mathbb{R}$, and $\wp_\Lambda$ maps V and $-V$ into $\mathbb{R}$ [**7**], so f maps $\mathbb{R}, V$, and $-V$ into L whenever $b \in L$. $\square$

We have therefore reduced the problem to one on L; moreover $f|_L : L \to J$, where J is the compact line segment from $e_2 + c_2 \in L$ to $e_3 + c_2 \in L$.

THEOREM 6.4. *If the basin of attraction of $c_3 \in L$ contains $[0, \lambda_1/2] + i\lambda_2/2$, then $J(f)$ is a Cantor set.*

PROOF. The symmetry of $\wp_\Lambda$ about critical points gives the proof. $\square$

We can now give the result from [**13**], stated here in the current context.

THEOREM 6.5 (Stability theorem 4). *If $\Lambda = \Lambda(g_2, g_3)$ is real rectangular, and satisfies $\left[-g_3 \pm \left(\frac{g_2}{3}\right)^{3/2}\right]^{\frac{1}{2}} < 1$, then $J(f_m)$ is a Cantor set, for $m = (\Lambda, 1, c_3 - e_3)$ and f_m is stable.*

PROOF. By (2.10), Λ real rectangular implies, that $g_2^3 - 27g_3^2 > 0$; equivalently we have

$$\left(\frac{g_2}{3}\right)^3 > g_3^2,$$

so taking either square root, $\left[-g_3 \pm \left(\frac{g_2}{3}\right)^{3/2}\right]^{\frac{1}{2}}$ is real. Then Proposition 2.8 implies that $|\wp_\Lambda{}'|$ restricted to L is always strictly less than 1, so the entire line L, or equivalently the interval $[e_2, e_3]$ is attracted to c_3 under iteration of f. An application of Corollary 3.5 gives the result. □

7. Hyperbolic components of $\mathfrak{M}$ with connected Julia sets

There are many results on $\wp_\Lambda$ and $J(\wp_\Lambda)$ that imply stability for small values of b (e.g., [**9**–**11**]); in other words, $J(a\wp_\Lambda + b)$ moves holomorphically in a neighborhood of $(\Lambda, 1, 0)$. For the examples we consider in this section, $J(\wp_\Lambda)$ is connected. The connectivity of $J(\wp_\Lambda)$ was studied in [**11**] where the following was proved.

THEOREM 7.1. *$J(\wp_\Lambda)$ is connected if one (or more) of the following holds:*

(1) *Each critical value e_1, e_2, e_3 lies in its own Fatou component;*
(2) *There are 3 distinct attracting cycles.*
(3) *Every periodic Fatou component is completely contained in one fundamental period of Λ.*

These are sufficient but not necessary conditions; for Λ square or triangular, we always have that $J(\wp_\Lambda)$ is connected even if there are no attracting cycles or no Fatou components, both of which can occur ([**3**, **4**, **9**, **11**]). In the square lattice case, since $e_3 = 0$ is a pole, small perturbations can easily move e_3 to a component of the Fatou set as shown in Theorem 4.1. However, Theorem 7.1(2) will yield stable maps.

PROPOSITION 7.2. *If $\Lambda = k[1, \tau]$ and $\wp_\Lambda$ has 3 attracting cycles, then $m = (\Lambda, 1, 0)$ lies in a hyperbolic component of moduli space $\mathfrak{M}$.*

PROOF. There are 3 distinct critical values and each attracting cycle must have at least one critical value in its immediate attracting basin. Since there are no free critical values and the attracting cycles persist under small perturbations, the result follows. □

Figure 11 shows $J(\wp_\Lambda)$ for Λ real triangular, and Figure 15 shows $J(a\wp_\Lambda)$ for a (different) real triangular Λ, both with three distinct superattracting fixed points; they are discussed below.

7.1. Triangular lattices with hyperbolic maps $a\wp_\Lambda + b$ near $b = 0$. The symmetry in the lattice and the resulting dynamics of $\wp_\Lambda$ when Λ is a triangular lattice lead to many hyperbolic maps of the form $\wp_\Lambda$, and these lie in hyperbolic components of the family $a\wp_\Lambda + b$. We describe the setting briefly here.

In ([**9**], Theorem 8.3) the authors showed that there are infinitely many values of $g_3 \in \mathbb{R}$ corresponding to triangular lattices with $\wp_\Lambda$ having three superattracting fixed points, $p_1, p_2 = e^{2\pi i/3}p_1$, and $p_3 = e^{4\pi i/3}p_1$. It is clear that these maps will be in hyperbolic components of $\mathfrak{M}$ since the fixed points and their multipliers move holomorphically near $(\Lambda, 1, 0)$. We show a typical Julia set in Figure 11; there are bifurcations to stable maps with higher period attracting cycles as shown in Figures 12 (with $g_3 \approx 3.082$), coming from the period 2 limb of the Mandelbrot-like

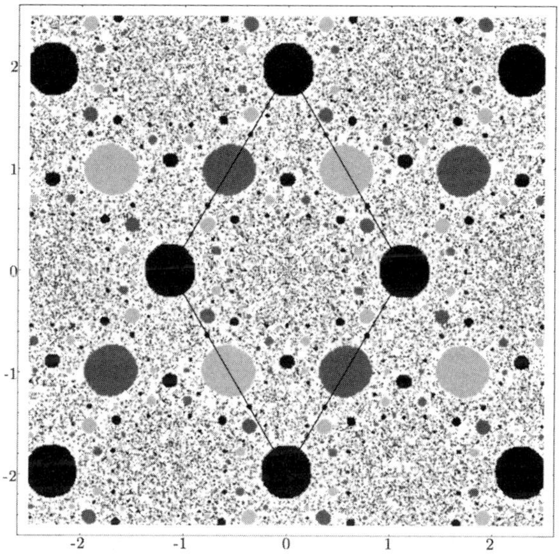

FIGURE 11. $J(\wp_{\Lambda_0})$ for Λ_0 a triangular lattice chosen according to ([**10**], Thm 8.3). There are 3 superattracting fixed points which move stably in Λ, a, and b for $a\wp_\Lambda + b$ near $(\Lambda_0, 1, 0)$. The different fixed point basins are colored distinct shades of gray.

set shown in Figure 14, and a perturbation in $\mathfrak{M}$ shown in Figure 13. Examples of the type shown in Figures 11 and 12 were first shown in [**9**].

We also have the following result which holds more generally in the triangular lattice setting. An example illustrating Theorem 7.3 appears in Figure 15.

THEOREM 7.3. *Suppose Λ is any triangular lattice. If c_1, c_2, c_3 are critical points with corresponding critical values e_1, e_2, e_3, then choosing $a = c_j/e_j$ yields a hyperbolic map of the form $a\wp_\Lambda$, with three superattracting fixed points.*

PROOF. We set $\omega = \exp(2\pi i/3)$, and label the residue classes of critical points to be the smallest critical points, and such that c_1 makes the smallest positive angle with the real axis. We then proceed counterclockwise to label $c_2 = \omega c_1$ and $c_3 = \omega^2 c_1$. We also have that $e_j = (g_3/4)^{1/3}$, where the roots are labelled by the critical points. Since $\wp_\Lambda(c_j) = e_j$, fix say $j = 1$ and choose $a = c_1/e_1$; then $a\wp_\Lambda(c_1) = ae_1 = c_1$. For this value of a, we have $a\wp_\Lambda(c_2) = (c_1/e_1)e_2$, but $e_2/e_1 = \omega$, so $a\wp_\Lambda(c_2) = \omega_1 c_1 = c_2$.

Similarly $a\wp_\Lambda(c_3) = (c_1/e_1)e_3 = \omega^2 c_1 = c_3$. Therefore we have three superattracting fixed points so by Proposition 7.2 (2), $f(z) = a\wp_\Lambda + b$ is stable in a neighborhood of $(\Lambda, a, 0)$. □

7.2. Toral bands for hyperbolic maps $a\wp_\Lambda + b$ near $b = 0$. A *toral band* is any Fatou component that is not completely contained in one period parallelogram; compare this definition with Definition 3.1.

In Section 3 we discussed how double toral bands give rise to hyperbolic components in $\mathfrak{M}$; this is also the case for certain toral bands with connected Julia sets. Depending on the nature of the toral band, these types of Fatou components can persist in hyperbolic components of $\mathfrak{M}$. We show in Figure 16 an example of

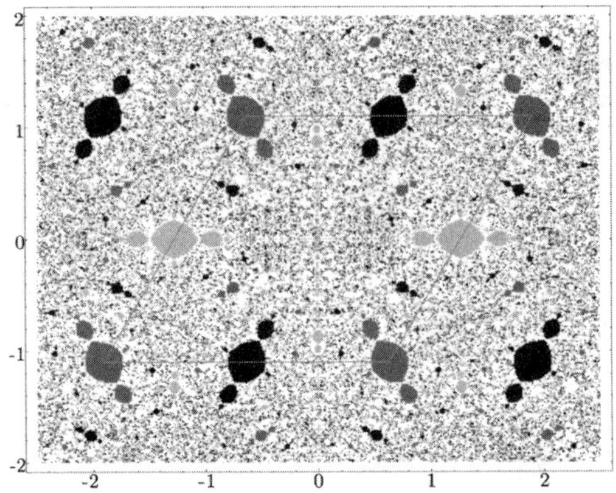

FIGURE 12. $J(\wp_\Lambda)$ for Λ a triangular lattice and $\wp_\Lambda$ a hyperbolic map with three attracting period two orbits, each a different shade of gray.

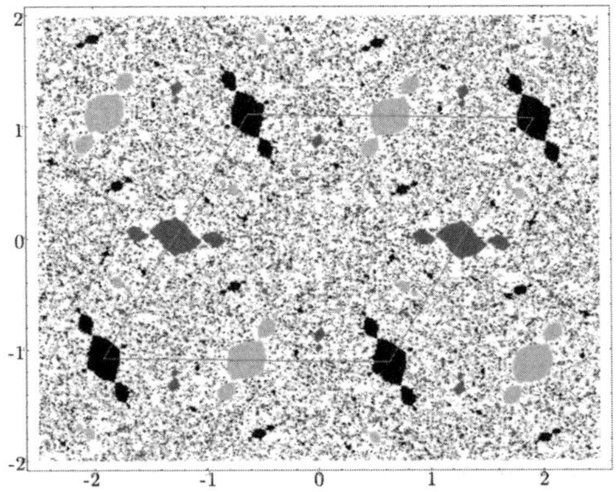

FIGURE 13. $J(\wp_{\Lambda'} + b)$, with Λ' a non-real lattice near Λ in Fig. 12 and b close to 0. The attracting period two orbits persist.

a lattice Λ with the following properties: (1) $\wp_\Lambda$ has a toral band; (2) $\wp_\Lambda$ has two attracting fixed points, one with a toral band in its immediate attracting basin, and another with small basins of attraction. (3) The Julia set is connected [**11**], and the map is hyperbolic. We show the Julia set of a perturbation of $\wp_\Lambda$ in $\mathfrak{M}$ to the right of $J(\wp_\Lambda)$.

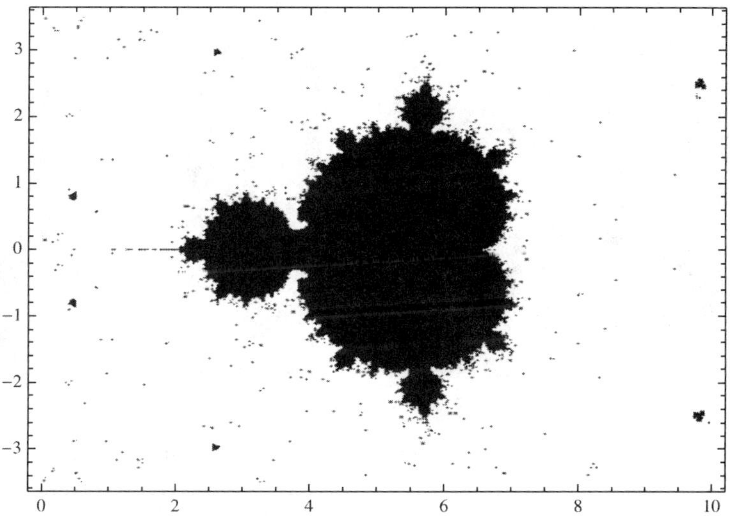

FIGURE 14. We show g_3 space for triangular lattices [**10**]. The periods of attracting cycles roughly match those in the Mandelbrot set; this shows that there are many hyperbolic components as the cycles move homomorphically for $a\wp_\Lambda + b$ near $(\Lambda_0, 1, 0)$ so the cycles persist.

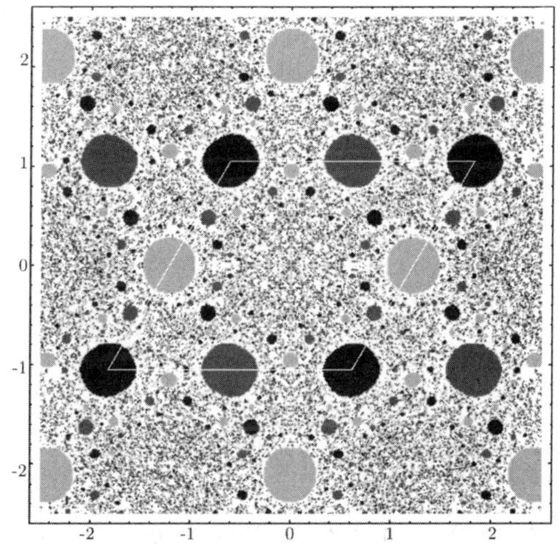

FIGURE 15. $J(a\wp_\Lambda)$ for Λ a "random" real triangular lattice, then scaled by $a = c_j/e_j$ for any of $j = 1, 2$, or 3. This results in 3 superattracting fixed points which move stably in Λ, a, and b for $a\wp_\Lambda + b$ near $(\Lambda, a, 0)$. The different fixed points and their attracting basins are colored distinct shades of gray.

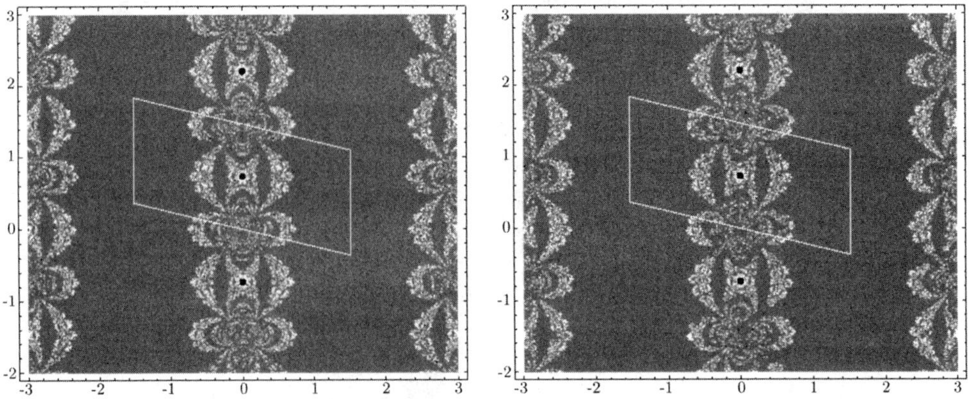

FIGURE 16. On the left, $J(\wp_\Lambda)$ for Λ a real rhombic lattice, and on the right, $J(a\wp_\Lambda + b)$ for $a = 1 + .02i$, $b = .1i$, show the stability. The maps each have an attracting fixed point with a black basin, and an attracting fixed point whose gray basin is a toral band.

References

[1] I. N. Baker, J. Kotus, and L. Yinian, *Iterates of meromorphic functions. I*, Ergodic Theory Dynam. Systems **11** (1991), no. 2, 241–248, DOI 10.1017/S014338570000612X. MR1116639

[2] W. Bergweiler, *Iteration of meromorphic functions*, Bull. Amer. Math. Soc. (N.S.) **29** (1993), no. 2, 151–188, DOI 10.1090/S0273-0979-1993-00432-4. MR1216719

[3] J. J. Clemons, *Connectivity of Julia sets for Weierstrass elliptic functions on square lattices*, Proc. Amer. Math. Soc. **140** (2012), no. 6, 1963–1972, DOI 10.1090/S0002-9939-2011-11079-7. MR2888184

[4] J. J. Clemons and L. Koss, *Higher order elliptic functions with connected Julia sets*, Topology Proc. **53** (2019), 57–72. MR3800419

[5] R. L. Devaney and L. Keen, *Dynamics of tangent*, Dynamical systems (College Park, MD, 1986), Lecture Notes in Math., vol. 1342, Springer, Berlin, 1988, pp. 105–111, DOI 10.1007/BFb0082826. MR970550

[6] R. L. Devaney and L. Keen, *Dynamics of meromorphic maps: maps with polynomial Schwarzian derivative*, Ann. Sci. École Norm. Sup. (4) **22** (1989), no. 1, 55–79. MR985854

[7] P. Du Val, *Elliptic functions and elliptic curves*, Cambridge University Press, London-New York, 1973. London Mathematical Society Lecture Note Series, No. 9. MR0379512

[8] J. Hawkins, *Smooth Julia sets of elliptic functions for square rhombic lattices*, Topology Proc. **30** (2006), no. 1, 265–278. MR2280672

[9] J. Hawkins and L. Koss, *Ergodic properties and Julia sets of Weierstrass elliptic functions*, Monatsh. Math. **137** (2002), no. 4, 273–300, DOI 10.1007/s00605-002-0504-1. MR1947915

[10] J. Hawkins and L. Koss, *Parametrized dynamics of the Weierstrass elliptic function*, Conform. Geom. Dyn. **8** (2004), 1–35, DOI 10.1090/S1088-4173-04-00103-1. MR2060376

[11] J. Hawkins and L. Koss, *Connectivity properties of Julia sets of Weierstrass elliptic functions*, Topology Appl. **152** (2005), no. 1-2, 107–137, DOI 10.1016/j.topol.2004.08.018. MR2160809

[12] J. Hawkins, L. Koss, and J. Kotus, *Elliptic functions with critical orbits approaching infinity*, J. Difference Equ. Appl. **16** (2010), no. 5-6, 613–630, DOI 10.1080/10236190903203895. MR2642469

[13] J. Hawkins and M. Moreno Rocha, *Dynamics and Julia set of iterated elliptic functions*, New York J. Math. **24** (2018), 947–979. MR3874958

[14] L. Keen and J. Kotus, *Ergodicity of some classes of meromorphic functions*, Ann. Acad. Sci. Fenn. Math. **24** (1999), no. 1, 133–145. MR1670876

[15] L. Koss, *Examples of parametrized families of elliptic functions with empty Fatou sets*, New York J. Math. **20** (2014), 607–625. MR3262023

[16] L. Koss and K. Roy, *Dynamics of vertical real rhombic Weierstrass elliptic functions*, Involve **10** (2017), no. 3, 361–378, DOI 10.2140/involve.2017.10.361. MR3583871

[17] J. Kotus, *Elliptic functions with critical points eventually mapped onto infinity*, Monatsh. Math. **149** (2006), no. 2, 103–117, DOI 10.1007/s00605-005-0373-5. MR2264577

[18] J. Kotus and M. Urbański, *Hausdorff dimension and Hausdorff measures of Julia sets of elliptic functions*, Bull. London Math. Soc. **35** (2003), no. 2, 269–275, DOI 10.1112/S0024609302001686. MR1952406

[19] R. Mañé, P. Sad, and D. Sullivan, *On the dynamics of rational maps*, Ann. Sci. École Norm. Sup. (4) **16** (1983), no. 2, 193–217. MR732343

[20] C. T. McMullen, *Complex dynamics and renormalization*, Annals of Mathematics Studies, vol. 135, Princeton University Press, Princeton, NJ, 1994. MR1312365

[21] L. M. Milne-Thomson, *Jacobian elliptic function tables*, Dover Publications, Inc., New York, N. Y., 1950. MR0088071

[22] J. Milnor, *Dynamics in one complex variable*, 3rd ed., Annals of Mathematics Studies, vol. 160, Princeton University Press, Princeton, NJ, 2006. MR2193309

[23] M. Moreno Rocha and P. Pérez Lucas, *A class of even elliptic functions with no Herman rings*, Topology Proc. **48** (2016), 151–162. MR3355212

Department of Mathematics, CB #325), University of North Carolina at Chapel Hill, Chapel Hill, NC 27599

Email address: jmh@math.unc.edu

Contemporary Mathematics
Volume **736**, 2019
https://doi.org/10.1090/conm/736/14848

Pressure and escape rates for random subshifts of finite type

Kevin McGoff

ABSTRACT. In this work we consider several aspects of the thermodynamic formalism in a randomized setting. Let X be a non-trivial mixing shift of finite type, and let $f : X \to \mathbb{R}$ be a Hölder continuous potential with associated Gibbs measure μ. Further, fix a parameter $\alpha \in (0,1)$. For each $n \geq 1$, let $\mathcal{F}_n$ be a random subset of words of length n, where each word of length n that appears in X is included in $\mathcal{F}_n$ with probability $1 - \alpha$ (and excluded with probability α), independently of all other words. Then let $Y_n = Y(\mathcal{F}_n)$ be the random subshift of finite type obtained by forbidding the words in $\mathcal{F}_n$ from X. In our first main result, for α sufficiently close to 1 and n tending to infinity, we show that the pressure of f on Y_n converges in probability to the value $P_X(f)+\log(\alpha)$, where $P_X(f)$ is the pressure of f on X. Additionally, let $H_n = H(\mathcal{F}_n)$ be the random hole in X consisting of the union of the cylinder sets of the words in $\mathcal{F}_n$. For our second main result, for α sufficiently close to one and n tending to infinity, we show that the escape rate of μ-mass through H_n converges in probability to the value $-\log(\alpha)$ as n tends to infinity.

1. Introduction

Random subshifts of finite type were introduced in [**29**], and they have subsequently been studied in [**3, 30, 31**]. Let us quickly recall their definition. Let X be a non-trivial mixing subshift of finite type (SFT), and let $B_n(X)$ be the set of words of length n that appear in X. For a fixed parameter $\alpha \in (0,1)$ and $n \in \mathbb{N}$, let $\mathcal{F}_n$ be the randomly selected subset of $B_n(X)$ formed by including each word from $B_n(X)$ in $\mathcal{F}_n$ with probability $1-\alpha$ (and excluding it from $\mathcal{F}_n$ with probability α), independently of all other words. Then let $Y_n = Y(\mathcal{F}_n)$ be the set of points in X that do not contain any word from $\mathcal{F}_n$. We refer to Y_n as a random SFT. In order to study random SFTs, we fix the ambient system X and the parameter α. Then we seek to describe the properties of Y_n that have probability tending to one as n tends to infinity. This framework gives a precise way to describe the behavior of "typical" SFTs within the ambient system X.

Previous work on random SFTs [**3, 29–31**] has established the existence of at least one critical value α_c such that the typical behavior of Y_n changes abruptly as α crosses this value. Indeed, when $\alpha < \alpha_c$, there is a positive limit for the probability that Y_n is empty, and Y_n has zero entropy with probability tending to one. On the other hand, for $\alpha > \alpha_c$, the probability that Y_n is empty tends to zero, and the entropy of Y_n converges in probability to the value $h(X) + \log(\alpha)$, where $h(X)$

2010 *Mathematics Subject Classification.* Primary: 37B10.

is the entropy of X. (Note that this value is positive for $\alpha > \alpha_c$.) Furthermore, for α close enough to one, Y_n contains a unique "giant component," which is itself a mixing SFT with full entropy, as well as a random number of isolated periodic orbits. See [**29**] for details.

In the present work, we study some aspects of the thermodynamic formalism for random SFTs in the super-critical regime ($\alpha > \alpha_c$). In our first main result (Theorem 1.1), we describe the distribution of the pressure of random SFTs for a fixed potential function, and in our second main result (Theorem 1.4), we describe the distribution of the escape rate of mass of Gibbs measures through random holes. Although these two topics may not at first appear to be related, they are in fact quite closely connected, as demonstrated by Proposition 7.1.

1.1. Pressure of random SFTs. Suppose that $f : X \to \mathbb{R}$ is a fixed Hölder continuous potential function. We seek to identify the limiting behavior of the pressure of f restricted to the random SFT Y_n in the limit as n tends to infinity. For notation, let $P_Y(f)$ denote the topological pressure of f restricted to any subshift $Y \subset X$ (see Section 2 for definitions).

THEOREM 1.1. *Let X be a non-trivial mixing SFT and $f : X \to \mathbb{R}$ Hölder continuous. Then there exists $\gamma_0 \in (0,1)$ such that for each $\alpha \in (\gamma_0, 1]$ and for each $\epsilon > 0$, there exists $\rho > 0$ such that for all large enough n,*

$$\mathbb{P}_\alpha\Big(\big|P_{Y_n}(f) - \big(P_X(f) + \log(\alpha)\big)\big| \geq \epsilon \Big) < e^{-\rho n}.$$

In other words, the pressure of f on the random SFT Y_n converges in probability to the value $P_X(f) + \log(\alpha)$. Note that when $f \equiv 0$, we recover the result of [**29**] regarding the entropy of random SFTs.

REMARK 1.2. The particular value of γ_0 that we use in our proof is given in Definition 2.1. It is possible that this value of γ_0 is not optimal, in the sense that for some choices of X and f, the statement might remain true with a smaller value of γ_0. However, one may check that for $f \equiv 0$, our definition of γ_0 is equal to α_c, which is optimal.

REMARK 1.3. The proof of Theorem 1.1 appears in Section 3. The broad outline of this proof is similar to the outline of the proof of [**29**, Theorem 1.3] concerning the entropy of random SFTs. However, the core technical results in the proof, which appear in Section 4, require new ideas to handle the fact that f may not be zero. In particular, we must estimate the μ measure of the appearance of certain types of repeated patterns, where μ is a Gibbs measure but not necessarily a measure of maximal entropy, and this generalization requires substantially new ideas.

1.2. Escape rate through random holes. Our second main result involves thinking of the random set of forbidden words $\mathcal{F}_n$ as a hole in the ambient system X, creating an open dynamical system. For an introduction to open systems, see [**14**] and references therein. Previous work on open systems has focused largely on the existence and properties of the escape rate of mass through the hole, as well as the existence and properties of conditionally invariant distributions (conditional on avoiding the hole); for an incomplete sampling of the literature on open systems, see [**4–13, 15–25, 27, 28, 33**]. In this work we focus on the escape rate of mass through the hole, which we define below. Let $\sigma : X \to X$ denote the left-shift map on X. For a Borel probability measure μ on X and a hole $H \subset X$, the escape rate of μ through H is defined to be $-\varrho(\mu : H)$, where

$$\varrho(\mu : H) = \lim_m \frac{1}{m} \log \mu\Big(\{x \in X : \forall k \in \{0, \dots, m-1\}, \sigma^k(x) \notin H\}\Big),$$

whenever the limit exists.

In this work we consider random holes H_n, constructed from the set of forbidden words $\mathcal{F}_n$ as follows:

$$H_n = \bigcup_{w \in \mathcal{F}_n} \{x \in X : x_0 \dots x_{n-1} = w\}.$$

Our goal is to describe the escape rate of mass of Gibbs measures through H_n.

THEOREM 1.4. *Let X be a non-trivial mixing SFT and let μ be the Gibbs measure associated to the Hölder continuous potential function $f : X \to \mathbb{R}$. Then there exists $\gamma_0 \in (0, 1)$ such that for each $\alpha \in (\gamma_0, 1]$ and for each $\epsilon > 0$, there exists $\rho > 0$ such that for all large enough n,*

$$\mathbb{P}\Big(\big|\varrho(\mu : H_n) - \log(\alpha)\big| \geq \epsilon\Big) < e^{-\rho n}.$$

Thus the escape rate of μ through the randomly selected hole H_n converges in probability to $-\log(\alpha)$.

REMARK 1.5. With probability tending to one, the hole H_n consists of the union of approximately $(1-\alpha)|B_n(X)|$ cylinder sets of length n. One may think of H_n as typically consisting of the union of many small holes spread randomly throughout the state space. Furthermore, the expected value of the μ-measure of H_n is $1 - \alpha$, which remains bounded away from zero as n tends to infinity. In this sense, the holes considered here differ substantially from the "small holes" studied in some previous work [**5**, **16**, **19**, **23**].

REMARK 1.6. In [**5**], the authors prove that in the deterministic setting, the escape rate depends on both the size (measure) of the hole and its precise location in state space. In contrast, Theorem 1.4 shows that for random holes, the escape rate is well approximated by a function that depends only on the expected measure of the hole. Indeed, the expected measure of the hole is $(1 - \alpha)$ (see Remark 2.3), so the expected measure of its complement is α. Then Theorem 1.4 yields that the escape rate converges in probability to the value $-\log(\alpha)$.

REMARK 1.7. A naive approximation of the hitting time for the hole is given by a geometrically distributed random variable τ with probability of success $p = 1-\alpha$. Since $\mathbb{P}(\tau = k) = \alpha^{k-1}(1-\alpha)$, we have that

$$\lim_k \frac{1}{k} \log \mathbb{P}(\tau > k) = \log \alpha.$$

From this perspective, Theorem 1.4 may be interpreted as giving precise meaning to the statement that the escape rate through the random hole is approximately the same as one would obtain if the hitting time of the hole were geometrically distributed with probability of success equal to the measure of the hole.

As a consequence of Theorem 1.4, we can also estimate the escape rate of mass of Gibbs measures for Axiom A diffeomorphisms through randomly selected Markov holes. As the proof relies solely on Theorem 1.4 and the well-known relationship between Markov partitions for Axiom A diffeomorphisms and SFTs (see [**2**]), we omit the proof.

COROLLARY 1.8. *Let $T : M \to M$ be an Axiom A diffeomorphism such that the restriction of T to its non-wandering set is topologically mixing, and let $f : M \to \mathbb{R}$ be a Hölder continuous potential with Gibbs measure ν. Let ξ be a finite Markov partition of M with diameter small enough that the symbolic dynamics is well-defined, and let $\xi^n = \bigvee_{k=0}^{n-1} T^{-k}\xi$. Then there exists $\gamma_0 \in (0,1)$ such that for each $\alpha \in (\gamma_0, 1]$, the following holds. Let H_n be the randomly selected hole obtained by including each cell of ξ^n independently with probability $1-\alpha$. Then for each $\epsilon > 0$, there exists $\rho > 0$ such that for large enough n,*

$$\mathbb{P}\Big(\big|\varrho(\nu : H_n) - \log(\alpha)\big| \geq \epsilon \Big) < e^{-\rho n}.$$

1.3. Outline of the paper. The following section collects some background definitions and results that are used elsewhere in the paper. In Section 3, we present the proof of Theorem 1.1 with the help of several technical lemmas. These technical lemmas are then proved in Sections 4, 5, and 6. Finally, in Section 7, we establish Proposition 7.1, which relates escape rates to pressure, and then we prove Theorem 1.4.

2. Background and notation

2.1. Symbolic dynamics. Let $\mathcal{A}$ be a finite set, which we call the alphabet. We let $\Sigma = \mathcal{A}^{\mathbb{Z}}$ denote the full-shift, and we let $\sigma : \Sigma \to \Sigma$ be the left-shift map, $\sigma(x)_n = x_{n+1}$. We endow Σ with the product topology from the discrete topology on $\mathcal{A}$, which makes σ a homeomorphism. We define the metric $d(\cdot,\cdot)$ on Σ by the rule $d(x,y) = 2^{-n(x,y)}$, where $n(x,y)$ is the infimum of all $|m|$ such that $x_m \neq y_m$.

A subset $X \subset \Sigma$ is a *subshift* if it is closed and $\sigma(X) = X$. In the context of a subshift X, we also use the symbol σ to denote the restriction of the left-shift to X. A *word* on $\mathcal{A}$ is an element of $\mathcal{A}^m$ for some $m \geq 1$. We also refer to the empty word as a word. If $u = u_1 \dots u_m$ is in $\mathcal{A}^m$, then we say that u has length m, and we let u_i^j denote the subword $u_i \dots u_j$. Further, we let $B_m(X)$ denote the set of words of length m that appear in some point in X. For any word w in $B_m(X)$, we let $[w]$ denote the set of points $x \in X$ such that $x_0 \dots x_{m-1} = w$. Also, for $x \in X$ and $i \leq j$, we let $x[i,j]$ denote the set of points $y \in X$ such that $y_i \dots y_j = x_i \dots x_j$.

A subset $X \subset \Sigma$ is a *subshift of finite type* (SFT) if there exists a natural number m and a collection of words $\mathcal{F} \subset \mathcal{A}^m$ such that X is exactly the set of points in Σ that contain no words from $\mathcal{F}$. We say that an SFT is non-trivial if it contains at least two points. The SFT X is mixing if there exists N such that for all points $x, y \in X$, there exists a point $z \in X$ such that $x(-\infty,0] = z(-\infty,0]$ and $y[N,\infty) = z[N,\infty)$.

For any subshift $X \subset \Sigma$, we let $M(X,\sigma)$ denote the set of Borel probability measures μ on X such that $\mu(\sigma^{-1}A) = \mu(A)$ for all Borel sets $A \subset X$. Suppose $\mu \in M(X,\sigma)$. When it will not cause confusion, we write $\mu(w)$ to denote the measure of the cylinder set $[w]$, where $w \in B_m(X)$ for some $m \geq 1$.

For any measure $\mu \in M(X, \sigma)$, one may define the entropy of μ as

$$h(\mu) = \lim_m \frac{1}{m} \sum_{w \in B_m(X)} -\mu(w) \log \mu(w),$$

where the limit exists by subadditivity.

2.2. Pressure and equilibrium states. Let Y be a subshift, and let $f : Y \to \mathbb{R}$ be continuous. For $m \geq 1$ and w in $B_m(Y)$, let

$$S_m f(w) = \sup_{x \in Y \cap [w]} \sum_{k=0}^{m-1} f \circ \sigma^k(x).$$

Then let

$$\Lambda_m(Y) = \sum_{w \in B_m(Y)} e^{S_m f(w)}.$$

Finally, the (topological) pressure of f on Y is defined as

$$P_Y(f) = \lim_{m \to \infty} \frac{1}{m} \log \Lambda_m(Y),$$

where the limit exists by subadditivity.

The well-known Variational Principle (see [**34**]) states that

$$P_Y(f) = \sup\left\{ \int f \, d\mu + h(\mu) : \mu \in M(Y, \sigma) \right\}.$$

For a subshift Y, this supremum must be realized, and any measure that attains the supremum is known as an equilibrium state for f on Y.

Now suppose that X is a mixing SFT and $f : X \to \mathbb{R}$ is Hölder continuous. In this case, it is known that there is a unique equilibrium state $\mu \in M(X, \sigma)$ for f, and furthermore μ satisfies the following Gibbs property: there exists $K > 1$ such that for all $n \geq 1$ and $x \in X$,

$$K^{-1} \leq \frac{\mu(x[0, n-1])}{\exp\left(-P_X(f) \cdot n + \sum_{k=0}^{n-1} f \circ \sigma^k(x)\right)} \leq K. \tag{2.1}$$

We may now give a definition for the parameter γ_0 that appears in Theorems 1.1 and 1.4.

DEFINITION 2.1. Let X be a non-empty mixing SFT, and let $f : X \to \mathbb{R}$ be a Hölder continuous potential with associated Gibbs measure μ. Then let $\gamma_0 = \gamma_0(X, f)$ be defined by

$$\gamma_0 = \inf\left\{ \gamma > 0 : \exists n_0, \forall m \geq n_0, \forall u \in B_m(X), \, \mu(u) \leq \gamma^m \right\}.$$

Note that by [**1**, Lemma 5], if X is non-trivial, then $\gamma_0 < 1$. We also make the following remark. Suppose $1 \geq \gamma > \gamma_0$, and fix n_0 such that $\mu(u) \leq \gamma^{|u|}$ whenever $|u| \geq n_0$. Then for any word u we have $\mu(u) \leq \gamma^{|u|-n_0}$; indeed, if $|u| \geq n_0$, then it follows from the choice of n_0, and if $|u| \leq n_0$, then $\mu(u) \leq 1 \leq \gamma^{|u|-n_0}$.

It is well-known (see, *e.g.*, [**2**, Proof of Proposition 1.14]) that μ satisfies a mixing property called ψ-mixing, from which a variety of mixing-type estimates may be deduced. The bounds required for the present work are summarized in the following lemma, which we state without proof.

LEMMA 2.2. *Let X be a non-trivial mixing SFT with Hölder continuous potential $f : X \to \mathbb{R}$ and associated Gibbs measure μ. Then there exist constants $K > 0$ and $g_0 \geq 1$ such that:*

- *the Gibbs property (2.1) holds;*
- *for all $m, n \geq 1$ and for all $u \in B_m(X)$ and $v \in B_n(X)$ such that $uv \in B_{m+n}(X)$, we have*
$$\mu(uv) \leq K\mu(u)\mu(v);$$
- *for all $m, n \geq 1$ and for all $u \in B_m(X)$ and $v \in B_n(X)$ such that $uv \in B_{m+n}(X)$, we have*
$$\mu(\sigma^{-m}[v] \mid [u]) \leq K\mu([v]);$$
- *for $g \geq g_0$, for all $m, n \geq 1$ and for all $u \in B_m(X)$ and $v \in B_n(X)$, we have*
$$\mu\Big([u] \bigcap \sigma^{-g+m}[v]\Big) \geq K^{-1}\mu([u])\mu([v]).$$

2.3. Basics of random SFTs. Let X be a non-trivial mixing SFT. Fix $\alpha \in (0, 1)$. Recall that $\mathcal{F}_n$ denotes the random subset of $B_n(X)$ formed by including each word with probability $1-\alpha$, independently of all other words, and $Y_n = Y(\mathcal{F}_n)$ is the random SFT formed by forbidding the words $\mathcal{F}_n$ from X. Here we establish some notation and basic facts for random SFTs.

Let $u \in B_k(X)$ for some $k \geq n$. We let $W_n(u)$ denote the set of all words of length n that appear in u:

$$W_n(u) = \big\{v \in B_n(X) : \exists j \in \{1, \dots, k-n+1\},\, u_j^{j+n-1} = v\big\}.$$

Then let ξ_u denote the indicator function of the event that u contains no words from $\mathcal{F}_n$, *i.e.*,

$$\xi_u = \begin{cases} 1, & \text{if } W_n(u) \cap \mathcal{F}_n = \varnothing \\ 0, & \text{otherwise.} \end{cases}$$

Since each word in $W_n(u)$ is excluded from $\mathcal{F}_n$ with probability α, independently of all other words, we have that

$$\mathbb{E}\big[\xi_u\big] = \mathbb{P}\big(W_n(u) \cap \mathcal{F}_n = \varnothing\big) = \alpha^{|W_n(u)|}. \tag{2.2}$$

Furthermore, for $u, v \in B_k(X)$, the covariance of ξ_u and ξ_v is given by

$$\begin{aligned} \mathbb{E}\Big[\big(\xi_u - \mathbb{E}[\xi_u]\big)\big(\xi_v - \mathbb{E}[\xi_v]\big)\Big] &= \mathbb{E}\big[\xi_u\xi_v\big] - \mathbb{E}\big[\xi_u\big]\mathbb{E}\big[\xi_v\big] \\ &= \alpha^{|W_n(u)\cup W_n(v)|} - \alpha^{|W_n(u)|+|W_n(v)|} \\ &= \alpha^{|W_n(u)\cup W_n(v)|}\big(1 - \alpha^{|W_n(u)\cap W_n(v)|}\big). \end{aligned} \tag{2.3}$$

REMARK 2.3. Let X be a non-trivial mixing SFT, and let $\mu \in M(X, \sigma)$. Then the expected value of the μ-measure of the hole H_n is $1 - \alpha$, since

$$\begin{aligned} \mathbb{E}\big[\mu(H_n)\big] &= \mathbb{E}\Bigg[\sum_{u\in B_n(X)} \mu(u)(1 - \xi_u)\Bigg] \\ &= \sum_{u\in B_n(X)} \mu(u)(1 - \mathbb{E}\big[\xi_u\big]) = (1-\alpha)\,\mu(B_n(X)) = 1 - \alpha. \end{aligned}$$

2.4. Repeats and repeat covers. We use interval notation to denote intervals in $\mathbb{Z}$. For example, $[1,3] = \{1,2,3\}$ and $[0,5) = \{0,1,2,3,4\}$. Furthermore, for a set $F \subset \mathbb{Z}$ and $t \in \mathbb{Z}$, we let $t + F = \{t + s : s \in F\}$. For n in $\mathbb{N}$ and $F \subset \mathbb{Z}$, we let $\mathcal{C}_n(F)$ denote the set of intervals of length n contained in F:

$$\mathcal{C}_n(F) = \{t + [0,n) : t + [0,n) \subset F\}.$$

We will also consider sets of pairs of intervals; that is, we consider sets $\mathcal{R} \subset \mathcal{C}_n(F) \times \mathcal{C}_n(F)$. We let $\mathrm{Power}(\mathcal{C}_n(F) \times \mathcal{C}_n(F))$ denote the power set of $\mathcal{C}_n(F) \times \mathcal{C}_n(F)$. For such $\mathcal{R}$, we let $|\mathcal{R}|$ denote the number of pairs in $\mathcal{R}$, and we let

$$A(\mathcal{R}) = \bigcup_{(I_1,I_2)\in\mathcal{R}} I_2.$$

Now we define *repeats* and *repeat covers*, which were used implicitly in [**29**] and then defined explicitly in [**31**].

DEFINITION 2.4. Let $\mathcal{A}$ be a finite set. Let $F \subset \mathbb{Z}$, and let $u \in \mathcal{A}^F$. A pair (I_1, I_2) in $\mathcal{C}_n(F) \times \mathcal{C}_n(F)$ is an *n-repeat* (or just a *repeat*) for u if $u_{I_1} = u_{I_2}$ and I_1 is the lexicographically minimal occurrence of the word u_{I_1} in u. In that case, the word u_{I_1} is called a repeated word for u. Furthermore, a set $\mathcal{R} \subset \mathcal{C}_n(F) \times \mathcal{C}_n(F)$ is a *repeat cover* for u if

(1) each pair $(I_1, I_2) \in \mathcal{R}$ is a repeat for u, and
(2) for each repeat (I_1, I_2) for u, we have $I_2 \subset A(\mathcal{R})$.

Note that every pattern $u \in \mathcal{A}^F$ has a repeat cover, which contains all repeats for u. When $\mathcal{R}$ is a repeat cover for u, we may refer to $A(\mathcal{R})$ as the repeat area for u (and we note that it is independent of the choice of repeat cover for u). In many cases, we seek to find efficient repeat covers, by which we mean repeat covers $\mathcal{R}$ such that $|\mathcal{R}|$ is small enough for our purposes. In this paper, we only require the bound supplied by the following lemma, which is a slightly weaker version of Lemma 3.8 in [**31**].

LEMMA 2.5. *Let $F \subset \mathbb{Z}$ be a finite union of intervals of length n, and suppose $u \in \mathcal{A}^F$. Then there exists an n-repeat cover $\mathcal{R}$ for u such that*

$$|\mathcal{R}| \leq 4|F|/n.$$

Additionally, our proofs make use of the following estimate relating the number of unique words of length n in $u \in B_k(X)$ and the cardinality of the repeat area for u.

LEMMA 2.6. *Let $F \subset \mathbb{Z}$ be a finite union of intervals of length n, and let $a = |\{s : s + [0,n) \subset F\}|$. Suppose that $u \in \mathcal{A}^F$ satisfies $|W_n(u)| = j < a$. Then for any repeat cover $\mathcal{R}$ for u,*

$$|A(\mathcal{R})| \geq a + n - j - 1.$$

PROOF. Let $r = a - j$, which is the number of repeats for u. The lexicographically minimal repeat for u contributes n elements to $A(\mathcal{R})$, and each of the other $r - 1$ repeats must contribute at least one element. Altogether, we must have $|A(\mathcal{R})| \geq n + r - 1 = n + a - j - 1$. □

In many of the proofs in Section 4, we decompose words into alternating blocks of repeated regions (*i.e.*, regions contained in $A(\mathcal{R})$ for some repeat cover $\mathcal{R}$) and non-repeated regions. The following definition standardizes some notation that is useful for such decompositions. We endow $\mathbb{Z}$ with the standard graph structure, in which two nodes $x, y \in \mathbb{Z}$ are connected by an edge whenever $|x-y| = 1$. We then endow all subsets of $\mathbb{Z}$ with the induced subgraph structure, and references to connected components refer to this subgraph structure. Furthermore, we give $\mathbb{Z}$ the standard ordering, and if I and J are disjoint subsets of $\mathbb{Z}$, then we let $I < J$ whenever $x < y$ for all $x \in I$ and $y \in J$.

DEFINITION 2.7. Let $A \subset [0,k)$ be a union of intervals of length n such that $0 \notin A$. Then the *interval decomposition* of $[0,k)$ induced by A consists of $\big((I_m)_{m=1}^{N+1}, (J_m)_{m=1}^{N}\big)$, where

- each J_m is a non-empty maximal connected component of A, and $\bigcup_m J_m = A$;
- each I_m is a maximal connected component of $[0,k) \setminus A$, and $\bigcup_m I_m = [0,k) \setminus A$;
- only I_{N+1} may be empty;
- for each $m = 1, \ldots, N$, we have $I_m < J_m < I_{m+1}$.

Now suppose $b \in \mathcal{A}^k$ and $\mathcal{R} \subset \mathcal{C}_n([0,k)) \times \mathcal{C}_n([0,k))$. Let $A = A(\mathcal{R})$, and let $\big((I_m)_{m=1}^{N+1}, (J_m)_{m=1}^{N}\big)$ be the interval decomposition of $[0,k)$ induced by A. For each m, we let $u_m = b|_{I_m}$ and $v_m = b|_{J_m}$. We refer to $\big((u_m)_{m=1}^{N+1}, (v_m)_{m=1}^{N}\big)$ as the *block decomposition* of b. If $\mathcal{R}$ is a repeat cover for b, then we refer to $\big((u_m)_{m=1}^{N+1}, (v_m)_{m=1}^{N}\big)$ as the *repeat block decomposition* of b. Note that $N \leq |\mathcal{R}|$.

When μ is a Gibbs measure associated to a Hölder continuous potential, the following lemma, which is used several times in Section 4, gives an estimate of the μ-measure of any word b in terms of a block decomposition.

LEMMA 2.8. *Let X be a non-trivial mixing SFT with Hölder continuous potential $f : X \to \mathbb{R}$ and associated Gibbs measure μ. Let $K > 0$ satisfy the conclusions of Lemma 2.2. Let $b \in B_k(X)$, and suppose that $\big((u_m)_{m=1}^{N+1}, (v_m)_{m=1}^{N}\big)$ is a block decomposition of b. Then*

$$\mu(b) \leq K^{2N} \prod_{m=1}^{N+1} \mu(u_m) \prod_{m=1}^{N} \mu(v_m).$$

PROOF. Let $\big((I_m)_{m=1}^{N+1}, (J_m)_{m=1}^{N}\big)$ be a block decomposition of b. Let s_m be the minimal element of the corresponding interval I_m, and let t_m be the minimal element of the interval J_m. In order to avoid confusion, in this proof we use proper cylinder set notation: for $u \in B_m(X)$, we let $[u]$ denote the set of points x in X such that $x_0 \ldots x_{m-1} = u$. Using conditional probabilities, we have

$$\begin{aligned} \mu([b]) = \mu([u_1]) &\prod_{m=1}^{N} \mu\big(\sigma^{-t_m}[v_m] \mid [u_1 v_1 \ldots v_{m-1} u_m]\big) \\ &\times \prod_{m=1}^{N} \mu\big(\sigma^{-s_{m+1}}[u_{m+1}] \mid [u_1 v_1 \ldots u_m v_m]\big). \end{aligned}$$

Then by our choice of K, we have

$$\mu([b]) \leq K^{2N} \prod_{m=1}^{N+1} \mu([u_m]) \prod_{m=1}^{N} \mu([v_m]),$$

as desired. □

3. Pressure of random SFTs

In this section we give a proof of Theorem 1.1. The broad outline of the proof involves finding upper and lower bounds on the pressure in terms of some other random variables, followed by a second moment argument showing that these auxiliary random variables each converge in probability to $P_X(f) + \log \alpha$. For the sake of exposition, we present the argument here and defer the proofs of the many technical lemmas to later sections of the paper. We hope that this presentation helps clarify the main argument and also motivate the technical lemmas. Note that at the beginning of this proof we define some notation and choose some parameters, including the random variables $\phi_{n,k}$ and $\psi_{n,k}$, and we make frequent reference to both the notation and the parameters throughout Sections 4 - 6 in the technical lemmas.

PROOF OF THEOREM 1.1. Let X be a non-trivial mixing SFT. Let $f : X \to \mathbb{R}$ be a Hölder continuous potential with associated Gibbs measure μ. Choose $\gamma_0 = \gamma_0(X, f)$ as in Definition 2.1, and note that $\gamma_0 < 1$. Let $\alpha \in (\gamma_0, 1]$, and let $\epsilon > 0$. Furthermore, fix K and g_0 as in Lemma 2.2.

We begin by selecting a variety of parameters for our proof. Since $\alpha > \gamma_0$, there exists γ in the interval (γ_0, α). According to the definition of γ_0, since $\gamma > \gamma_0$, there exists n_0 such that for all $m \geq n_0$, for all $u \in B_m(X)$, we have $\mu(u) \leq \gamma^m$. We assume throughout that $n \geq n_0$. Choose $\delta > 0$ such that $\delta < \log(\alpha\gamma^{-1})/4$. Fix a sequence $k = k(n)$ such that $n/k \to 0$ and $k = o(n^2/\log n)$. (For example, one may choose $k = [n^{1+\nu}]$ for any $0 < \nu < 1$.) Now let $\ell = k - n + 1$, which is the number of positions s in $[0, k)$ such that $s + [0, n) \subset [0, k)$.

Having made these parameter choices, we now proceed to define our upper and lower bounds on $P_{Y_n}(f)$. First, for all $m \geq n$, for all $u \in B_m(X)$, recall from Section 2.3 that ξ_u is the random variable that is one if u is allowed (*i.e.* $W_n(u) \cap \mathcal{F}_n = \varnothing$) and zero otherwise. Then define

$$\phi_{n,k} = \sum_{u \in B_k(X)} e^{S_k f(u)} \xi_u.$$

By Lemma 6.1, $\phi_{n,k}$ may be used to provide an upper bound on $P_{Y_n}(f)$:

$$P_{Y_n}(f) \leq \frac{1}{k} \log \phi_{n,k}. \tag{3.1}$$

Now we turn towards the lower bound on $P_{Y_n}(f)$. Recall that we have already defined $\delta > 0$ above. Consider the set of words of length n that are entropy-typical for μ with tolerance δ:

$$E_n = \left\{ u \in B_n(X) : \left| -\frac{1}{n} \log \mu(u) - h(\mu) \right| < \delta \right\}.$$

Then let $G_{n,k}$ be the set of words of length k that begin and end with the same word of length n from E_n:

$$G_{n,k} = \left\{ u \in B_k(X) : u_1^n = u_\ell^k \text{ and } u_1^n \in E_n \right\}.$$

Next we define the random variable

$$\psi_{n,k} = \frac{1}{|E_n|} \sum_{u \in G_{n,k}} e^{S_k f(u)} \xi_u.$$

By Lemma 6.2, we see that $\psi_{n,k}$ may be used to bound $P_{Y_n}(f)$ from below: for all large enough n,

$$P_{Y_n}(f) \geq \frac{1}{k} \log \psi_{n,k} - \epsilon/2. \tag{3.2}$$

By Lemmas 5.1, 5.2, 5.3, and 5.4, we have the following asymptotic results on the expectation and variance of $\phi_{n,k}$ and $\psi_{n,k}$:

(I) $\lim_n \frac{1}{k} \log \mathbb{E}\left[\phi_{n,k}\right] = P_X(f) + \log(\alpha)$;
(II) $\lim_n \frac{1}{k} \log \mathbb{E}\left[\psi_{n,k}\right] = P_X(f) + \log(\alpha)$;
(III) there exists $\rho_1 > 0$ such that for all large enough n,

$$\frac{\operatorname{Var}\left[\phi_{n,k}\right]}{\mathbb{E}\left[\phi_{n,k}\right]^2} \leq e^{-\rho_1 n};$$

(IV) there exists $\rho_2 > 0$ such that for all large enough n,

$$\frac{\operatorname{Var}\left[\psi_{n,k}\right]}{\mathbb{E}\left[\psi_{n,k}\right]^2} \leq e^{-\rho_2 n}.$$

The first two properties indicate that we expect $\phi_{n,k}$ and $\psi_{n,k}$ to be on the correct exponential order of magnitude, while the third and fourth properties show that these random variables are well concentrated around their expected values. Combining these properties with Chebyshev's inequality, we are able to finish the proof as follows.

By the monotonicity of $\mathbb{P}$ under inclusion, the union bound, and the displays (3.1) and (3.2), for all large enough n, we have

(3.3)

$$\begin{aligned}
&\mathbb{P}\Big(|P_{Y_n}(f) - (P_X(f) + \log \alpha)| \geq \epsilon \Big) \\
&\leq \mathbb{P}\Big(P_{Y_n}(f) \geq P_X(f) + \log \alpha + \epsilon \Big) + \mathbb{P}\Big(P_{Y_n}(f) \leq P_X(f) + \log \alpha - \epsilon \Big) \\
&\leq \mathbb{P}\Big(\frac{1}{k} \log \phi_{n,k} \geq P_X(f) + \log \alpha + \epsilon \Big) + \mathbb{P}\Big(\frac{1}{k} \log \psi_{n,k} - \epsilon/2 \leq P_X(f) + \log \alpha - \epsilon \Big) \\
&= \mathbb{P}\Big(\phi_{n,k} \geq e^{k(P_X(f)+\log \alpha + \epsilon)} \Big) + \mathbb{P}\Big(\psi_{n,k} \leq e^{k(P_X(f)+\log \alpha - \epsilon/2)} \Big).
\end{aligned}$$

We proceed to bound the two terms on the right-hand side separately. For the first, Chebyshev gives

$$\begin{aligned}
&\mathbb{P}\Big(\phi_{n,k} \geq e^{k(P_X(f)+\log \alpha+\epsilon)} \Big) \\
&= \mathbb{P}\Big(\phi_{n,k} - \mathbb{E}[\phi_{n,k}] \geq e^{k(P_X(f)+\log \alpha+\epsilon)} - \mathbb{E}[\phi_{n,k}] \Big) \\
&= \mathbb{P}\Big(\phi_{n,k} - \mathbb{E}[\phi_{n,k}] \geq \operatorname{Var}[\phi_{n,k}]^{1/2} \frac{\mathbb{E}[\phi_{n,k}]}{\operatorname{Var}[\phi_{n,k}]^{1/2}} \Big(e^{k(P_X(f)+\log \alpha+\epsilon)}/\mathbb{E}[\phi_{n,k}] - 1 \Big) \Big) \\
&\leq \frac{\operatorname{Var}[\phi_{n,k}]}{\mathbb{E}[\phi_{n,k}]^2} \Big(e^{k(P_X(f)+\log \alpha+\epsilon)}/\mathbb{E}[\phi_{n,k}] - 1 \Big)^{-2}.
\end{aligned}$$

Then by properties (I) and (III), there exists $\rho_3 > 0$ such that for all large n,

$$\mathbb{P}\left(\phi_{n,k} \geq e^{k(P_X(f)+\log\alpha+\epsilon)} \right) < e^{-\rho_3 n}. \tag{3.4}$$

Similarly, for the second term in the last line of (3.3), Chebyshev gives

$$\begin{aligned}
&\mathbb{P}\left(\psi_{n,k} \leq e^{k(P_X(f)+\log\alpha-\epsilon/2)} \right) \\
&= \mathbb{P}\left(\psi_{n,k} - \mathbb{E}[\psi_{n,k}] \leq e^{k(P_X(f)+\log\alpha-\epsilon/2)} - \mathbb{E}[\psi_{n,k}] \right) \\
&= \mathbb{P}\left(\psi_{n,k} - \mathbb{E}[\psi_{n,k}] \leq \mathrm{Var}[\psi_{n,k}]^{1/2} \frac{\mathbb{E}[\psi_{n,k}]}{\mathrm{Var}[\psi_{n,k}]^{1/2}} \left(e^{k(P_X(f)+\log\alpha-\epsilon/2)}/\mathbb{E}[\psi_{n,k}] - 1 \right) \right) \\
&\leq \frac{\mathrm{Var}[\psi_{n,k}]}{\mathbb{E}[\psi_{n,k}]^2} \left(e^{k(P_X(f)+\log\alpha-\epsilon/2)}/\mathbb{E}[\psi_{n,k}] - 1 \right)^{-2}.
\end{aligned}$$

Then by properties (II) and (IV), there exists $\rho_4 > 0$ such that for all large n,

$$\mathbb{P}\left(\psi_{n,k} \leq e^{k(P_X(f)+\log\alpha-\epsilon/2)} \right) < e^{-\rho_4 n}. \tag{3.5}$$

Combining the inequalities in (3.3), (3.4), and (3.5), we obtain the desired result. □

4. Repeat probabilities

In this section we bound the μ measure of sets of words that have repeated subwords. By the well-known result of Ornstein and Weiss [**32**], the first return time of a μ-typical point x to its initial block of length n is approximately $e^{h(\mu)n}$. Then for μ-typical words of polynomial length in n, one would expect to find no repeated words of length n at all. However, to control the expectation and variance of $\phi_{n,k}$ and $\psi_{n,k}$, it is important to give more precise estimates on just how unlikely it is that a word of length k will have exactly j distinct subwords of length n, for each $1 \leq j \leq k - n + 1$.

Throughout this section we use the same environment (notation, parameters, and assumptions) laid out at the beginning of the proof of Theorem 1.1. The results of this section are used in the following section to establish properties (I) - (IV) from the proof of Theorem 1.1.

We begin by considering some sets of words that have exactly j distinct subwords of length n. For $1 \leq j \leq \ell$, we define the following sets:

$$\begin{aligned}
B^j_{n,k} &= \{u \in B_k(X) : |W_n(u)| = j\} \\
G^j_{n,k} &= \{u \in G_{n,k} : |W_n(u)| = j\}.
\end{aligned}$$

Furthermore, for $1 \leq j \leq 2\ell$, we let

$$\begin{aligned}
D^j_{n,k} &= \{(u,v) \in B_k(X) \times B_k(X) : W_n(u) \cap W_n(v) \neq \varnothing,\ |W_n(u) \cup W_n(v)| = j\} \\
Q_{n,k} &= \{(u,v) \in G_{n,k} \times G_{n,k} : W_n(u) \cap W_n(v) \neq \varnothing\} \\
Q^j_{n,k} &= \{(u,v) \in Q_{n,k} : |W_n(u) \cup W_n(v)| = j\}.
\end{aligned}$$

In Lemmas 4.1 and 4.2 we find bounds on the μ measure of the sets $B^j_{n,k}$ and $G_{n,k}$. In subsequent lemmas (Lemmas 4.3 - 4.5), we also find bounds on the $\mu \times \mu$ measure of $D^j_{n,k}$, $Q_{n,k}$, and $Q^j_{n,k}$. These estimates are used in the following section to bound the expectation and variance of $\phi_{n,k}$ and $\psi_{n,k}$.

LEMMA 4.1. *There exists a polynomial* $p_1(x)$ *such that for all large enough* n, *for each* $1 \leq j \leq \ell$, *we have*

$$\mu(B^j_{n,k}) \leq p_1(n)^{k/n}\gamma^{k-j}.$$

PROOF. Consider $n \geq n_0$ and and $1 \leq j \leq \ell$. We define a map $\varphi : B^j_{n,k} \to \{(\mathcal{R}, w) : \mathcal{R} \subset \mathcal{C}_n([0,k)) \times \mathcal{C}_n([0,k)),\ w \in \mathcal{A}^{[0,k)\setminus A(\mathcal{R})}\}$ as follows. Let b be in $B^j_{n,k}$. Let $\mathcal{R}$ be a repeat cover of b such that $|\mathcal{R}| \leq 4k/n$, which exists by Lemma 2.5. Let $\big((u_m)_{m=1}^{N+1}, (v_m)_{m=1}^{N}\big)$ be a repeat block decomposition of b (as in Definition 2.7). Then set $\varphi(b) = (\mathcal{R}, (v_m)_{m=1}^N)$. Furthermore, note that by Lemma 2.8,

$$\mu(b) \leq K^{2N} \prod_{m=1}^{N} \mu(v_m) \prod_{m=1}^{N+1} \mu(u_m).$$

Since each block v_m has length at least n (it contains at least one repeated word of length n from b) and $n \geq n_0$, we have that $\mu(v_m) \leq \gamma^{|v_m|}$. Then

$$\mu(b) \leq K^{2N}\gamma^{\sum_m |v_m|} \prod_{m=1}^{N+1} \mu(u_m) = K^{2N}\gamma^{|A(\mathcal{R})|} \prod_{m=1}^{N+1} \mu(u_m)$$

Using that $N \leq |\mathcal{R}| \leq 4k/n$ and $|A(\mathcal{R})| \geq k - j$ (by Lemma 2.6), we see that

$$\begin{aligned} \mu(b) &\leq \big(K^8\big)^{k/n}\gamma^{|A(\mathcal{R})|} \prod_{m=1}^{N+1} \mu(u_m) \\ &\leq \big(K^8\big)^{k/n}\gamma^{k-j} \prod_{m=1}^{N+1} \mu(u_m). \end{aligned} \tag{4.1}$$

Now define the projection map $\pi : \varphi(B^j_{n,k}) \to \mathrm{Power}(\mathcal{C}_n(k) \times \mathcal{C}_n(k))$, given by $\pi((\mathcal{R}, w)) = \mathcal{R}$. Let $S = \pi \circ \varphi(B^j_{n,k})$. Note that $|\mathcal{C}_n([0,k))| \leq k$, and therefore $|\mathcal{C}_n([0,k)) \times \mathcal{C}_n([0,k))| \leq k^2$. Furthermore, since each $\mathcal{R}$ in $\pi \circ \varphi(B^j_{n,k})$ satisfies $|\mathcal{R}| \leq 4k/n$, we have that $|S| \leq |\mathcal{C}_n([0,k)) \times \mathcal{C}_n([0,k))|^{4k/n} \leq \big(k^2\big)^{4k/n} = \big(k^8\big)^{k/n}$.

Having established these bounds, we may now estimate the μ measure of $B^j_{n,k}$ as follows. By rearranging the sum, we have

$$\begin{aligned} \mu(B^j_{n,k}) &= \sum_{b \in B^j_{n,k}} \mu(b) \\ &= \sum_{\mathcal{R} \in S} \sum_{(\mathcal{R},(u_m)) \in \pi^{-1}(\mathcal{R})} \sum_{b \in \varphi^{-1}(\mathcal{R},(u_m))} \mu(b). \end{aligned}$$

Then by (4.1), we get

$$\begin{aligned} \mu(B^j_{n,k}) &\leq \sum_{\mathcal{R} \in S} \sum_{(\mathcal{R},(u_m)) \in \pi^{-1}(\mathcal{R})} \sum_{b \in \varphi^{-1}(\mathcal{R},(u_m))} \big(K^8\big)^{k/n}\gamma^{k-j} \prod_{m=1}^{N+1} \mu(u_m) \\ &= \big(K^8\big)^{k/n}\gamma^{k-j} \sum_{\mathcal{R} \in S} \sum_{(\mathcal{R},(u_m)) \in \pi^{-1}(\mathcal{R})} \prod_{m=1}^{N+1} \mu(u_m). \end{aligned}$$

Since μ is a probability measure, the sum of $\mu(u_m)$ over any set of words u_m of the same length is less than or equal to one. Then

$$\begin{aligned}\mu(B^j_{n,k}) &\leq \left(K^8\right)^{k/n}\gamma^{k-j}\sum_{\mathcal{R}\in S}\ \sum_{(\mathcal{R},(u_m))\in\pi^{-1}(\mathcal{R})}\ \prod_{m=1}^{N+1}\mu(u_m)\\ &\leq \left(K^8\right)^{k/n}\gamma^{k-j}|S|\\ &\leq \left(K^8k^8\right)^{k/n}\gamma^{k-j},\end{aligned}$$

where we have used that $|S| \leq \left(k^8\right)^{k/n}$ (established in the previous paragraph).

Recall that $k = o(n^2/\log(n))$. Therefore for large enough n, we have $k \leq n^2$. Let $p_1(x) = K^8x^{16}$. Then for large enough n, for all $1 \leq j \leq \ell$, the previous display yields that $\mu(B^j_{n,k}) \leq p_1(n)^{k/n}\gamma^{k-j}$, as desired. □

The following lemma gives both upper and lower bounds on the μ measure of $G_{n,k}$.

LEMMA 4.2. *There exists $\rho_0 > 0$ such that for all large enough n,*

$$K^{-1}e^{-(h(\mu)+\delta)n}(1-e^{-\rho_0 n}) \leq \mu(G_{n,k}) \leq Ke^{-(h(\mu)-\delta)n}.$$

PROOF. For $u \in E_n$, let $G_{n,k}(u) = \{v \in G_{n,k} : v_1^n = u\}$. Then by our choice of K, for large enough n, we have

$$\mu(G_{n,k}(u)) = \sum_{v\in G_{n,k}(u)}\mu(v) \geq K^{-1}\mu(u)^2.$$

Also, note that $G_{n,k} = \bigsqcup_{u\in E_n} G_{n,k}(u)$. Then

$$\mu(G_{n,k}) = \sum_{u\in E_n}\mu(G_{n,k}(u)) \geq K^{-1}\sum_{u\in E_n}\mu(u)^2 \geq K^{-1}e^{-(h(\mu)+\delta)n}\mu(E_n),$$

where the last inequality results from the fact that $\min_{u\in E_n}\mu(u) \geq e^{-(h(\mu)+\delta)n}$. Additionally, using the Gibbs property (2.1) and the large deviations results for Gibbs measures [**35**], one may check that there exists $\rho_0 > 0$ such that $\mu(E_n) \geq 1 - e^{-\rho_0 n}$ for all large enough n. Combining this fact with the above inequalities yields the desired lower bound.

For the upper bound, for all large enough n and for each $u \in E_n$, we have that

$$\mu(G_{n,k}(u)) = \sum_{v\in G_{n,k}(u)}\mu(v) \leq K\mu(u)^2.$$

Then

$$\mu(G_{n,k}) = \sum_{u\in E_n}\mu(G_{n,k}(u)) \leq K\sum_{u\in E_n}\mu(u)^2 \leq Ke^{-(h(\mu)-\delta)n},$$

where we have used that $\max_{u\in E_n}\mu(u) \leq e^{-(h(\mu)-\delta)n}$. □

In the following three lemmas, we estimate the $\mu\times\mu$ measure of sets of pairs of words with various repeat properties. The outline of these proofs is similar to the outline of the proof of Lemma 4.1, but each proof requires some arguments that are specific to the particular repeat structure of interest.

LEMMA 4.3. *There exists a polynomial $p_2(x)$ such that for all large enough n, for all $1 \leq j \leq 2\ell - 1$,*

$$(\mu\times\mu)(D^j_{n,k}) \leq p_2(n)^{k/n}\gamma^{2\ell-j+n}.$$

PROOF. Consider $n \geq n_0$, and let $1 \leq j \leq 2\ell - 1$. First define the set $F = [0,k) \sqcup [k+1,2k]$. Then define a map $\varphi : D^j_{n,k} \to \{(\mathcal{R}, w) : \mathcal{R} \subset \mathcal{C}_n(F) \times \mathcal{C}_n(F), w \in \mathcal{A}^{F \setminus A(\mathcal{R})}\}$ as follows. Let $(a,b) \in D^j_{n,k}$. We use the notation $a \sqcup b$ to denote the element in $\mathcal{A}^F$ such that $(a \sqcup b)|_{[0,k)} = a$ and $(a \sqcup b)|_{[k+1,2k]} = b$. Let $\mathcal{R}$ be a repeat cover of $a \sqcup b$ such that $|\mathcal{R}| \leq 4|F|/n = 8k/n$, which exists by Lemma 2.5. Then let $\left((u_m)_{m=1}^{N_1+1}, (v_m)_{m=1}^{N_1}\right)$ be the repeat block decomposition of a induced by the set $A(\mathcal{R}) \cap [0,k)$, and let $\left((y_m)_{m=1}^{N_2+1}, (z_m)_{m=1}^{N_2}\right)$ be the repeat block decomposition of b induced by the set $A(\mathcal{R}) \cap [k+1,2k]$. Finally, we define $\varphi(a,b) = (\mathcal{R}, (u_m)_{m=1}^{N_1+1}, (y_m)_{m=1}^{N_2+1})$.

By Lemma 2.8, note that

$$\mu(a) \leq K^{2N_1} \prod_{m=1}^{N_1} \mu(v_m) \prod_{m=1}^{N_1+1} \mu(u_m),$$

and

$$\mu(b) \leq K^{2N_2} \prod_{m=1}^{N_2} \mu(z_m) \prod_{m=1}^{N_2+1} \mu(y_m).$$

Furthermore, since each of the blocks v_m and z_m has length at least n, for all large enough n, we have that $\mu(v_m) \leq \gamma^{|v_m|}$ and $\mu(z_m) \leq \gamma^{|z_m|}$. Therefore for all large enough n, we have

$$\mu(a) \leq K^{2N_1} \gamma^{\sum_m |v_m|} \prod_{m=1}^{N_1+1} \mu(u_m),$$

and

$$\mu(a) \leq K^{2N_2} \gamma^{\sum_m |z_m|} \prod_{m=1}^{N_2+1} \mu(y_m).$$

Using that $N_1 + N_2 \leq |\mathcal{R}| \leq 8k/n$ and $\sum_m |v_m| + \sum_m |z_m| = |A(\mathcal{R})| \geq 2k - j - n$ (by Lemma 2.6), we obtain

$$\mu(a)\mu(b) \leq \left(K^{16}\right)^{k/n} \gamma^{2k-j-n} \prod_{m=1}^{N_1+1} \mu(u_m) \prod_{m=1}^{N_2+1} \mu(y_m). \tag{4.2}$$

Now define the projection $\pi : \varphi(D^j_{n,k}) \to \text{Power}(\mathcal{C}_n(F) \times \mathcal{C}_n(F))$, given by $\pi(\mathcal{R}, (u_m), (y_m)) = \mathcal{R}$. Let $S = \pi \circ \varphi(D^j_{n,k})$. Note that $|\mathcal{C}_n(F)| \leq 2k$, and so $|\mathcal{C}_n(F) \times \mathcal{C}_n(F)| \leq (2k)^2$. Since each $\mathcal{R} \in \pi \circ \varphi(D^j_{n,k})$ satisfies $|\mathcal{R}| \leq 8k/n$, we then have that $|S| = |\pi \circ \varphi(D^j_{n,k})| \leq |\mathcal{C}_n(F) \times \mathcal{C}_n(F)|^{8k/n} \leq (2k)^{16k/n}$.

Let us now estimate $(\mu \times \mu)(D^j_{n,k})$. By rearranging the sum, we get

$$\begin{aligned}(\mu \times \mu)(D^j_{n,k}) &= \sum_{(a,b) \in D^j_{n,k}} \mu(a)\mu(b) \\ &= \sum_{\mathcal{R} \in S} \sum_{(\mathcal{R},(u_m),(y_m)) \in \pi^{-1}(\mathcal{R})} \sum_{(a,b) \in \varphi^{-1}(\mathcal{R},(u_m),(y_m))} \mu(a)\mu(b).\end{aligned}$$

Applying (4.2) to each term in the sum, we get

$$(\mu \times \mu)(D^j_{n,k})$$
$$\leq \left(K^{16}\right)^{k/n} \gamma^{2k-j-n} \sum_{\mathcal{R} \in S} \sum_{(\mathcal{R},(u_m),(y_m)) \in \pi^{-1}(\mathcal{R})} \prod_{m=1}^{N_1+1} \mu(u_m) \prod_{m=1}^{N_2+1} \mu(y_m).$$

Then since the sum of $\mu(u_m)$ over any set of words u_m of the same length is less than or equal to one, we see that

$$(\mu \times \mu)(D^j_{n,k}) \leq \left(K^{16}\right)^{k/n} \gamma^{2k-j-n} |S|.$$

Combining this estimate with the bound on $|S|$ established in the previous paragraph, we obtain

$$(\mu \times \mu)(D^j_{n,k}) \leq \left(2^{16} K^{16} k^{16}\right)^{k/n} \gamma^{2k-j-n}.$$

Recall that $k = o(n^2/\log(n))$. Then for all large enough n, we have $k \leq n^2$. Let $p_2(x) = (2K)^{16} x^{32}$. Then by the previous display, for all large enough n, we obtain that $(\mu \times \mu)(D^j_{n,k}) \leq p_2(n)^{k/n} \gamma^{2k-j-n}$. □

LEMMA 4.4. *There exists a polynomial $p_3(x)$ such that for all large enough n,*

$$(\mu \times \mu)(Q_{n,k}) \leq p_3(n) e^{-2n(h(\mu)-\delta)} \gamma^n.$$

PROOF. Consider $n \geq n_0$. First define the set $F = [0, k) \sqcup [k+1, 2k]$. Then we define a map $\varphi : Q_{n,k} \to \{(\mathcal{R}, w) : \mathcal{J} \subset \mathcal{C}_n(S) \times \mathcal{C}_n(S), w \in \mathcal{A}^{S \setminus A(\mathcal{R})}\}$ as follows. Let $(a, b) \in Q_{n,k}$. We let $a \sqcup b$ denote the element of $\mathcal{A}^F$ such that $(a \sqcup b)|_{[0,k)} = a$ and $(a \sqcup b)|_{[k+1,2k]} = b$. Since $W_n(a) \cap W_n(b) \neq \varnothing$, there exists $I \in \mathcal{C}_n([0, k))$ and $J \in \mathcal{C}_n([k+1, 2k])$ such that $(a \sqcup b)|_I = (a \sqcup b)|_J$, and we assume that J is (lexicographically) minimal among all such intervals. We partition $Q_{n,k}$ into $\tilde{Q}_{n,k}$ and $\hat{Q}_{n,k}$, where $\tilde{Q}_{n,k}$ consists of all pairs (a, b) such that $J \cap [2\ell - n, 2k] = \varnothing$, and $\hat{Q}_{n,k}$ contains the remaining pairs. Our definition of $\varphi(a, b)$ will depend on whether (a, b) is in $\tilde{Q}_{n,k}$ or $\hat{Q}_{n,k}$.

First suppose that $(a, b) \in \tilde{Q}_{n,k}$. Let $\mathcal{R} \subset \mathcal{C}_n(F) \times \mathcal{C}_n(F)$ be the following set containing three pairs of intervals: $\{([0, n), [\ell - 1, k)), (I, J), ([k+1, k+n], [2k-n+1, 2k])\}$. Note that the block decomposition of a induced by $A(\mathcal{R}) \cap [0, k)$ has the form (u_1, v_1), where $v_1 = a|_{[\ell-1,k)}$. Similarly, the block decomposition of b induced by $A(\mathcal{R}) \cap [k+1, 2k]$ has the form $((y_m)_{m=1}^2, (z_m)_{m=1}^2)$, where $z_1 = (a \sqcup b)|_J$ and $z_2 = (a \sqcup b)|_{[2k-n+1,2k]}$. Finally, we define $\varphi(a, b) = (\mathcal{R}, u_1, (y_m)_{m=1}^2)$. Furthermore, we note that

$$\mu(a) = \mu(u_1 v_1) \leq K \mu(u_1) \mu(v_1),$$

and

$$\mu(b) = \mu(y_1 z_1 y_2 z_2) \leq K^3 \mu(y_1) \mu(y_2) \mu(z_1) \mu(z_2).$$

Since $a, b \in G_{n,k}$, we must have that $v_1, z_2 \in E_n$, and therefore $\mu(v_1) \leq e^{-(h(\mu)-\delta)n}$ and $\mu(z_2) \leq e^{-(h(\mu)-\delta)n}$. Also, since z_1 has length $n \geq n_0$, we have that $\mu(z_1) \leq \gamma^n$. Putting these estimates together, we obtain

$$\mu(a)\mu(b) \leq K^4 e^{-2n(h(\mu)-\delta)} \gamma^n \mu(u_1) \mu(y_1) \mu(y_2). \tag{4.3}$$

Now suppose that $(a, b) \in \hat{Q}_{n,k}$. In this case we let $\mathcal{R} \subset \mathcal{C}_n(F) \times \mathcal{C}_n(F)$ be a different set of three pairs of intervals: $\mathcal{R} = \{([0, n), [\ell - 1, k)), (I, J), ([2k -$

$n+1,2k],[k+1,k+n])\}$. Note that the third pair of intervals listed is not in lexicographical order. Let (u_1,v_1) be the repeat block decomposition of a induced by $A(\mathcal{R})\cap[0,k)$, and let $((z_m)_{m=1}^2,(y_m)_{m=1}^2)$ be the block decomposition induced by $A(\mathcal{R})|_{[k+1,2k]}$, by which we mean that $(a\sqcup b)|_{[k+1,2k]}=z_1y_1z_2y_2$, where z_1 and z_2 have length n. In this case, we define $\varphi(a,b)=(\mathcal{R},u_1,(y_m)_{m=1}^2)$. Note that

$$\mu(a)=\mu(u_1v_1)\leq K\mu(u_1)\mu(v_1),$$

and

$$\mu(b)=\mu(z_1y_1z_2y_2)\leq K^3\mu(y_1)\mu(y_2)\mu(z_1)\mu(z_2).$$

Since $a,b\in G_{n,k}$, we have that $v_1,z_1\in E_n$, and thus $\mu(v_1)\leq e^{-(h(\mu)-\delta)n}$ and $\mu(z_1)\leq e^{-(h(\mu)-\delta)n}$. Also, since z_2 has length $n\geq n_0$, we have that $\mu(z_2)\leq\gamma^n$. Combining these estimates, we see that

$$\mu(a)\mu(b)\leq K^4e^{-2n(h(\mu)-\delta)}\gamma^n\mu(u_1)\mu(y_1)\mu(y_2). \tag{4.4}$$

Now define the projection map $\pi:\varphi(Q_{n,k})\to\mathrm{Power}(\mathcal{C}_n(F)\times\mathcal{C}_n(F))$, given by $\pi(\mathcal{R},w)=\mathcal{R}$. Let $S=\pi\circ\varphi(Q_{n,k})$. Since $|\mathcal{C}_n(F)|\leq(2k)$, we get $|\mathcal{C}_n(F)\times\mathcal{C}_n(F)|\leq(2k)^2$. Moreover, since each $\mathcal{R}$ in $\pi\circ\varphi(Q_{n,k})$ satisfies $|\mathcal{R}|=3$, we get $|S|=|\pi\circ\varphi(Q_{n,k})|\leq|\mathcal{C}_n(F)\times\mathcal{C}_n(F)|^3\leq(2k)^6$.

By rearranging the sum, we find

$$\begin{aligned}(\mu\times\mu)(Q_{n,k})&=\sum_{(a,b)\in Q_{n,k}}\mu(a)\mu(b)\\&=\sum_{\mathcal{R}\in S}\ \sum_{(\mathcal{R},u_1,(y_m)_{m=1}^2)\in\pi^{-1}(\mathcal{R})}\ \sum_{(a,b)\in\varphi^{-1}(\mathcal{R},u_1,(y_m)_{m=1}^2)}\mu(a)\mu(b).\end{aligned}$$

Then by applying the estimates (4.3) and (4.4) to each term, we get

$$(\mu\times\mu)(Q_{n,k})\leq\sum_{\mathcal{R}\in S}\ \sum_{(\mathcal{R},u_1,(y_m)_{m=1}^2)\in\pi^{-1}(\mathcal{R})}K^4e^{-2n(h(\mu)-\delta)}\gamma^n\mu(u_1)\mu(y_1)\mu(y_2).$$

Summing over all u_1, y_1, and y_2, we obtain

$$(\mu\times\mu)(Q_{n,k})\leq K^4e^{-2n(h(\mu)-\delta)}\gamma^n|S|\leq K^4(2k)^6e^{-2n(h(\mu)-\delta)}\gamma^n,$$

where the second inequality uses the bound on $|S|$ established in the previous paragraph

Recall that $k=o(n^2/\log(n))$, and hence for all large enough n, we have $k\leq n^2$. Let $p_3(x)=2^6K^4x^{12}$. Then by the previous inequality, for all large enough n, we see that $(\mu\times\mu)(Q_{n,k})\leq p_3(n)e^{-2n(h(\mu)-\delta)}\gamma^n$. □

LEMMA 4.5. *There exists a polynomial $p_4(x)$ such that for all large enough n and $1\leq j\leq 2\ell$,*

$$(\mu\times\mu)(Q_{n,k}^j)\leq p_4(k)^{k/n}e^{-2n(h(\mu)-\delta)}\gamma^{2\ell-j}.$$

PROOF. Consider $n\geq n_0$ and $1\leq j\leq 2\ell$. Let $F=[0,k)\cup[k+1,2k]$. We begin by defining a map $\varphi:Q_{n,k}^j\to\{(\mathcal{R},w):\mathcal{R}\subset\mathcal{C}_n(S)\times\mathcal{C}_n(S),w\in\mathcal{A}^{F\setminus A(\mathcal{R})}\}$ as follows. Let $(a,b)\in Q_{n,k}^j$. We let $(a\sqcup b)$ denote the element of $\mathcal{A}^F$ such that $(a\sqcup b)|_{[0,k)}=a$ and $(a\sqcup b)|_{[k+1,2k]}=b$. Let $\mathcal{R}$ be a repeat cover for $a\sqcup b$ such that $|\mathcal{R}|\leq 4|F|/n=8k/n$, which exists by Lemma 2.5. Then let $((u_m)_{m=1}^{N_1+1},(v_m)_{m=1}^{N_1})$ be the repeat block decomposition of a induced by the set $A(\mathcal{R})\cap[0,k-n)$, and let $((y_m)_{m=1}^{N_2+1},(z_m)_{m=1}^{N_2})$ be the repeat block decomposition for b induced by the set

$A(\mathcal{R}) \cap [k+1, 2k-n]$. Additionally, let $v_{N_1+1} = a_{[k-n,k)}$ and $z_{N_2+1} = b_{[2k-n+1,2k]}$. Set $\varphi(a,b) = (\mathcal{R}, (u_m)_{m=1}^{N_1+1}, (y_m)_{m=1}^{N_2+1})$. Furthermore, by Lemma 2.8, we have that

$$\mu(a) \le K^{2N_1+1} \prod_{m=1}^{N_1+1} \mu(v_m) \prod_{m=1}^{N_1+1} \mu(u_m),$$

and

$$\mu(b) \le K^{2N_2+1} \prod_{m=1}^{N_2+1} \mu(z_m) \prod_{m=1}^{N_2+1} \mu(y_m).$$

Since $v_{N_1+1}, z_{N_2+1} \in E_n$, we have $\mu(v_{N_1+1}) \le e^{-n(h(\mu)-\delta)}$ and $\mu(z_{N_2+1}) \le e^{-n(h(\mu)-\delta)}$. Also, for $m = 1, \ldots, N_1 - 1$, the length of v_m is at least n, and we get $\mu(v_m) \le \gamma^{|v_m|}$. For v_{N_1}, we always have $\mu(v_{N_1}) \le \gamma^{|v_{N_1}|-n_0}$. Similarly, for $m = 1, \ldots, N_2 - 1$, the length of z_m is at least n, and we get $\mu(z_m) \le \gamma^{|z_m|}$. As for z_{N_2}, we always have $\mu(z_{N_2}) \le \gamma^{|z_{N_2}|-n_0}$. Then for large enough n, we have

$$\mu(a) \le K^{2N_1+1} \gamma^{\sum_m |v_m| - n_0} e^{-n(h(\mu)-\delta)} \prod_{m=1}^{N_1+1} \mu(u_m),$$

and

$$\mu(a) \le K^{2N_2+1} \gamma^{\sum_m |z_m| - n_0} e^{-n(h(\mu)-\delta)} \prod_{m=1}^{N_2+1} \mu(y_m).$$

Using that $N_1 + N_2 \le |\mathcal{R}| \le 8k/n$ and $\sum_m |v_m| + \sum_m |z_m| = |A(\mathcal{R})| \ge 2\ell - j - 2$ (by Lemma 2.6), we obtain

$$\mu(a)\mu(b) \le \left(K^{32}\right)^{k/n} \gamma^{2\ell - j - 2 - 2n_0} e^{-2n(h(\mu)-\delta)} \prod_{m=1}^{N_1+1} \mu(u_m) \prod_{m=1}^{N_2+1} \mu(y_m). \tag{4.5}$$

Now we define the projection map $\pi : \varphi(Q_{n,k}^j) \to \mathrm{Power}(\mathcal{C}_n(F) \times \mathcal{C}_n(F))$, given by $\pi(\mathcal{R}, (u_m), (y_m)) = \mathcal{R}$. Let $S = \pi \circ \varphi(Q_{n,k}^j)$. Since $|\mathcal{C}_n(F)| \le 2k$, we see that $|\mathcal{C}_n(F) \times \mathcal{C}_n(F)| \le (2k)^2$. Moreover, since each $\mathcal{R}$ in S satisfies $|\mathcal{R}| \le 8k/n$, we estimate $|S| = |\pi \circ \varphi(Q_{n,k}^j)| \le |\mathcal{C}_n(F) \times \mathcal{C}_n(F)|^{8k/n} \le (2^{16}k^{16})^{k/n}$.

Let us now estimate $(\mu \times \mu)(Q_{n,k}^j)$. By rearranging the sum, we get

$$\begin{aligned}(\mu \times \mu)(Q_{n,k}^j) &= \sum_{(a,b) \in Q_{n,k}^j} \mu(a)\mu(b) \\ &= \sum_{\mathcal{R} \in S} \sum_{(\mathcal{R},(u_m),(y_m)) \in \pi^{-1}(\mathcal{R})} \sum_{(a,b) \in \varphi^{-1}(\mathcal{R},(u_m),(y_m))} \mu(a)\mu(b).\end{aligned}$$

By applying (4.5) to each term in the sum, we see that

$$\begin{aligned}&(\mu \times \mu)(Q_{n,k}^j) \\ &\le \sum_{\mathcal{R} \in S} \sum_{(\mathcal{R},(u_m),(y_m)) \in \pi^{-1}(\mathcal{R})} (K^{32}\gamma^{-2n_0-2})^{k/n} \gamma^{2\ell-j} e^{-n(h(\mu)-\delta)} \prod_{m=1}^{N_1+1} \mu(u_m) \prod_{m=1}^{N_2+1} \mu(y_m),\end{aligned}$$

and then summing over all u_m and y_m gives

$$(\mu \times \mu)(Q_{n,k}^j) \le (K^{32}\gamma^{-2n_0-2})^{k/n} e^{-n(h(\mu)-\delta)} \gamma^{2\ell-j} |S|. \tag{4.6}$$

Combining this estimate with the bound on $|S|$ from the previous paragraph, we obtain

$$(\mu \times \mu)(Q^j_{n,k}) \leq \left(2^{32} K^{32} \gamma^{-2n_0-2} k^{16}\right)^{k/n} e^{-n(h(\mu)-\delta)} \gamma^{2\ell-j}.$$

Recall that $k = o(n^2/\log(n))$, and hence for all large enough n, we have $k \leq n^2$. Let $p_4(x) = (2K)^{32}\gamma^{-2n_0-2}x^{32}$. Then by the previous displayed inequality, for all large enough n and all $1 \leq j \leq 2\ell$, we have $(\mu \times \mu)(Q^j_{n,k}) \leq p_4(n)^{k/n} e^{-n(h(\mu)-\delta)} \gamma^{2\ell-j}$. □

5. Moment bounds

In this section we prove properties (I)-(IV) concerning the expectation and variance of $\phi_{n,k}$ and $\psi_{n,k}$, which are used in the proof of Theorem 1.1. Throughout this section, we use the same environment (notation, parameters, and assumptions) as in the proof of Theorem 1.1.

LEMMA 5.1. *For all $n \geq 1$, the expectation of $\phi_{n,k}$ satisfies*

$$\mathbb{E}\big[\phi_{n,k}\big] \geq K^{-1}\alpha^\ell e^{Pk}.$$

Furthermore,

$$\lim_n \frac{1}{k} \log \mathbb{E}\big[\phi_{n,k}\big] = P + \log(\alpha).$$

PROOF. Let $n \geq 1$. Then by (2.2) and our choice of K, we have

$$\begin{aligned}
\mathbb{E}\big[\phi_{n,k}\big] &= \sum_{u \in B_k(X)} e^{S_k f(u)} \alpha^{|W_n(u)|} \\
&= \sum_{j=1}^{\ell} \alpha^j \sum_{u \in B^j_{n,k}} e^{S_k f(u)} \\
&\geq K^{-1} e^{Pk} \sum_{j=1}^{\ell} \alpha^j \mu(B^j_{n,k}).
\end{aligned}$$

Using that $\alpha^j \geq \alpha^\ell$ for all $j \leq \ell$ and $\sum_j \mu(B^j_{n,k}) = \mu(B_n(X)) = 1$, we get

$$\mathbb{E}\big[\phi_{n,k}\big] \geq K^{-1}\alpha^\ell e^{Pk},$$

which establishes the first conclusion of the lemma.

Now we consider letting n tend to infinity. By the first conclusion of the lemma, we have that

$$\liminf_n \frac{1}{k} \log \mathbb{E}\big[\phi_{n,k}\big] \geq P + \log \alpha.$$

Also, for any n, our choice of K yields

$$\mathbb{E}\big[\phi_{n,k}\big] = \sum_{j=1}^{k-n+1} \alpha^j \sum_{u \in B^j_{n,k}} e^{S_k f(w)} \leq K e^{Pk} \sum_{j=1}^{k-n+1} \alpha^j \mu(B^j_{n,k}).$$

Recall that $\alpha > \gamma$, and therefore $\alpha\gamma^{-1} > 1$. By Lemma 4.1, there exists a polynomial $p_1(x)$ such that for large enough n, we have

$$\begin{aligned}
\mathbb{E}\big[\phi_{n,k}\big] &\leq Ke^{Pk}\sum_{j=1}^{\ell}\alpha^j\mu(B^j_{n,k}) \\
&\leq Ke^{Pk}\sum_{j=1}^{\ell}\alpha^j p_1(n)^{k/n}\gamma^{k-j} \\
&\leq Ke^{Pk}p_1(n)^{k/n}\gamma^k\sum_{j=1}^{\ell}(\alpha\gamma^{-1})^j \\
&\leq Ke^{Pk}p_1(n)^{k/n}\gamma^k(\alpha\gamma^{-1})^{\ell}\frac{1}{1-(\alpha^{-1}\gamma)} \\
&= Ke^{Pk}p_1(n)^{k/n}\alpha^{\ell}\gamma^{n}\frac{1}{1-(\alpha^{-1}\gamma)}.
\end{aligned}$$

Since $n/k \to 0$ and $n^{-1}\log p_1(n) \to 0$, we obtain that

$$\limsup_n \frac{1}{k}\log\mathbb{E}\big[\phi_{n,k}\big] \leq P + \log\alpha,$$

which finishes the proof. □

LEMMA 5.2. *For all $n \geq 1$, the expectation of $\psi_{n,k}$ satsifies*

$$\mathbb{E}\big[\psi_{n,k}\big] \geq |E_n|^{-1}K^{-1}\alpha^{\ell}e^{Pk}\mu\big(G_{n,k}\big).$$

Furthermore,

$$\lim_n \frac{1}{k}\log\mathbb{E}\big[\psi_{n,k}\big] = P + \log(\alpha).$$

PROOF. Let $n \geq 1$. Then by (2.2) and our choice of K, we have

$$\begin{aligned}
\mathbb{E}\big[\psi_{n,k}\big] &= \frac{1}{|E_n|}\sum_{u\in G_{n,k}} e^{S_k f(u)}\alpha^{|W_n(u)|} \\
&= \frac{1}{|E_n|}\sum_{j=1}^{\ell}\alpha^j\sum_{u\in G^j_{n,k}} e^{S_k f(u)} \\
&\geq |E_n|^{-1}K^{-1}e^{Pk}\sum_{j=1}^{\ell}\alpha^j\mu\big(G^j_{n,k}\big).
\end{aligned}$$

Since $\alpha^j \geq \alpha^{\ell}$ for all $j \leq \ell$ and $\sum_j \mu(G^j_{n,k}) = \mu(G_{n,k})$, we get

$$\mathbb{E}\big[\psi_{n,k}\big] \geq |E_n|^{-1}K^{-1}\alpha^{\ell}e^{Pk}\mu\big(G_{n,k}\big),$$

which establishes the first conclusion of the lemma.

Now we consider letting n tend to infinity. By the first conclusion of the lemma, we have that

$$\liminf_n \frac{1}{k}\log\mathbb{E}\big[\psi_{n,k}\big] \geq P + \log\alpha + \liminf_n \frac{1}{k}\log|E_n|^{-1} + \frac{1}{k}\log\mu\big(G_{n,k}\big).$$

Note that $|E_n| \le |\mathcal{A}|^n$, and by Lemma 4.2, for large enough n, we have $\mu(G_{n,k}) \ge 2^{-1}K^{-1}e^{-n(h(\mu)+\delta)}$. Therefore

$$\liminf_n \frac{1}{k}\log \mathbb{E}\big[\psi_{n,k}\big] \ge P + \log\alpha + \liminf_n \left(\frac{n}{k}\log|\mathcal{A}|^{-1} + \frac{n}{k}\log(e^{-(h(\mu)+\delta)})\right).$$

Finally, using that $n/k \to 0$, we obtain

$$\liminf_n \frac{1}{k}\log \mathbb{E}\big[\psi_{n,k}\big] \ge P + \log\alpha.$$

Also, by Lemma 5.1 and the fact that $\psi_{n,k} \le \phi_{n,k}$, we have

$$\limsup_n \frac{1}{k}\log \mathbb{E}\big[\psi_{n,k}\big] \le \limsup_n \frac{1}{k}\log \mathbb{E}\big[\phi_{n,k}\big] \le P + \log\alpha.$$

Taken together, the previous two inequalities yield that

$$\lim_n \frac{1}{k}\log \mathbb{E}\big[\psi_{n,k}\big] = P + \log\alpha,$$

as desired. □

LEMMA 5.3. *There exists $\rho_1 > 0$ such that for all large enough n,*

$$\frac{\operatorname{Var}\big[\phi_{n,k}\big]}{\mathbb{E}\big[\phi_{n,k}\big]^2} \le e^{-\rho_1 n};$$

PROOF. Using the fact that the variance of a sum is the sum of the covariances, (2.3), and our choice of K, we have

$$\begin{aligned}
\operatorname{Var}\big[\phi_{n,k}\big] &= \sum_{u,v\in B_k(X)} \alpha^{|W_n(u)\cup W_n(v)|}\Big(1-\alpha^{|W_n(u)\cap W_n(v)|}\Big)e^{S_kf(u)+S_kf(v)} \\
&\le \sum_{j=1}^{2\ell-1}\alpha^j \sum_{(u,v)\in D^j_{n,k}} e^{S_kf(u)+S_kf(v)} \\
&\le K^2e^{2Pk}\sum_{j=1}^{2\ell-1}\alpha^j(\mu\times\mu)\big(D^j_{n,k}\big).
\end{aligned}$$

Let $C = 1/(1-(\alpha^{-1}\gamma))$. Then by the lower bound on $\mathbb{E}[\phi_{n,k}]$ from Lemma 5.1 and the upper bound on $(\mu\times\mu)(D^j_{n,k})$ from Lemma 4.3, there exists a polynomial $p_2(x)$ such that for large enough n, we have

$$\begin{aligned}
\frac{\operatorname{Var}\big[\phi_{n,k}\big]}{\mathbb{E}\big[\phi_{n,k}\big]^2} &\le \frac{K^2e^{2Pk}\sum_{j=1}^{2\ell-1}\alpha^j(\mu\times\mu)\big(D^j_{n,k}\big)}{K^{-2}\alpha^{2\ell}e^{2Pk}} \\
&= K^4\alpha^{-2\ell}\sum_{j=1}^{2\ell-1}\alpha^j(\mu\times\mu)\big(D^j_{n,k}\big) \\
&\le K^4\alpha^{-2\ell}\sum_{j=1}^{2\ell-1}\alpha^j p_2(n)^{k/n}\gamma^{2\ell+n-j} \\
&\le K^4p_2(n)^{k/n}\alpha^{-2\ell}\gamma^{2\ell+n}\sum_{j=1}^{2\ell-1}(\alpha\gamma^{-1})^j \\
&\le K^4p_2(n)^{k/n}\alpha^{-2\ell}\gamma^{2\ell+n}(\alpha\gamma^{-1})^{2\ell}C.
\end{aligned}$$

Rewriting this estimate, we find

$$\begin{aligned}\frac{\operatorname{Var}\big[\phi_{n,k}\big]}{\mathbb{E}\big[\phi_{n,k}\big]^2} &= \exp\Big(4\log K + \log C + \frac{k}{n}\log p_2(n) + n\log\gamma\Big)\\ &\le \exp\Big(n\Big(q\frac{k\log n}{n^2} + \log\gamma + \frac{4}{n}\log K + \frac{1}{n}\log C\Big)\Big),\end{aligned}$$

where $p_2(x) \le x^q$ for all large enough x. Since $k = o(n^2/\log n)$ and $\log\gamma < 0$, we obtain the desired bound. □

LEMMA 5.4. *There exists $\rho_2 > 0$ such that for all large enough n,*

$$\frac{\operatorname{Var}\big[\psi_{n,k}\big]}{\mathbb{E}\big[\psi_{n,k}\big]^2} \le e^{-\rho_2 n}.$$

PROOF. Let $b = 2\ell - n$. Using the fact that the variance of a sum is the sum of the covariances, (2.3), and our choice of K, we have

$$\begin{aligned}\operatorname{Var}\big[\psi_{n,k}\big] &= \frac{1}{|E_n|^2}\sum_{u,v\in G_{n,k}} \alpha^{|W_n(u)\cup W_n(v)|}\Big(1-\alpha^{|W_n(u)\cap W_n(v)|}\Big)e^{S_kf(u)+S_kf(v)}\\ &\le \frac{1}{|E_n|^2}\sum_{j=1}^{2\ell}\alpha^j \sum_{(u,v)\in Q^j_{n,k}} e^{S_kf(u)+S_kf(v)}\\ &\le K^2\frac{e^{2Pk}}{|E_n|^2}\sum_{j=1}^{2\ell}\alpha^j(\mu\times\mu)\big(Q^j_{n,k}\big).\end{aligned}$$

Dividing by $\mathbb{E}[\psi_{n,k}]$ and using the lower bound on $\mathbb{E}[\psi_{n,k}]$ in Lemma 5.2 and the lower bound on $\mu(G_{n,k})$ in Lemma 4.2, we see that

$$\begin{aligned}\frac{\operatorname{Var}\big[\psi_{n,k}\big]}{\mathbb{E}\big[\psi_{n,k}\big]^2} &\le \frac{K^2e^{2Pk}\sum_{j=1}^{2\ell-1}\alpha^j(\mu\times\mu)\big(Q^j_{n,k}\big)}{K^{-2}\alpha^{2\ell}e^{2Pk}\mu(G_{n,k})^2}\\ &\le \frac{K^4\alpha^{-2\ell}\sum_{j=1}^{2\ell-1}\alpha^j(\mu\times\mu)\big(Q^j_{n,k}\big)}{(2K)^{-2}e^{-2n(h(\mu)+\delta)}}\\ &= (2K)^6\alpha^{-2\ell}e^{2n(h(\mu)+\delta)}\left(\sum_{j=1}^{b-1}\alpha^j(\mu\times\mu)\big(Q^j_{n,k}\big) + \sum_{j=b}^{2\ell-1}\alpha^j(\mu\times\mu)\big(Q^j_{n,k}\big)\right).\end{aligned}$$

Let $C = 1/(1-(\alpha^{-1}\gamma))$. Note that $\alpha^j \le \alpha^b$ for $j \ge b$. Applying this fact and the upper bounds on $(\mu\times\mu)(Q_{n,k})$ and $(\mu\times\mu)(Q^j_{n,k})$ from Lemmas 4.4 and 4.5, respectively, we get that there are polynomials $p_3(x)$ and $p_4(x)$ such that for all large enough n,

$$\begin{aligned}\frac{\mathrm{Var}\left[\psi_{n,k}\right]}{\mathbb{E}\left[\psi_{n,k}\right]^2} &\leq (2K)^6\alpha^{-2\ell}e^{2n(h(\mu)+\delta)} \\ &\quad\times\left(\sum_{j=1}^{b-1}\alpha^j p_4(n)^{k/n}e^{-2n(h(\mu)-\delta)}\gamma^{2\ell-j}+\alpha^b p_3(n)e^{-2n(h(\mu)-\delta)}\gamma^n\right) \\ &\leq (2K)^6\alpha^{-2\ell}e^{4n\delta}\left(p_4(n)^{k/n}\gamma^{2\ell}\sum_{j=1}^{b-1}(\alpha\gamma^{-1})^j+\alpha^b\gamma^n p_3(n)\right) \\ &\leq (2K)^6\alpha^{-2\ell}e^{4n\delta}\left(p_4(n)^{k/n}\gamma^{2\ell}C(\alpha\gamma^{-1})^b+\alpha^b\gamma^n p_3(n)\right) \\ &= (2K)^6 e^{4n\delta}\gamma^n\alpha^{-n}\left(Cp_4(n)^{k/n}+p_3(n)\right).\end{aligned}$$

Rewriting this estimate, we have

$$\frac{\mathrm{Var}\left[\psi_{n,k}\right]}{\mathbb{E}\left[\psi_{n,k}\right]^2} \leq \exp\left(n\left(\log(\gamma\alpha^{-1})+4\delta+\frac{6}{n}\log(2K)+\frac{1}{n}\log\left(Cp_4(n)^{k/n}+p_3(n)\right)\right)\right).$$

Let $q>1$ be such that for all large enough x, we have $Cp_4(x)^{k/n}+p_3(x)\leq x^{qk/n}$. Then for all large enough n, we get

$$\frac{\mathrm{Var}\left[\psi_{n,k}\right]}{\mathbb{E}\left[\psi_{n,k}\right]^2} \leq \exp\left(n\left(\log(\gamma\alpha^{-1})+4\delta+\frac{6}{n}\log(2K)+\frac{qk}{n^2}\log n\right)\right).$$

Since $k=o(n^2/\log(n))$, and $\log(\gamma\alpha^{-1})+4\delta<0$ (by our choice of δ in the proof of Theorem 1.1), we obtain the desired bound. □

6. Bounds on pressure

In this section we work with the same notation, parameters, and assumptions as in the proof of Theorem 1.1.

LEMMA 6.1. *For each n,*

$$P_{Y_n}(f)\leq\frac{1}{k}\log\phi_{n,k}.$$

PROOF. By subadditivity in the definition of pressure, we have that for all $m\geq 1$,

$$P_{Y_n}(f)\leq\frac{1}{m}\log\sum_{u\in B_m(Y_n)}e^{S_m f(u)}.$$

We apply this inequality with $m=k$. Also, since $B_k(Y_n)\subset\{u\in B_k(X):\xi_u=1\}$, we have

$$\begin{aligned}\frac{1}{k}\log\sum_{u\in B_k(Y_n)}e^{S_m f(u)} &\leq \frac{1}{k}\log\sum_{u\in B_k(X)}e^{S_k f(u)}\xi_u \\ &= \frac{1}{k}\log\phi_{n,k}.\end{aligned}$$

Combining the two previous inequalities yields the desired conclusion. □

LEMMA 6.2. *For any $\epsilon > 0$, for all large enough n,*

$$\frac{1}{k}\log \psi_{n,k} - \epsilon/2 \leq P_{Y_n}(f).$$

PROOF. Let $\mathcal{F} = \mathcal{F}_n$ and $Y = Y_n$. For $v \in E_n$, and $m \geq n$, we let

$$Z_m(v) = \Big\{ u \in B_m(X) : W_n(u) \cap \mathcal{F} = \varnothing, \text{ and } \forall q \in \{0, \dots, \lfloor m/\ell - 1 \rfloor\},\ u_{q\ell+1}^{q\ell+n} = v \Big\}.$$

Note that $\psi_{n,k}$ may be viewed as an average over the set E_n:

$$\psi_{n,k} = \frac{1}{|E_n|} \sum_{v \in E_n} \sum_{u \in Z_k(v)} e^{S_k f(u)}.$$

Since the average over a finite set is always less than or equal to the maximum, there exists $v \in E_n$ such that

$$\psi_{n,k} \leq \sum_{u \in Z_k(v)} e^{S_k f(u)}.$$

For the sake of this proof, if $u \in B_m(X)$, then we let $\underline{S}_m f(u) = \inf_{x \in [u]} \sum_{j=0}^{m-1} f \circ \sigma^j(x)$. Observe that elements of $Z_\ell(v)$ can be arbitrarily concatenated to form words in Y. Hence, for any $q \in \mathbb{N}$, we note that $Z_{q\ell}(v) \subset B_{q\ell}(Y)$, and then we have

$$\begin{aligned} \sum_{u \in B_{q\ell}(Y)} e^{S_{q\ell} f(u)} &\geq \sum_{u \in Z_{q\ell}(v)} e^{S_{q\ell} f(u)} \\ &\geq \sum_{u \in Z_{q\ell}(v)} e^{\underline{S}_{q\ell} f(u)}. \end{aligned}$$

Then by our choice of K, we get

$$\begin{aligned} \sum_{u \in B_{q\ell}(Y)} e^{S_{q\ell} f(u)} &\geq K^{-q} \sum_{u_0 \dots u_{q-1} \in Z_{q\ell}(v)} e^{\sum_{i=0}^{q-1} \underline{S}_\ell f(u_i)} \\ &= K^{-q} \sum_{u_0 \in Z_\ell(v)} \cdots \sum_{u_{q-1} \in Z_\ell(v)} e^{\sum_{i=0}^{q-1} \underline{S}_\ell f(u_i)} \\ &= K^{-q} \Big(\sum_{u \in Z_\ell(v)} e^{\underline{S}_\ell(u)} \Big)^q \\ &\geq K^{-2q} \Big(\sum_{u \in Z_\ell(v)} e^{S_\ell(u)} \Big)^q \\ &\geq K^{-2q} \Big(\sum_{u \in Z_k(v)} e^{S_k(u)} \Big)^q e^{-\|f\|_\infty nq}, \end{aligned}$$

where $\|f\|_\infty = \sup_{x \in X} |f(x)|$. Now take logarithm, divide by $q\ell$, and let q tend to infinity:

$$P_Y(f) \geq \frac{1}{\ell} \log \sum_{u \in Z_k(v)} e^{S_k f(u)} - \frac{2q \log K}{\ell} - \|f\|_\infty \frac{n}{\ell}.$$

Then

$$P_Y(f) \geq \frac{1}{k} \log \psi_{n,k} - \frac{2q \log K}{\ell} - \|f\|_\infty \frac{n}{\ell}.$$

Finally, since $n/\ell \to 0$, we may choose n large enough that

$$\frac{2q \log K}{\ell} + \|f\|_\infty \frac{n}{\ell} < \epsilon/2,$$

which finishes the proof of the lemma. □

7. Connection between pressure and escape rate

Here we relate the notions of pressure and escape rate. For a hole H in an SFT X, we define the *survivor set* to be the set of points that never fall into the hole (in either forward or backward time):

$$Y = X \setminus \left(\bigcup_{m \in \mathbb{Z}} \sigma^{-m}(H) \right).$$

For an SFT (X, σ), a hole H consisting of a finite union of cylinder sets, and an equilibrium state μ associated to a Hölder continuous potential function f, the following proposition relates the escape rate of μ through the hole H to the pressure of f on the survivor set Y. Although various versions of this result appear to be well-known (see, *e.g.*, [**7**, **13**]), we could not find an explicit reference for it, and we include a proof for completeness. For analogous results in various smooth settings, see the discussion of the *escape rate formula* in [**4**] and references therein.

PROPOSITION 7.1. *Let X be a non-trivial mixing SFT, $f : X \to \mathbb{R}$ a Hölder continuous potential, and μ the Gibbs measure associated to f. Further, let H be a finite union of cylinder sets in X, and let Y be the survivor set of the open system (X, σ, H). Then*

$$-\varrho(\mu : H) = P_X(f) - P_Y(f).$$

PROOF. Let K satisfy the conclusions of Lemma 2.2 for X, f, and μ. We suppose without loss of generality that H is the union of cylinder sets corresponding to words of length n. For $k \geq n$, let $B_k(X, H)$ denote the set of $w \in B_k(X)$ such that w contains no subword in H, and let $P = P_X(f)$. Let $\ell = \ell(k) = k - n + 1$. Recall that

$$M_\ell = \left\{ x \in X : \forall j \in \{0, \ldots, \ell - 1\}, \, \sigma^j(x) \notin H \right\},$$

so that we have $\mu(M_\ell) = \mu(B_k(X, H))$. Note that since n is fixed in this context, we have $\lim_{k \to \infty} \ell/k = 1$.

By our choice of K and the fact that $B_k(Y) \subset B_k(X, H)$, we have that

$$\begin{aligned}
\mu(B_k(X, H)) &= \sum_{w \in B_k(X,H)} \mu(w) \\
&\geq \sum_{w \in B_k(X,H)} K^{-1} e^{-Pk + S_k f(w)} \\
&= K^{-1} e^{-Pk} \sum_{w \in B_k(X,H)} e^{S_k f(w)} \\
&\geq K^{-1} e^{-Pk} \sum_{w \in B_k(Y)} e^{S_k f(w)}.
\end{aligned}$$

It follows that

$$\frac{1}{k} \log \mu(B_k(X, H)) \geq -P + \frac{1}{k} \log \Lambda_k(Y) - \frac{1}{k} \log K,$$

and letting k tend to infinity, we see that

$$\liminf_k \frac{1}{k}\log\mu(M_\ell) \geq -P + \liminf_{k\to\infty}\frac{1}{k}\log\Lambda_k(Y) = -P + P_Y(f). \tag{7.1}$$

Similarly, we have the following upper bound:

$$\limsup_k \frac{1}{k}\log\mu(M_\ell) \leq -P + \limsup_{k\to\infty}\frac{1}{k}\log\left(\sum_{w\in B_k(X,H)} e^{S_k f(w)}\right). \tag{7.2}$$

Comparing the bounds in (7.1) and (7.2) , we see that in order to finish the proof, it suffices to show that

$$\limsup_{k\to\infty}\frac{1}{k}\log\left(\sum_{w\in B_k(X,H)} e^{S_n f(w)}\right) \leq P_Y(f). \tag{7.3}$$

To get this inequality, we use ideas from [**26**] to find an invariant measure ν supported on Y such that $h(\nu) + \int f d\nu$. The measure ν is obtained as follows.

For $k \geq 1$, suppose $B_k(X,H) = \{w_1^k, \dots, w_{m_k}^k\}$. Let x_i^k be in $[w_i^k]$ such that $S_k f(w_i^k) = S_m f(x_i^k)$ (which exists by compactness and continuity). Then let

$$\mu_k = \frac{\sum_{j=1}^m e^{S_k f(x_j^k)}\delta_{x_j^k}}{\sum_{j=1}^m e^{S_k f(x_j^k)}}$$

$$\nu_k = \frac{1}{k}\sum_{j=0}^{k-1} S^j \mu_k.$$

Since the space of Borel probability measures on X is weak* compact, there exists a Borel probability measure ν on X and a subsequence (ν_{k_j}) of (ν_k) such that $\nu_{k_j} \to \nu$ and along which the lim sup in (7.3) is obtained. Note that ν is in $M(X,S)$. Furthermore, we have that

$$\begin{aligned}\log\left(\sum_{w\in B_k(X,H)} e^{S_k f(w)}\right) &= \log\left(\sum_{j=1}^m e^{S_k f(x_j^k)}\right)\\ &= H_{\mu_k}(\xi^k) + \int f d\nu_k,\end{aligned}$$

where ξ is the natural partition of X according to the symbol in the zero coordinate and $\xi^k = \bigvee_{j=0}^{k-1}\sigma^{-j}\xi$. Arguing as in Proposition 3.6 of [**26**], we obtain that

$$\limsup_{k\to\infty}\frac{1}{k}\log\left(\sum_{w\in B_k(X,H)} e^{S_k f(w)}\right) \leq h(\nu) + \int f d\nu. \tag{7.4}$$

Now we claim that ν is supported on Y. Let $[w]$ be a cylinder set in X such that $Y \cap [w] = \varnothing$ and w has length N. We show that $\nu([w]) = 0$. Since $Y \cap [w] = \varnothing$ and since X is compact, there must exist k_0 such that for all $k \geq k_0$ and for all u in $B_k(X,H)$, it holds that w is not a subword of u. Then for u in $B_k(X,H)$, x in $[u]$, and $j = 0,\dots,k-N$, we have that $S^j(x) \notin [w]$. Hence $S^j\mu_k(w) = 0$ for $j = 0,\dots,k-N$, and therefore

$$\nu_k(w) = \frac{1}{k}\sum_{j=0}^{k-1} S^j\mu_k(w) = \frac{1}{k}\sum_{j=k-N+1}^{k-1} S^j\mu_k(w) \leq \frac{N}{k}.$$

Letting k tend to infinity along the subsequence (k_j), we obtain that $\nu([w]) = 0$, as desired. Hence ν is supported on Y.

Then by (7.4) and the variational principle for $P_Y(f)$, we have that

$$\limsup_{k\to\infty} \frac{1}{k} \log\left(\sum_{w\in B_k(X,H)} e^{S_k f(w)} \right) \le h(\nu) + \int f d\nu \le P_Y(f),$$

which establishes (7.3) and finishes the proof. □

7.1. Proof of Theorem 1.4. Having established Proposition 7.1, we are now in a position to prove Theorem 1.4. The proof simply uses Proposition 7.1 to reduce Theorem 1.4 to Theorem 1.1.

Proof of Theorem 1.4. Let X be a non-trivial mixing SFT, $f : X \to \mathbb{R}$ a Hölder continuous potential with associated Gibbs measure μ, and $\gamma_0 = \gamma_0(X, f)$ as in Theorem 1.1. Let $\alpha \in (\gamma_0, 1]$. Let $\epsilon > 0$. By Theorem 1.1, there exists $\rho > 0$ such that for all large enough n,

$$\mathbb{P}\Big(\big| P_{Y_n}(f) - (P_X(f) + \log(\alpha)) \big| \ge \epsilon \Big) < e^{-\rho n}.$$

Observe that Y_n is the survivor set of the open system (X, σ, H_n). Then by Proposition 7.1, we have

$$-\varrho(\mu : H_n) = P_X(f) - P_{Y_n}(f).$$

Then for all large enough n, we see that

$$\mathbb{P}\Big(\big| \varrho(\mu : H_n) - \log(\alpha) \big| \ge \epsilon \Big) = \mathbb{P}\Big(\big| P_{Y_n}(f) - (P_X(f) + \log(\alpha)) \big| \ge \epsilon \Big) < e^{-\rho n},$$

as was to be shown. □

Acknowledgments

The author would like to thank the anonymous referee for helpful comments and suggestions. This work was supported by the National Science Foundation through the grant DMS 1613261.

References

[1] Miguel Abadi, *Sharp error terms and necessary conditions for exponential hitting times in mixing processes*, Ann. Probab. **32** (2004), no. 1A, 243–264, DOI 10.1214/aop/1078415835. MR2040782

[2] Rufus Bowen, *Equilibrium states and the ergodic theory of Anosov diffeomorphisms*, Lecture Notes in Mathematics, Vol. 470, Springer-Verlag, Berlin-New York, 1975. MR0442989

[3] Ryan Broderick, *Finite orbits in random subshifts of finite type*, Qual. Theory Dyn. Syst. **16** (2017), no. 3, 531–545, DOI 10.1007/s12346-017-0224-5. MR3703513

[4] Henk Bruin, Mark Demers, and Ian Melbourne, *Existence and convergence properties of physical measures for certain dynamical systems with holes*, Ergodic Theory Dynam. Systems **30** (2010), no. 3, 687–728, DOI 10.1017/S0143385709000200. MR2643708

[5] Leonid A. Bunimovich and Alex Yurchenko, *Where to place a hole to achieve a maximal escape rate*, Israel J. Math. **182** (2011), 229–252, DOI 10.1007/s11856-011-0030-8. MR2783972

[6] N. Chernov and R. Markarian, *Anosov maps with rectangular holes. Nonergodic cases*, Bol. Soc. Brasil. Mat. (N.S.) **28** (1997), no. 2, 315–342, DOI 10.1007/BF01233396. MR1479506

[7] N. Chernov and R. Markarian, *Ergodic properties of Anosov maps with rectangular holes*, Bol. Soc. Brasil. Mat. (N.S.) **28** (1997), no. 2, 271–314, DOI 10.1007/BF01233395. MR1479505

[8] N. Chernov, R. Markarian, and S. Troubetzkoy, *Conditionally invariant measures for Anosov maps with small holes*, Ergodic Theory Dynam. Systems **18** (1998), no. 5, 1049–1073, DOI 10.1017/S0143385798117492. MR1653291

[9] N. Chernov, R. Markarian, and S. Troubetzkoy, *Invariant measures for Anosov maps with small holes*, Ergodic Theory Dynam. Systems **20** (2000), no. 4, 1007–1044, DOI 10.1017/S0143385700000560. MR1779391

[10] H. van den Bedem and N. Chernov, *Expanding maps of an interval with holes*, Ergodic Theory Dynam. Systems **22** (2002), no. 3, 637–654, DOI 10.1017/S0143385702000329. MR1908547

[11] Pierre Collet, Servet Martínez, and Bernard Schmitt, *The Yorke-Pianigiani measure and the asymptotic law on the limit Cantor set of expanding systems*, Nonlinearity **7** (1994), no. 5, 1437–1443. MR1294552

[12] Pierre Collet, Servet Martínez, and Bernard Schmitt, *Quasi-stationary distribution and Gibbs measure of expanding systems*, Instabilities and nonequilibrium structures, V (Santiago, 1993), Nonlinear Phenom. Complex Systems, vol. 1, Kluwer Acad. Publ., Dordrecht, 1996, pp. 205–219, DOI 10.1007/978-94-009-0239-8_19. MR1406590

[13] Pierre Collet, Servet Martínez, and Bernard Schmitt, *The Pianigiani-Yorke measure for topological Markov chains*, Israel J. Math. **97** (1997), 61–70, DOI 10.1007/BF02774026. MR1441238

[14] Mark F. Demers and Lai-Sang Young, *Escape rates and conditionally invariant measures*, Nonlinearity **19** (2006), no. 2, 377–397, DOI 10.1088/0951-7715/19/2/008. MR2199394

[15] Mark F. Demers and Bastien Fernandez, *Escape rates and singular limiting distributions for intermittent maps with holes*, Trans. Amer. Math. Soc. **368** (2016), no. 7, 4907–4932, DOI 10.1090/tran/6481. MR3456165

[16] Mark Demers, Paul Wright, and Lai-Sang Young, *Escape rates and physically relevant measures for billiards with small holes*, Comm. Math. Phys. **294** (2010), no. 2, 353–388, DOI 10.1007/s00220-009-0941-y. MR2579459

[17] Mark F. Demers, *Markov extensions and conditionally invariant measures for certain logistic maps with small holes*, Ergodic Theory Dynam. Systems **25** (2005), no. 4, 1139–1171, DOI 10.1017/S0143385704000963. MR2158400

[18] Mark F. Demers, *Markov extensions for dynamical systems with holes: an application to expanding maps of the interval*, Israel J. Math. **146** (2005), 189–221, DOI 10.1007/BF02773533. MR2151600

[19] Mark F. Demers, *Dispersing billiards with small holes*, Ergodic theory, open dynamics, and coherent structures, Springer Proc. Math. Stat., vol. 70, Springer, New York, 2014, pp. 137–170, DOI 10.1007/978-1-4939-0419-8_8. MR3213499

[20] Mark F. Demers, Christopher J. Ianzano, Philip Mayer, Peter Morfe, and Elizabeth C. Yoo, *Limiting distributions for countable state topological Markov chains with holes*, Discrete Contin. Dyn. Syst. **37** (2017), no. 1, 105–130, DOI 10.3934/dcds.2017005. MR3583472

[21] Mark F. Demers and Paul Wright, *Behaviour of the escape rate function in hyperbolic dynamical systems*, Nonlinearity **25** (2012), no. 7, 2133–2150, DOI 10.1088/0951-7715/25/7/2133. MR2947939

[22] Mark F. Demers, Paul Wright, and Lai-Sang Young, *Entropy, Lyapunov exponents and escape rates in open systems*, Ergodic Theory Dynam. Systems **32** (2012), no. 4, 1270–1301, DOI 10.1017/S0143385711000344. MR2955314

[23] Andrew Ferguson and Mark Pollicott, *Escape rates for Gibbs measures*, Ergodic Theory Dynam. Systems **32** (2012), no. 3, 961–988, DOI 10.1017/S0143385711000058. MR2995652

[24] Gary Froyland and Ognjen Stancevic, *Escape rates and Perron-Frobenius operators: Open and closed dynamical systems*, Discrete Contin. Dyn. Syst. Ser. B **14** (2010), no. 2, 457–472, DOI 10.3934/dcdsb.2010.14.457. MR2660868

[25] Gerhard Keller, *Rare events, exponential hitting times and extremal indices via spectral perturbation*, Dyn. Syst. **27** (2012), no. 1, 11–27, DOI 10.1080/14689367.2011.653329. MR2903242

[26] François Ledrappier and Peter Walters, *A relativised variational principle for continuous transformations*, J. London Math. Soc. (2) **16** (1977), no. 3, 568–576, DOI 10.1112/jlms/s2-16.3.568. MR0476995

[27] Carlangelo Liverani and Véronique Maume-Deschamps, *Lasota-Yorke maps with holes: conditionally invariant probability measures and invariant probability measures on the survivor*

set (English, with English and French summaries), Ann. Inst. H. Poincaré Probab. Statist. **39** (2003), no. 3, 385–412, DOI 10.1016/S0246-0203(02)00005-5. MR1978986

[28] Artur Lopes and Roberto Markarian, *Open billiards: invariant and conditionally invariant probabilities on Cantor sets*, SIAM J. Appl. Math. **56** (1996), no. 2, 651–680, DOI 10.1137/S0036139995279433. MR1381665

[29] Kevin McGoff, *Random subshifts of finite type*, Ann. Probab. **40** (2012), no. 2, 648–694, DOI 10.1214/10-AOP636. MR2952087

[30] Kevin McGoff and Ronnie Pavlov, *Factor maps and embeddings for random $\mathbb{Z}^d$ shifts of finite type*, Israel J. Math., to appear.

[31] Kevin McGoff and Ronnie Pavlov, *Random $\mathbb{Z}^d$-shifts of finite type*, J. Mod. Dyn. **10** (2016), 287–330, DOI 10.3934/jmd.2016.10.287. MR3538865

[32] Donald Samuel Ornstein and Benjamin Weiss, *Entropy and data compression schemes*, IEEE Trans. Inform. Theory **39** (1993), no. 1, 78–83, DOI 10.1109/18.179344. MR1211492

[33] Giulio Pianigiani and James A. Yorke, *Expanding maps on sets which are almost invariant. Decay and chaos*, Trans. Amer. Math. Soc. **252** (1979), 351–366, DOI 10.2307/1998093. MR534126

[34] Peter Walters, *An introduction to ergodic theory*, Graduate Texts in Mathematics, vol. 79, Springer-Verlag, New York-Berlin, 1982. MR648108

[35] Lai-Sang Young, *Large deviations in dynamical systems*, Trans. Amer. Math. Soc. **318** (1990), no. 2, 525–543, DOI 10.2307/2001318. MR975689

9201 University City Blvd., Charlotte, NC 28223
Email address: kmcgoff1@uncc.edu

Contemporary Mathematics
Volume **736**, 2019
https://doi.org/10.1090/conm/736/14833

On the complexity function for sequences which are not uniformly recurrent

Nic Ormes and Ronnie Pavlov

ABSTRACT. We prove that every non-minimal transitive subshift X satisfying a mild aperiodicity condition satisfies $\limsup(c_n(X) - 1.5n) = \infty$, and give a class of examples which shows that the threshold of $1.5n$ cannot be increased. As a corollary, we show that any transitive X satisfying $\limsup(c_n(X) - n) = \infty$ and $\limsup(c_n(X) - 1.5n) < \infty$ must be minimal. We also prove some restrictions on the structure of transitive non-minimal X satisfying

$$\liminf(c_n(X) - 2n) = -\infty,$$

which imply unique ergodicity (for a periodic measure) as a corollary, which extends a result of Boshernitzan [**2**] from the minimal case to the more general transitive case.

1. Introduction and definitions

In this work, we describe some simple connections between the recurrence properties of a two-sided sequence x and the so-called **word complexity function** $c_n(x)$ which measures the number of words of length n appearing in x. One of the most fundamental results of this sort is the Morse-Hedlund theorem, which has slightly different statements in the one- and two-sided cases (see [**7**]).

THEOREM 1.1. (Morse-Hedlund Theorem) *Suppose that A is a finite alphabet, $x \in A^{\mathbb{N}}$ or $x \in A^{\mathbb{Z}}$, and there exists n such that the number of n-letter subwords of x is less than or equal to n. Then, if $x \in A^{\mathbb{Z}}$, x must be periodic, and if $x \in A^{\mathbb{N}}$, then x must be eventually periodic.*

One way to view this theorem is that it yields a lower bound on $c_n(x)$; if x is two-sided and not periodic, then $c_n(x) \geq n+1$ for all n. It is well-known that this bound is sharp; there exist aperiodic sequences called **Sturmian** sequences (see Chapter 6 of [**5**] for an introduction) for which $c_n(x) = n + 1$ for all n. There are also other examples in the literature ([**1**]) with $1 < \liminf(c_n(x)/n) < \limsup(c_n(x)/n) < 1 + \epsilon$ for arbitrarily small ϵ. All of these examples are **uniformly recurrent** sequences, meaning that for every subword w, there exists N so that every N-letter subword contains w. Equivalently, a sequence is uniformly recurrent whenever the

2010 *Mathematics Subject Classification.* Primary: 37B10; Secondary: 05A05, 37B20.

Key words and phrases. Symbolic dynamics, word complexity, transitive, minimal, uniquely ergodic.

The second author gratefully acknowledges the support of NSF grant DMS-1500685.

shift map acting on its orbit closure forms a minimal topological dynamical system [**3**, Ch. 2].

There are also fairly simple examples of sequences which are not uniformly recurrent and yet have $c_n(x) < n + k$ for all n and some constant k, given by any x which is not periodic but is eventually periodic in both directions. For example, if $x = \ldots 121212344444\ldots$ then $c_n(x) = n + 3$ for all $n \geq 1$, and x is clearly not uniformly recurrent since the word 1234 occurs just once.

This leads to a natural question: must a sequence with complexity function "close to n" be either uniformly recurrent or eventually periodic in both directions? Our main result shows that this is indeed the case.

THEOREM 1.2. *If x is not uniformly recurrent and it is not true that x is eventually periodic in both directions, then* $\limsup(c_n(x) - 1.5n) = \infty$.

This gives a large gap in the complexity functions achievable by sequences which are not uniformly recurrent; any such complexity is either below $n+k$ for all n and some constant k, or has a subsequence along which $c_n(x) - 1.5n$ approaches infinity. In particular, this means that some interesting examples from the literature ([**1**], [**6**]) with $1 < \limsup(c_n(x)/n) < 1.5$ can only be achieved by uniformly recurrent sequences. We also show that Theorem 1.2 is tight in the sense that the threshold of $1.5n$ cannot be meaningfully increased.

THEOREM 1.3. *For any nondecreasing $g : \mathbb{N} \to \mathbb{R}$ with $\lim g(n) = \infty$, there exists an x which is not uniformly recurrent and for which it is not true that x is eventually periodic in both directions where $c_n(x) < 1.5n + g(n)$ for sufficiently large n.*

Our proof is an analysis by cases, and in most of the cases, the much stronger bound $\liminf(c_n(x) - 2n) > -\infty$ holds. We can then prove a fairly strong structure on those x for which it does not.

THEOREM 1.4. *If x is not uniformly recurrent, it is not true that x is eventually periodic in both directions, and* $\liminf(c_n(x) - 2n) = -\infty$, *then there exist a constant k and periodic orbit M with the following property: for every N, there exists $m > N$ so that every $(3m + k)$-letter subword of x contains an m-letter subword of a point in M.*

Informally, the conclusion of Theorem 1.4 says that x can be partitioned, at arbitrarily large "scales," into long (possibly infinite on one side) pieces of the periodic orbit M and pieces not in M which are not much longer. Unsurprisingly, this structure is quite similar to the structure of the examples proving Theorem 1.3, as we will see in Section 3.

Theorem 1.4 implies a useful corollary which extends a result of Boshernitzan. He proved in [**2**] that if X is minimal and $\liminf(c_n(X) - 2n) = -\infty$, then X is uniquely ergodic, i.e. there is only one shift-invariant Borel probability measure on X. The following result uses the same complexity hypothesis, but applies to non-minimal transitive systems.

THEOREM 1.5. *If $X = \overline{\mathcal{O}(x)}$, x is not uniformly recurrent, it is not true that x is eventually periodic in both directions, and* $\liminf(c_n(X) - 2n) = -\infty$, *then X is uniquely ergodic, with unique shift-invariant measure supported on a periodic orbit.*

(We would like to note that the proof in [**2**] could theoretically be applied to transitive systems with very few changes, and so the main new content in our result is the triviality of the measure in the non-minimal case.)

Cyr and Kra ([**4**]) recently generalized a different result of Boshernitzan's, proving that under no assumption on X whatsoever, for any $k \in \mathbb{N}$, $\liminf(c_n(X)/n) < k$ implies that X has fewer than k nonatomic shift-invariant measures which have so-called generic points. Theorem 1.5 applies only to the case $k = 2$ and assumes transitivity of X and some aperiodicity of x, but uses a weaker complexity hypothesis and implies that X cannot have multiple shift-invariant measures at all, rather than only forbidding multiple nonatomic shift-invariant measures.

The authors would like to thank the referee of this paper for numerous helpful comments.

2. Definitions

Let A denote a finite set, which we will refer to as our alphabet.

DEFINITION 2.1. *A bi-infinite sequence $x \in A^{\mathbb{Z}}$ is* ***periodic*** *if there exists $n \neq 0$ so that $x(k) = x(k+n)$ for all $k \in \mathbb{Z}$. A one-sided sequence $x \in A^{\mathbb{N}}$ is* ***eventually periodic*** *if there exist $n, N \in \mathbb{N}$ so that $x(k) = x(k+n)$ for all $k > N$; the definition is analogous for $x \in A^{-\mathbb{N}}$. A bi-infinite sequence x is* ***eventually periodic in both directions*** *if $x(0)x(1)x(2)\ldots$ and $\ldots x(-2)x(-1)$ are each eventually periodic.*

DEFINITION 2.2. *A* ***subshift*** *X on an alphabet A is any subset of $A^{\mathbb{Z}}$ which is invariant under the left shift map σ and closed in the product topology.*

DEFINITION 2.3. *A subshift X is* ***transitive*** *if it can be written as $\overline{\mathcal{O}(x)}$ for some $x \in A^{\mathbb{Z}}$, where $\mathcal{O}(x) := \{\sigma^n x \ : \ n \in \mathbb{Z}\}$.*

DEFINITION 2.4. *A subshift X is* ***minimal*** *if it contains no proper nonempty subshift; equivalently, if $X = \overline{\mathcal{O}(x)}$ for all $x \in X$.*

A routine application of Zorn's Lemma shows that every nonempty subshift contains a nonempty minimal subshift.

DEFINITION 2.5. *A* ***word*** *over A is a member of A^n for some $n \in \mathbb{N}$. For $w \in A^n$ we call n the* ***length*** *of w and denote it by $|w|$. A word w is called a* ***subword*** *of a longer word or infinite or bi-infinite sequence u if there exists i so that $u(i+j) = w(j)$ for all $1 \leq j \leq |w|$.*

DEFINITION 2.6. *A sequence $x \in A^{\mathbb{Z}}$ is* ***recurrent*** *if every subword of x appears infinitely many times within x, and* ***uniformly recurrent*** *if, for every $w \in W(x)$, there exists N so that every N-letter subword of x contains w as a subword.*

DEFINITION 2.7. *For any words $v \in A^n$ and $w \in A^m$, we define the* ***concatenation*** *vw to be the word in A^{n+m} whose first n letters are the letters forming v and whose next m letters are the letters forming w.*

DEFINITION 2.8. *For a word $u \in A^n$, if u can be written as the concatenation of two words $u = vw$ then we say that v is a* ***prefix*** *of u and that w is a* ***suffix*** *of u.*

DEFINITION 2.9. *For any infinite or bi-infinite sequence x, we denote by $W(x)$ the set of all subwords of x and, for any $n \in \mathbb{N}$, define $W_n(x) = W(x) \cap A^n$, the set of subwords of x with length n. For a subshift X, we define $W(X) = \bigcup_{x \in X} W(x)$ and $W_n(X) = \bigcup_{x \in X} W_n(x)$.*

DEFINITION 2.10. *For any infinite or bi-infinite sequence x, $c_n(x) := |W_n(x)|$ is the* ***word complexity function*** *of x; for a subshift X, $c_n(X)$ is similarly defined.*

DEFINITION 2.11. *A word w is* ***right-special*** *within a subshift X if there exist $a \neq b \in A$ so that $wa, wb \in W(X)$.*

We note that for every subshift X and $w \in W(X)$, there exists at least one letter a so that $wa \in W(X)$. Therefore, for any n, $c_{n+1}(X) - c_n(X)$ is greater than or equal to the number of right-special words in $W_n(X)$.

DEFINITION 2.12. *A* ***sliding block code with anticipation*** *a* ***and memory*** *m is a function ϕ defined on a subshift X where $\phi(X)$ is a subshift and $(\phi(x))(i)$ depends only on $x(i-m)x(i-m+1)\dots x(i+a-1)x(i+a)$ for all $x \in X$ and $i \in \mathbb{Z}$. For a sliding block code, we define the* ***window size*** *to be $a+m+1$, the length of $x(i-m)x(i-m+1)\dots x(i+a-1)x(i+a)$.*

All of the sliding block codes we will construct in this paper will have memory 0. For such a sliding block code ϕ with window size k even though ϕ technically is defined on X, it induces an obvious action on words in $W_n(X)$ for $n \geq k$; for any such w, one can define $\phi(w) \in W_{n-k+1}(\phi(X))$ to be $(\phi(x))(0)\dots(\phi(x))(n-k)$ for any x with $x(0)\dots x(n-1) = w$. (This is independent of choice of x by the definition of sliding block code.) This induces a surjection from $W_n(X)$ to $W_{n-k+1}(\phi(X))$, and so for any such ϕ and $n \geq k$, $c_n(X) \geq c_{n-k+1}(\phi(X))$.

3. Proofs

3.1. Proof of Theorem 1.2. Throughout, x will represent a bi-infinite sequence and X will represent its orbit closure, $X = \overline{\mathcal{O}(x)}$. Note that then $W_n(X)$ is just the set of words of length n appearing as subwords of x, and $c_n(X)$ is the number of such words, i.e., $W_n(X) = W_n(x)$ and $c_n(X) = c_n(x)$. We assume throughout that x is not uniformly recurrent and that it is not true that x is eventually periodic in both directions, and will now break into various cases and give lower bounds on $c_n(x)$ in each.

3.1.1. *x is non-recurrent.*

LEMMA 3.1. *If x is non-recurrent and it is not true that x is eventually periodic in both directions, then* $\liminf(c_n(X) - 2n) > -\infty$.

PROOF. Since x is not recurrent, there exists a word v which appears in x only finitely many times. We can then write $x = \ell w r$ where w contains all occurrences of v in x; then w occurs only once in x. Then ℓ and r do not contain w, and one of ℓ or r is not eventually periodic. We treat only the r case here, as the ℓ case is similar. Since r is not eventually periodic, by Theorem 1.1, it contains at least $n+1$ distinct n-letter subwords for every n, and none of these contain w as a subword. In addition, $x = \ell w r$ contains $n - |w| + 1$ subwords of length n which contain w, which are all distinct since they contain w exactly once at different locations. Therefore, $c_n(X) \geq (n+1) + (n - |w| + 1) = 2n - |w| + 2$ for all n, which implies $c_n(X) - 2n$ is bounded below by $2 - |w|$, i.e., $\liminf(c_n(X) - 2n) > -\infty$. □

3.1.2. *x is recurrent and not uniformly recurrent.*

Recall that the shift map on $X = \overline{\mathcal{O}(x)}$ is minimal if and only if x is uniformly recurrent. Therefore, if x is not uniformly recurrent, X must properly contain some minimal subshift.

LEMMA 3.2. *If X properly contains an infinite minimal subshift M, then* $\liminf(c_n(X) - 2n) > -\infty$.

PROOF. Suppose that X, M are as in the theorem. Since $\overline{\mathcal{O}(x)} = X \neq M$, x contains a subword not in $W(M)$, let's call it w. By shifting x if necessary, we may assume that $w = x(0)\dots x(|w|-1)$. By recurrence, x contains infinitely many occurrences of w. However, since $\overline{\mathcal{O}(x)} = X \supset M$, x contains arbitrarily long subwords in $W(M)$, none of which may contain w. Choose any $n \geq |w|$, and consider a subword of x of length n which does not contain w; take it to be $x(k)\dots x(k+n-1)$, and for now assume that $k > 0$. Now, the word $x(0)\dots x(k+n-1)$ contains w at least once (as a prefix), so we may define the rightmost occurrence of w within it; say this happens at $x(j)\dots x(j+|w|-1)$. Note that since $x(k)\dots x(k+n-1)$ contains no occurrences of w, we know that $j < k$.

Finally, consider the n-letter subwords of x defined by $u_i = x(i)\dots x(i+n-1)$, where $j-n+|w| \leq i \leq j$. Each contains the occurrence of w at $x(j)\dots x(j+|w|-1)$, and no occurrence of w to the right, by definition of j. Therefore, all are distinct, and so x contains $n - |w| + 1$ subwords of length n, which each contain w.

On the other hand, since M is an infinite minimal subshift, it is aperiodic, and so by Theorem 1.1, $W_n(M)$ contains at least $n + 1$ subwords of length n, none of which contain w since $w \notin W(M)$. Since $M \subset X$, $W_n(M) \subset W_n(X)$, and so $c_n(X) > (n+1) + (n - |w| + 1) = 2n - |w| + 2$ for all $n \geq |w|$, completing the proof when $k > 0$. Since the complexity function is unaffected by reflecting x (and w) about the origin, the same holds when $k < 0$, completing the proof. □

We now only need treat the case where X contains only finite minimal subshifts (i.e. periodic orbits), and will first deal with the case where it contains more than one.

LEMMA 3.3. *If X contains two minimal subshifts and it is not true that x is eventually periodic in both directions, then* $\liminf(c_n(X) - 2n) > -\infty$.

PROOF. Denote by M and M' two different minimal subshifts of X; by definition of minimality, M and M' are disjoint closed sets. If either M or M' is infinite, then we are done by Lemma 3.2. So, assume that both are finite, and therefore periodic orbits. Let p denote the product of the cardinalities of M and M'. Then every point $z \in M \cup M'$ satisfies $\sigma^p(z) = z$, or $z(i) = z(i + p)$ for all $i \in \mathbb{Z}$.

Suppose $k > p$. We claim $W_k(M) \cap W_k(M') = \varnothing$. To see this, note that if $W_k(M) \cap W_k(M')$ is nonempty then by considering a prefix of length p, we have a word $w \in W_p(M) \cap W_p(M')$. But then by infinitely concatenating w, we obtain a sequence $\dots wwwww \dots$ which is in both M and M', a contradiction.

We may assume without loss of generality that x is not eventually periodic to the right. Thus we can find arbitrarily large indices i such that $x(i) \neq x(i + p)$. Since $\overline{\mathcal{O}(x)}$ contains M, there are arbitrarily long words from $W(M)$ in x (similarly for M'). It follows that for all $n > p$, there exists ℓ so that $x(\ell)\dots x(\ell+n-1) \in W(M)$ and $x(\ell+n-p)\dots x(\ell+n) \notin W(M)$. Similarly, there exists m so that $x(m)\dots x(m+n-1) \in W(M')$ and $x(m+n-p)\dots x(m+n) \notin W(M')$.

Define the n-letter words $u_i = x(\ell + i) \dots x(\ell + i + n - 1)$ and $v_j = x(m + j) \dots x(m + j + n - 1)$ for $0 \leq i, j < n - p$; clearly all are in $W_n(X)$. In each u_i, the leftmost $(p+1)$-letter word not in $W(M)$ is $u_i(n-i-p+1) \dots u_i(n-i+1) = x(\ell+n-p) \dots x(\ell+n)$, and so all u_i are distinct. The same argument (using $W(M')$) shows that all v_j are distinct. Finally, all u_i begin with a word in $W_{p+1}(M)$ and all v_j begin with a word in $W_{p+1}(M')$, and so the sets $\{u_i\}$ and $\{v_j\}$ are also disjoint. Therefore, $c_n(X) \geq 2n - 2p$ for $n > p$, completing the proof. □

The remaining case is that x is recurrent and that X properly contains a periodic orbit M, which is the only minimal subshift contained in X. For simplicity, we assume that M is a single fixed point, which we may do via the following lemma.

LEMMA 3.4. *Suppose that x is recurrent and $X = \overline{\mathcal{O}(x)}$ strictly contains a periodic orbit M which is the only minimal subshift contained in X. Then there is a sliding block code ϕ with the following properties: $\phi(X)$ has alphabet $\{0,1\}$, $\phi(X)$ strictly contains the unique minimal subshift $\{0^\infty\}$, and $\phi(w) = 0^i$ implies that $w \in W(M)$.*

PROOF. Choose such X and M, and choose any k greater than the period p of M. Define ϕ as follows: for every i, $(\phi(x))(i) = 0$ if $x(i) \dots x(i+k-1) \in W_k(M)$, and 1 otherwise. Trivially $\phi(X)$ has alphabet $\{0,1\}$. If $\phi(w) = 0^i$, then w has period p (since all words in $W_k(M)$ have period p) and begins with a k-letter word in $W(M)$, and is therefore itself in $W(M)$. Since $X \supsetneq M$, $\phi(X)$ contains points other than 0^∞. Finally, if $\phi(X)$ contained a minimal subshift not equal to $\{0^\infty\}$, then it would be disjoint from $\{0^\infty\}$, and so its preimage would contain a minimal subshift of X other than M, a contradiction. □

3.1.3. *x is recurrent, $x \in \{0,1\}^{\mathbb{Z}}$, $M = \{0^\infty\}$ is the only minimal subsystem of X.*

In this case, x must contain infinitely many 1s (by recurrence) and must contain 0^n as a subword for every n (since $0^\infty \in \overline{\mathcal{O}(x)} = X$). We will need the following slightly stronger fact.

LEMMA 3.5. *For x satisfying the conditions of this section, and for all n, 0^n1 and 10^n are subwords of x.*

PROOF. Choose any n. We know already that 0^n is a subword of x. If neither 0^n1 nor 10^n were subwords of x, then every occurrence of 0^n in x would force 0s on both sides, implying $x = 0^\infty$, a contradiction. Therefore, either 0^n1 or 10^n is a subword of x; assume without loss of generality that it is the former. Then by recurrence, 0^n1 appears twice as a subword of x, implying that x contains a subword of the form 0^n1w0^n1. Remove the terminal 1, and consider the rightmost 1 in the remaining word; it must be followed by 0^n, and so x also (in addition to 0^n1) contains 10^n as a subword. Since n was arbitrary, this completes the proof. □

By Lemma 3.5, for every n there exists a one-sided sequence y_n beginning with 1 so that 0^ny_n appears in x. By compactness, there exists a limit point y of the y_n (which begins with 1), and then since X is closed, $0^\infty y \in X$. Similarly, there exists a one-sided sequence z ending with 1 so that $z0^\infty \in X$. We first treat the case whether either y or z is not unique.

THEOREM 3.6. *For x satisfying the conditions of this section, if there exist either $y \neq y' \in \{0,1\}^{\mathbb{N}}$ beginning with 1 for which $0^\infty y, 0^\infty y' \in X$ or $z \neq z' \in \{0,1\}^{-\mathbb{N}}$ ending with 1 for which $z0^\infty, z'0^\infty \in X$, then* $\liminf(c_n(X) - 2n) > -\infty$.

PROOF. We prove only the statement for y, y', as the corresponding proof for z, z' is trivially similar. Assume that such y, y' exist. Since $y \neq y'$, there exists k so that $y(k) \neq y'(k)$.

For any $n \geq k$, define the n-letter words $u_i = 0^i y(1) \dots y(n-i)$ and $v_i = 0^i y'(1) \dots y'(n-i)$ for $0 \leq i \leq n-k$. First, note that u_i and v_i both begin with $0^i 1$ for every i, and since $0^i 1$ is never a prefix of $0^j 1$ for $i \neq j$, the sets $\{u_i, v_i\}$ and $\{u_j, v_j\}$ are disjoint whenever $i \neq j$. Finally, for every i, $u_i(i+k) = y(k) \neq y'(k) = v_i(i+k)$, so $u_i \neq v_i$. This yields $2n - 2k + 2$ words in $W_n(X)$ for $n \geq k$, or $c_n(X) - 2n \geq -2k + 2$, which completes the proof. □

We from now on assume that y and z are unique sequences beginning with 1 and ending with 1 respectively which satisfy $0^\infty y, z0^\infty \in X$.

THEOREM 3.7. *For x satisfying the conditions of the section, if either y or z contains only finitely many 1s, then* $\liminf(c_n(X) - 2n) > -\infty$.

PROOF. We again treat only the y case, as the proof for the z case is similar. Suppose that y contains only finitely many 1s. Then, we can write $y = w0^\infty$ for some w beginning and ending with 1, and $0^\infty y = 0^\infty w 0^\infty \in X$. By recurrence, x contains a subword v which contains more than $|w|$ 1s.

Again, by recurrence x contains v infinitely many times. Also, x contains the subword 0^n for all n, which never contains v. Therefore, for every n, there exists a word u of length n so that either vu or uv is a subword of x and contains v only once as a subword. We treat only the former case, as the latter is similar, and so suppose that $x(k) \dots x(k+n+|v|-1) = vu$.

For any $n \geq \max(|v|, |w|)$, consider the n-letter subwords of x given by $t_j = x(j) \dots x(j+n-1)$ for $k - n + |v| \leq j \leq k$. The rightmost occurrence of v within t_j begins at the $(k-j+1)$th letter of t_j, and so all t_j are distinct. This yields $n - |v| + 1$ words in $W_n(X)$ which each contain v. On the other hand, we can define $u_i = 0^i w 0^{n-|w|-i}$ for $0 \leq i \leq n - |w|$, each of which is contained in $0^\infty w 0^\infty \in X$. Each u_i contains w exactly once, beginning at the $(i+1)$th letter, and so all are distinct. In addition, each u_i contains at most $|w|$ 1s, and so none contains v, meaning no t_j and u_i can be equal.

Therefore, for $n \geq \max(|v|, |w|)$, $c_n(X) \geq (n - |v| + 1) + (n - |w| + 1) = 2n - |v| - |w| + 2$, completing the proof. □

THEOREM 3.8. *For x satisfying the conditions of the section, if the lengths of runs of 0s in y or z are bounded, then* $\liminf(c_n(X) - 2n) > -\infty$.

PROOF. As usual, we treat only the y case since the z case is similar. Suppose that there exists k so that 0^k is not a subword of y. Then, for any $n > k$, consider the n-letter words $u_i = 0^i y(1) \dots y(n-i)$, $k \leq i < n$, and $v_j = z(-j) \dots z(-1) 0^{n-j}$, $0 < j \leq n-k$. Each u_i begins with $0^i 1$, and $0^i 1$ is never a prefix of $0^{i'} 1$ for $i \neq i'$, so all u_i are distinct; a similar argument shows that all v_j are distinct. In addition, all v_j end with 0^k, and all u_i either have final k letters containing $y(1) = 1$ or end with a k-letter subword of y, and in either case do not end with 0^k. Therefore, no u_i and v_j can be equal, and so $c_n(X) \geq 2n - k$ for $n > k$, completing the proof. □

We finally arrive at the only case in which $\liminf(c_n(x)-2n)$ may be $-\infty$: y and z contain infinitely many 1s and arbitrarily long runs of 0s. In this case, we instead prove the weaker bound from the conclusion of Theorem 1.2, and interestingly only require the stated hypotheses on y.

THEOREM 3.9. *For x satisfying the conditions of the section, if y contains infinitely many 1s and contains 0^n as a subword for all n, then* $\limsup(c_n(x) - 1.5n) = \infty$.

PROOF. For every k, choose $m \geq 2k$ so that $y(1)\dots y(m)$ ends with 1 and contains exactly $2k$ 1s; note that $y(1)\dots y(m)$ does not contain 0^{m-2k+1} as a subword. Then, choose ℓ so that $y(1)\dots y(\ell)0^{m-2k+1}$ is a prefix of y and contains 0^{m-2k+1} only at the end, i.e. $y(\ell+1)\dots y(\ell+m-2k+1)$ is the first occurrence of 0^{m-2k+1} in y. Clearly, $\ell \geq m$ since $y(1)\dots y(m)$ ends with 1 and did not contain 0^{m-2k+1}. Also, by definition of ℓ, $y(1)\dots y(\ell)0^{m-2k} = y(1)\dots y(\ell+m-2k)$ does not contain 0^{m-2k+1}.

Now, consider the $(\ell+m-2k)$-letter words defined by $u_i = 0^i y(1)\dots y(\ell+m-2k-i)$, $0 \leq i < \ell+m-2k$, and $v_j = z(-j)\dots z(-1)0^{\ell+m-2k-j}$, $0 \leq j < \ell$. Again, since each u_i begins with 0^i1, all u_i are distinct; similarly, all v_j are distinct. In addition, all v_j end with 0^{m-2k+1}, and all u_i either have final $m-2k+1$ letters containing $y(1)=1$ or end with a subword of $y(1)\dots y(\ell+m-2k)$, and in either case do not end with 0^{m-2k+1}. Therefore, no u_i and v_j can be equal, and so $c_{\ell+m-2k}(X) \geq (\ell+m-2k)+\ell = 2\ell+m-2k$.

Recall that $\ell \geq m$; therefore, $2\ell+m-2k \geq 1.5\ell+1.5m-2k = 1.5(\ell+m-2k)+k$. In other words, for $n = \ell+m-2k$, $c_n(X) \geq 1.5n+k$. Since k was arbitrary, $c_n(x) - 1.5n$ is unbounded from above, completing the proof. □

We are now prepared to combine the results from the previous subsections to prove Theorem 1.2.

PROOF OF THEOREM 1.2. We assume that X is not minimal and that it is not the case that x is eventually periodic in both directions. By Lemma 3.1, if x is non-recurrent, then $\liminf(c_n(X)-2n) > -\infty$, implying that $\limsup(c_n(X)-1.5n) = \infty$. By Lemmas 3.2 and 3.3, if X contains either two minimal subsystems or an infinite minimal subsystem, then $\liminf(c_n(X)-2n) > -\infty$, implying that $\limsup(c_n(X)-1.5n) = \infty$.

So, we can assume that x is recurrent and that X properly contains a unique minimal subsystem, which is finite. Take the sliding block code ϕ (with window size k) guaranteed by Lemma 3.4. If we define $y = \phi(x)$ and $Y = \phi(X)$, then $Y = \overline{\mathcal{O}(y)}$ has alphabet $\{0,1\}$, strictly contains the unique minimal subshift $\{0^\infty\}$, and (since ϕ has window size k) satisfies $c_n(X) \geq c_{n-k+1}(Y)$ for all n.

By Theorems 3.6, 3.7, 3.8, and 3.9, $\limsup(c_n(Y)-1.5n) = \infty$, and since $c_n(X) \geq c_{n-k+1}(Y)$ for all n, it must be the case that $\limsup(c_n(X)-1.5n) = \infty$, completing the proof. □

3.2. Proof of Theorem 1.3. Fix any nondecreasing unbounded $g : \mathbb{N} \to \mathbb{R}$. Clearly there exist N and a nondecreasing unbounded $f : \mathbb{N} \to \mathbb{N}$ so that $f(n)+1 \leq g(n)$ for all $n > N$.

We will construct a point $x \in \{0,1\}^{\mathbb{Z}}$ of the following form

$$x = 0^\infty .\ 1\ 0^{g_1}\ 1\ 0^{g_2}\ 1\ 0^{g_3}\ 1\ 0^{g_4}\ 1\dots$$

where all $g_i \geq 1$. We will refer to these numbers $\{g_i\}$ as the *gaps* (between 1s in x). Next we describe how the gaps are defined.

We will construct an increasing sequence of natural numbers $n_0 < n_2 < n_3 < \cdots$, and for every i define

$$g_i = n_k \text{ if } i \text{ is the product of } 2^k \text{ and an odd natural number where } k \geq 0.$$

As such, x will have the form

$$x = 0^\infty.\ 1\ 0^{n_0}\ 1\ 0^{n_1}\ 1\ 0^{n_0}\ 1\ 0^{n_2}\ 1\ 0^{n_0}\ 1\ 0^{n_1}\ 1\ 0^{n_0}\ 1\ 0^{n_3}\ldots$$

Our goal is to show that the natural numbers $n_0 < n_1 < n_2 < \cdots$ may be chosen so that $c_n(x) < 1.5n + 1 + f(n)$ for all $n \in \mathbb{N}$; since $1.5n + 1 + f(n) \leq 1.5n + g(n)$ for $n > N$, we will then be done.

We will establish this by consideration of right-special words occurring in x of various lengths. First note that for any $j \geq 1$, 0^j is a right-special word: both 0^{j+1} and $0^j 1$ appear in x.

Set $w(0) = 0^{n_0}10^{n_0}$, and for $k \geq 1$, let $w(k)$ be the unique word in x of the form $w = 0^{n_k}1u10^{n_k}$ where u has no occurrence of 0^{n_k}. The uniqueness of $w(k)$ can be seen from noting that a gap of n_k or longer must correspond to the ith gap in x where i is multiple of 2^k. By the construction of x, the sequence of gaps that occur between any two consecutive gaps of n_k or more are always the same and are equal to $g_1, g_2, \ldots, g_{2^k-1}$. Thus

$$w(k) = 0^{n_k}10^{g_1}10^{g_2}1\cdots 10^{g_{2^k-1}}10^{n_k} = 0^{n_k}10^{n_0}10^{n_1}10^{n_0}10^{n_2}\cdots 0^{n_2}10^{n_0}10^{n_1}10^{n_0}10^{n_k}.$$

We make a series of claims about the words $w(k)$ for all $k \geq 0$.

Claim 1: Every $w(k)$ is right-special. Because there are gaps of exactly n_k, $w(k)1$ occurs in x, and because there are gaps larger than n_k, $w(k)0$ occurs in x.

Claim 2: Neither $0w(k)$ nor $1w(k)$ are right-special. Given two consecutive multiples of 2^k, one is the product of an odd natural number and 2^k and the other is a multiple of 2^{k+1}. Therefore, given two consecutive gaps of n_k or longer, one is exactly n_k and the other is strictly more than n_k. Therefore, neither $0w(k)0$ nor $1w(k)1$ occur in x, but both $0w(k)1$ and $1w(k)0$ occur in x.

Claim 3: Any right-special word w is a suffix of $w(k)$ for some $k > 0$. Clearly, $w = 0^n$ is a suffix of $w(k)$ for k large enough that $n_k > n$. Now assume w is a right-special word of the form $u10^n$ for some word u and some $n \geq 0$. Then $w1$ occurs in x, meaning that $n = n_k$ for some $k \geq 0$. Therefore, $w = u10^{n_k}$. If $|w| \leq |w(k)|$ then w is a suffix of $w(k)$ by the uniqueness of $w(k)$. Now assume $|w| > |w(k)|$. Then again by the uniqueness of $w(k)$, $w = vw(k)$ for some word v with $|v| > 1$. But since any suffix of a right-special word is right-special, this implies that either $0w(k)$ or $1w(k)$ is right-special, contradicting Claim 2.

Next we give a recursive formula for $|w(k)|$. Between any two consecutive multiples of 2^k, for $1 \leq j \leq k$, there are 2^{j-1} odd multiples of 2^{k-j}. Therefore, in $w(k)$ we have two runs of n_k 0s, 2^k 1s, and 2^{j-1} gaps of n_{k-j} for $0 < j \leq k$. This gives

$$|w(k)| = 2^k + 2n_k + \sum_{j=1}^{k} 2^{j-1}n_{k-j} = 2^k + 2n_k + \sum_{j=0}^{k-1} 2^{k-j-1}n_j.$$

In order to analyze $c_n(x)$, we consider the number of right-special words of length n. For all $n \geq 1$, we have 0^n and for any $n \in (n_k, |w(k)|]$, we have a suffix of $w(k)$ that contains at least one 1. In what follows, we will always recursively choose the sequence $\{n_k\}$ so that $n_k > |w(k-1)|$, implying that the intervals $(n_k, |w(k)|]$ are pairwise disjoint. Therefore, for some values of n we will have exactly one right-special word (0^n), and for n which are in $(n_k, |w(k)|]$ for some k, we have exactly two right-special words (0^n and the suffix of $w(k)$ of length n).

Let $R = \mathbb{N} \cap \bigcup_{k\geq 0}(n_k, |w(k)|]$. For $n \in R$ there are two right-special words of length n, and for $n \notin R$ there is just one right-special word of length n. This gives us the recursion formula

$$c_{n+1}(X) = c_n(X) + 1 + |\{n\} \cap R| .$$

From this, and the fact that $c_1(X) = 2$ it follows that

$$c_n(X) = n + 1 + |\{1, 2, \ldots, n-1\} \cap R| \tag{1}$$

for all $n \geq 1$.

It remains to show that the sequence $n_0 < n_1 < n_2 < \cdots$ can be chosen so that $|\{1, 2, \ldots, n-1\} \cap R| < 0.5n + f(n)$ for all $n \in \mathbb{N}$. First, we choose $n_0 = 1$, meaning that $|w(0)| = 2n_0 + 1 = 3$. Then clearly $|\{1, 2, \ldots, n-1\} \cap R| \leq 2 < 0.5n + f(n)$ for $n \leq 3 = |w(0)|$.

3.2.1. *Choice of n_k, $k \geq 1$.* Suppose $n_0, n_1, \ldots, n_{k-1}$ have been chosen so that $|\{1, 2, \ldots, n-1\} \cap R| < 0.5n + f(n)$ for all $n \leq |w(k-1)|$. Choose n_k so that $f(n_k)$ is greater than

$$0.5|w(k)| - n_k + |R \cap \{1, \ldots, |w(k-1)|\}| = 2^{k-1} + \sum_{j=0}^{k-1} 2^{k-2-j} n_j + |R \cap \{1, \ldots, |w(k-1)|\}|. \tag{2}$$

For $n \in (|w(k-1)|, n_k)$, we have

$$\begin{aligned} |\{1, 2, \ldots, n-1\} \cap R| &= |\{1, 2, \ldots, |w(k-1)|\} \cap R| \\ &< 0.5(|w(k-1)| + 1) + f(|w(k-1)| + 1) \leq 0.5n + f(n). \end{aligned}$$

For $n \in [n_k, |w(k)|)$ we have

$$\begin{aligned} |\{1, 2, \ldots, n-1\} \cap R| &= n - n_k + |\{1, 2, \ldots, |w(k-1)|\} \cap R| \\ &= 0.5n + (0.5n - n_k + |\{1, 2, \ldots, |w(k-1)|\} \cap R|) \\ &< 0.5n + (0.5|w(k)| - n_k + |\{1, 2, \ldots, |w(k-1)|\} \cap R|) \\ &< 0.5n + f(n_k) \leq 0.5n + f(n). \end{aligned}$$

(The second-to-last inequality came from (2).) We've shown that $|\{1, 2, \ldots, n-1\} \cap R| < 0.5n + f(n)$ for all n, and so by (1), $c_n(X) < 1.5n + f(n) + 1$ for all n. Since $f(n) + 1 \leq g(n)$ for $n > N$, this means that $c_n(X) < 1.5n + g(n)$ for $n > N$, completing the proof.

3.3. Proofs of Theorems 1.4 and 1.5.

Proof of Theorem 1.4. We first note that by Lemmas 3.1, 3.2, and 3.3, $\liminf(c_n(X) - 2n) = -\infty$ implies that x is recurrent and that X properly contains a periodic orbit M, which is the unique minimal subshift contained in X. We will for now assume that $M = \{0^\infty\}$ and that X has alphabet $\{0, 1\}$, and will then extend to the general case by Lemma 3.4.

By Lemma 3.5, $0^n 1 \in W(X)$ for all n, and so 0^n is right-special for all n. Therefore, for any n where there is another right-special word in $W_n(X)$, $c_{n+1}(X) - c_n(X) \geq 2$.

For any N, choose n so that $c_n(X) - 2n \leq -N$, and define $S = \{j < n : 0^j$ is the only right-special word in $W_j(X)\}$. Then $c_n(X) = \sum_{j=0}^{n-1}(c_{j+1}(X) - c_j(X)) \geq |S| + 2(n - |S|)$, and so $|S| \geq N$. Define m to be the maximal element of S; then 0^m is the only right-special word in $W_m(X)$, $m \geq N$, and $c_m(X) = c_n(X) - \sum_{j=m}^{n-1}(c_{j+1}(X) - c_j(X)) \leq 2n - 2(n-m) = 2m$.

We claim that every word in $W_{3m}(X)$ contains 0^m, and so that every subword of length $3m$ of x contains 0^m. Suppose for a contradiction that this is false, i.e. that there is $y \in X$ where $y(1)\dots y(3m)$ does not contain 0^m. Since $c_m(X) \leq 2m$, there exist $1 \leq i < j \leq 2m$ so that $y(i)\dots y(i+m-1) = y(j)\dots y(j+m-1)$. Also, all m-letter words $y(k)\dots y(k+m-1)$ for $i \leq k \leq j$ are not 0^m and so, since $m \in S$, are not right-special, i.e. there is only one letter that can follow each of them in a point of X. This means that $y(i)y(i+1)\dots$ is in fact periodic with period $j-i$. Since $y(i)\dots y(j+m-1)$ does not contain 0^m (as a subword of $y(1)\dots y(3m)$), $y(i)y(i+1)\dots$ cannot contain 0^m, a contradiction to 0^∞ being the only minimal subsystem of X. This means that the original claim was true, completing the proof in the case $M = \{0^\infty\}$.

Now suppose that M is an arbitrary periodic orbit, and take the sliding block code ϕ (with window size k) guaranteed by Lemma 3.4. As before, define $y = \phi(x)$ and $Y = \phi(X)$; then Y has alphabet $\{0,1\}$, strictly contains the unique minimal subshift $\{0^\infty\}$, and satisfies $c_n(X) \geq c_{n-k+1}(Y)$ for all n, implying that $\liminf(c_n(Y) - 2n) = -\infty$.

From the above proof, for all N, there exists $m \geq N$ so that every $3m$-letter subword of y contains 0^m. Every $(3m+3k)$-letter subword of x has image under ϕ which is a $(3m+2k)$-letter subword of y, and therefore contains 0^m. By definition of ϕ, x contains an $(m+k)$-letter word at the corresponding location which is in $W(M)$; since $m + k \geq m \geq N$, the proof is complete. □

The proof of Theorem 1.5 uses a few basic notions from ergodic theory, which we briefly and informally summarize here. Firstly, a (shift-invariant Borel probability) measure μ is called **ergodic** if every measurable set A with $A = \sigma A$ has $\mu(A) \in \{0,1\}$. Ergodic measures are valuable because of the **pointwise ergodic theorem**, which says that μ-almost every point x in X is **generic for** μ, which means that for every $f \in C(X)$, $\frac{1}{n}\sum_{i=0}^{n-1} f(\sigma^i x) \to \int f \, d\mu$. For the purposes of the proof below, we need only a very simple application of the ergodic theorem: for any generic point for μ, the frequency of 0 symbols is equal to $\mu([0])$, where $[0]$ is the set of $z \in X$ containing a 0 at the origin. Finally, the **ergodic decomposition theorem** states that every (shift-invariant Borel probability) measure is a sort of generalized convex combination of ergodic measures. Again, we need only a very simple corollary: if a subshift has only one ergodic measure, then it has only one (shift-invariant Borel probability) measure. For a more detailed introduction to ergodic theory, see [**8**].

Proof of Theorem 1.5. Assume that x is not uniformly recurrent, it is not the case that x is periodic in both directions, and $\liminf(c_n(X) - 2n) = -\infty$. Then as above, X properly contains a periodic orbit M, which is the unique minimal subshift contained in X. We again first treat the case where $M = \{0^\infty\}$. Assume for a contradiction that X has a (shift-invariant Borel probability) measure μ not equal to δ_{0^∞}. Since δ_{0^∞} is obviously ergodic, by ergodic decomposition we may assume without loss of generality that μ is ergodic.

Since the set $\{0^\infty\}$ is invariant, $\mu(\{0^\infty\})$ is 0 or 1. Since $\mu \neq \delta_{0^\infty}$, $\mu(\{0^\infty\}) = 0$ and so there exists j so that $\mu([0^j]) < 1/6$. Since μ is ergodic, by the pointwise ergodic theorem there is a point z which is generic for μ. Note that the only periodic orbit in X is $\{0^\infty\}$, and so z cannot be eventually periodic in both directions, since then it would be generic for δ_{0^∞}. In addition, note that $\overline{\mathcal{O}(z)}$ must contain $M = \{0^\infty\}$ (since M is the unique minimal subshift contained in X), and so z is not uniformly recurrent.

Finally, since $\liminf(c_n(X) - 2n) = -\infty$, we know that $\liminf(c_n(z) - 2n) = -\infty$, and so z satisfies the hypotheses of Theorem 1.4. We apply that theorem with $N = 2j$ to find $m \geq 2j$ for which every $3m$-letter subword of z contains 0^m. Then, the frequency of occurrences of 0^j in z is at least $\frac{m-j}{3m}$, which is greater than or equal to $1/6$ since $m \geq 2j$. By genericity of z for μ, $\mu([0^j]) \geq 1/6$, contradicting the definition of j and so the existence of μ.

Now, suppose that M is an arbitrary periodic orbit, define the sliding block code ϕ (with window size k) guaranteed by Lemma 3.4, and again define $y = \phi(x)$ and $Y = \phi(X)$. As usual, Y has alphabet $\{0, 1\}$, strictly contains the unique minimal subshift $\{0^\infty\}$, and satisfies $c_n(X) \geq c_{n-k+1}(Y)$ for all n. We note that y cannot be eventually periodic in both directions; if it were, then it would have to begin and end with infinitely many 0s, which would imply that x was eventually periodic in both directions, a contradiction.

Finally, since $c_n(X) \geq c_{n-k+1}(Y)$ for all n, $\liminf(c_n(Y) - 2n) = -\infty$. So, by the proof above in the $M = \{0^\infty\}$ case, Y has unique invariant measure δ_{0^∞}. Any invariant measure ν in X then must have pushforward δ_{0^∞} under ϕ, and so must have $\nu(M) = 1$. It is easily checked that there is only one such ν, namely the measure equidistributed over the points of M. □

References

[1] Ali Aberkane, *Exemples de suites de complexité inférieure à 2n* (French, with French summary), Bull. Belg. Math. Soc. Simon Stevin **8** (2001), no. 2, 161–180. Journées Montoises d'Informatique Théorique (Marne-la-Vallée, 2000). MR1838940

[2] Michael Boshernitzan, *A unique ergodicity of minimal symbolic flows with linear block growth*, J. Analyse Math. **44** (1984/85), 77–96, DOI 10.1007/BF02790191. MR801288

[3] Michael Brin and Garrett Stuck, *Introduction to dynamical systems*, Cambridge University Press, Cambridge, 2002. MR1963683

[4] Van Cyr and Bryna Kra, *Counting generic measures for a subshift of linear growth*, J. Eur. Math. Soc. (JEMS) **21** (2019), no. 2, 355–380, DOI 10.4171/JEMS/838. MR3896204

[5] N. Pytheas Fogg, *Substitutions in dynamics, arithmetics and combinatorics*, Lecture Notes in Mathematics, vol. 1794, Springer-Verlag, Berlin, 2002. Edited by V. Berthé, S. Ferenczi, C. Mauduit and A. Siegel. MR1970385

[6] Alex Heinis, *The $P(n)/n$-function for bi-infinite words*, Theoret. Comput. Sci. **273** (2002), no. 1-2, 35–46, DOI 10.1016/S0304-3975(00)00432-1. WORDS (Rouen, 1999). MR1872441

[7] Michael E. Paul, *Minimal symbolic flows having minimal block growth*, Math. Systems Theory **8** (1974/75), no. 4, 309–315, DOI 10.1007/BF01780578. MR0380760

[8] Peter Walters, *An introduction to ergodic theory*, Graduate Texts in Mathematics, vol. 79, Springer-Verlag, New York-Berlin, 1982. MR648108

Nic Ormes, Department of Mathematics, University of Denver, 2390 S. York St., Denver, CO 80208
Email address: normes@du.edu
URL: www.math.du.edu/~normes/

Ronnie Pavlov, Department of Mathematics, University of Denver, 2390 S. York St., Denver, CO 80208
Email address: rpavlov@du.edu
URL: www.math.du.edu/~rpavlov/

Contemporary Mathematics
Volume **736**, 2019
https://doi.org/10.1090/conm/736/14845

Definitions and properties of entropy and distance for regular languages

Austin J. Parker, Kelly B. Yancey, and Matthew P. Yancey

Abstract. This paper addresses a need that has arisen for constructing a practical and intuitive distance function over regular languages. Some of the previously constructed distance functions are not well-defined; and we construct examples showing when they fail to meet intuitive expectations. We present thorough mathematical analysis and general reasoning for how and why our definitions fix these issues.

As regular languages are sets, most proposed distances between regular languages L_1 and L_2 are based on the "size" of $L_1 \triangle L_2$. As a related issue, this paper also addresses the need to construct a practical and intuitive function that describes the size of a regular language, which we name language entropy.

There exists a well-known map from the space of regular languages to sofic shifts, but theorems about sofic shifts have only been pulled back to the space of regular languages under strong assumptions. One of the contributions of this paper is a new map from regular languages to sofic shifts; in this new map the entropy of a regular language equals the topological entropy of the associated sofic shift.

1. Introduction

Deterministic finite automata are known to be similar to sofic shifts from symbolic dynamics [**BP97**]. In this paper we study the distance between regular languages using ideas from symbolic dynamics, as well as extend the standard Jaccard distance to infinite sets. Throughout this paper we use the word distance in a nontechnical capacity that refers to similarity. When each distance function is described we give a technical description of that function, i.e. metric, pseudo-metric, etc. There are many motivations for developing a good distance function to measure similarity between regular languages. Activities in bioinformatics, copy-detection [**CDFI13**], and network defense sometimes require large numbers of regular expressions be managed. Metrics aid in indexing and management of those regular expressions [**CGR03**]. Other applications include determining a regular language that best matches given data and an analogue of Cauchy's convergence test for a sequence of regular languages (see [**NS08**] and [**CMR06**], respectively, for probabilistic versions of these applications). Further, understanding the distance between regular languages requires an investigation into the structure of regular languages

2010 *Mathematics Subject Classification.* Primary 37B10, 68Q45; Secondary 37B40.

Key words and phrases. Deterministic Finite Automata, Sofic shift, regular language, entropy.

that we hope eliminates the need for similar theoretical investigations in the future. An extended abstract of this paper appeared in the proceedings of the Mathematical Foundations of Computer Science Conference in 2017 [**PYY17**]. The short abstract version of this paper introduces the main ideas and describes the results, however background, examples, and proofs of the results appear in this paper.

In symbolic dynamics we often deal with languages $\mathcal{L}$ which are composed of finite words over a finite alphabet $\mathcal{A}$. For example, a shift of finite type defines a language $\mathcal{L}$ where a finite list of blocks are forbidden to appear as subwords of words in $\mathcal{L}$. Similarly, regular languages define a special class of subsets of $\mathcal{A}^*$, where $\mathcal{A}^*$ refers to the set of all finite words over the alphabet $\mathcal{A}$. *Regular languages* are those that are accepted by a deterministic finite automaton and can be expressed by a regular expression [**HU79**].

We seek to develop and understand a notion of similarity between two regular languages over a common alphabet. A natural definition of distance between regular languages that you might think of initially is a naive extension of the standard Jaccard distance. For two finite sets A and B, the *Jaccard distance* between A and B is defined to be $\frac{|A \triangle B|}{|A \cup B|}$. Let $W_n(L)$ denote the set of words in a language L of length exactly n and define the n *Jaccard distance* between two languages L_1 and L_2 to be

$$J'_n(L_1, L_2) = \frac{|W_n(L_1 \triangle L_2)|}{|W_n(L_1 \cup L_2)|}$$

if $|W_n(L_1 \cup L_2)| > 0$ and $J'_n(L_1, L_2) = 0$ otherwise. The natural extension for this distance function to infinite languages would then be $\lim_{n\to\infty} J'_n(L_1, L_2)$. However, this definition has a fundamental flaw: the limit does not always exist. Let $\mathcal{A} = \{a\}$ and consider the distance between the language composed of all words over $\mathcal{A}$ (given by a^*) and the language composed of words of even length over $\mathcal{A}$ (given by $(aa)^*$). When n is even, $J'_n(L_1, L_2) = 0$, while when n is odd $J'_n(L_1, L_2) = 1$. Thus, the limit given above is not well defined for those two languages. In Theorem 5.7 we give conditions on when this limit exists.

Another extension of Jaccard distance is to replace W_n with $W_{\leq n}$, where $W_{\leq n}$ denotes the set of words in L of length at most n. That is, consider strings up-to a given length, instead of strings of a given length exactly. To that end, define the $\leq n$ *Jaccard distance* between two regular languages, L_1 and L_2, by

$$J_n(L_1, L_2) = \frac{|W_{\leq n}(L_1 \triangle L_2)|}{|W_{\leq n}(L_1 \cup L_2)|}$$

if $|W_{\leq n}(L_1 \cup L_2)| > 0$ and $J_n(L_1, L_2) = 0$ otherwise. In 2009 Dassow, Reyes, and Vico [**DMV09**] studied $\lim_{n\to\infty} J_n(L_1, L_2)$ for cyclic unary regular languages (a special class of languages with one element alphabets); they gave a closed form for the distance.

One of the topics that we address in this paper is the fact that the limit based Jaccard distances discussed thus far do not always converge. A fundamental contribution of our work is a limit-based distance related to the above that (1) exists, (2) can be computed from the deterministic finite automaton for the associated regular languages, and (3) does not invalidate expectations about the distance between languages.

The core idea is two-fold: (1) to rely on the number of strings *up-to* a given length rather than strings *of* a given length, and (2) to use Cesáro averages to smooth out the behavior of the limit. To do this we define the *Cesáro Jaccard distance* between languages L_1 and L_2 by

$$J_C(L_1, L_2) = \lim_{n\to\infty} \frac{1}{n} \sum_{i=1}^{n} J_i(L_1, L_2).$$

We prove the following theorem:

THEOREM 1.1. *Let L_1 and L_2 be two regular languages. Then, $J_C(L_1, L_2)$ is well-defined. That is, $\lim_{n\to\infty} \frac{1}{n}\sum_{i=1}^{n} J_i(L_1, L_2)$ exists.*

Tied up in this discussion is the *entropy* of a regular language, which is again a concept whose common definition needs tweaking due to limit-related considerations. Shannon's *channel capacity* (equation 7 from [**CM58**]) is commonly called the entropy of a regular language and is defined by

$$\lim_{n\to\infty} \frac{\log |W_n(L)|}{n}.$$

The problem here is that this limit does not always exist; consider words of even length over a finite alphabet. In Chomsky and Miller's seminal paper on regular languages [**CM58**] they use this definition of entropy. Several works since then [**CSMS03, CDFI13, Kui70**] define entropy as Chomsky and Miller, but add the caveat that they use the upper limit when the limit does not exist.

Chomsky and Miller's technique was to develop a recursive formula for the number of words accepted by a regular language. That recursive formula comes from the characteristic polynomial of the adjacency matrix for an associated automaton. The eigenvalues of the adjacency matrix describe the growth of the language (we use the same technique, but apply stronger theorems from linear algebra that were discovered several decades after Chomsky and Miller's work). The recursive formula can also be used to develop a generating function to describe the growth of the language (see [**SS78**]). Bodirsky, Gärtner, Oertzen, and Schwinghammer [**BGvOS04**] used the generating functions to determine the growth of a regular language over alphabet $\mathcal{A}$ relative to $|\mathcal{A}|^n$, and Kozik [**Koz05**] used them to determine the growth of a regular language relative to a second regular language. Our approaches share significant details: they relate the growth of a regular language to the poles of its generating function—which are the zeroes of the corresponding recurrence relation—which are the eigenvalues of the associated adjacency matrix. Our technique establishes the "size" of a regular language independent of a reference alphabet or language.

We define the *language entropy* of a regular language L to be

$$h(L) = \lim_{n\to\infty} \frac{\log |W_{\le n}(L)|}{n}.$$

Here we provide foundation for those works mentioned in the previous paragraph by showing the upper limit (used in channel capacity) to be correct:

THEOREM 1.2. *Let L be a non-empty regular language over the alphabet $\mathcal{A}$. Then,*

$$h(L) = \lim_{n\to\infty} \frac{\log |W_{\le n}(L)|}{n} = \limsup_{n\to\infty} \frac{\log |W_n(L)|}{n}.$$

We also explore the relationship with topological entropy. In Section 3 we describe how to obtain the associated sofic shift from a regular language. Let h_t

denote the topological entropy of a dynamical system. We show that all forms of entropy are compatible.

THEOREM 1.3. *Let L be a non-empty regular language over the alphabet $\mathcal{A}$, and let X_G be the associated sofic shift (where the graph G is a presentation of the sofic shift). Then,*

$$h(L) = \lim_{n\to\infty} \frac{\log |W_{\leq n}(L)|}{n} = h_t(X_G).$$

Entropy can be used to develop a distance between regular languages that measures something different than the Jaccard distances previously mentioned. In Section 4 we develop and explore several distance functions based on entropy. Of particular note is the *entropy sum distance* defined as

$$H_S(L_1, L_2) = h(L_1 \cap \overline{L_2}) + h(\overline{L_1} \cap L_2)$$

where L_1 and L_2 are regular languages. We prove that this function is a pseudometric (Theorem 4.6) and is also granular (Theorems 4.7 and 4.8).

Language entropy and Cesáro Jaccard are mostly disjoint in what they measure. Specifically we prove:

THEOREM 1.4. *Let L_1, L_2 be two regular languages.*

(1) *If $h(L_1 \triangle L_2) \neq h(L_1 \cup L_2)$, then $J_C(L_1, L_2) = 0$.*
(2) *If $h(L_1 \cap L_2) \neq h(L_1 \cup L_2)$, then $J_C(L_1, L_2) = 1$.*
(3) *If $0 < J_C(L_1, L_2) < 1$, then the following equal each other: $h(L_1)$, $h(L_2)$, $h(L_1 \cap L_2)$, $h(L_1 \triangle L_2)$, $h(L_1 \cup L_2)$.*

Table 1 provides an overview of the distance functions that are discussed throughout this paper.

Notation	Name	Formula	Appearance
$J'_n(L_1, L_2)$	n Jaccard Distance	$\frac{\lvert W_n(L_1 \triangle L_2)\rvert}{\lvert W_n(L_1 \cup L_2)\rvert}$	
$J_n(L_1, L_2)$	$n_{\leq}$ Jaccard Distance	$\frac{\lvert W_{\leq n}(L_1 \triangle L_2)\rvert}{\lvert W_{\leq n}(L_1 \cup L_2)\rvert}$	**[CGR03, DMV09]**
$J_C(L_1, L_2)$	Cesàro Jaccard Distance	$\lim_{n\to\infty} \frac{1}{n} \sum_{i=1}^{n} J_i(L_1, L_2)$	New
$H(L_1, L_2)$	Entropy Distance	$\frac{h(L_1 \triangle L_2)}{h(L_1 \cup L_2)}$	**[CDFI13, Koz05]**
$H_S(L_1, L_2)$	Entropy Sum Distance	$h(L_1 \cap \overline{L_2}) + h(\overline{L_1} \cap L_2)$	New

TABLE 1. The distance functions considered in this paper are listed in this table.

This paper is structured as follows. Section 2 provides the necessary background on deterministic finite automata and symbolic dynamics, including discussion on operations on DFA and also topological entropy. In Section 3 we develop the theory of language entropy and prove Theorems 1.2 and 1.3. We discuss the entropy related distance functions in Section 4. In Section 5 we dive into various limit based Jaccard distances and prove Theorem 1.1. In that section we also discuss the relationship between language entropy and Cesáro Jaccard, proving Theorem 1.4. Finally, Section 6 provides a conclusion and details some potential future work.

2. Background

In this section we provide the necessary background on deterministic finite automata, shift spaces, and topological entropy, as well as describe the construction of a sofic shift from a DFA.

2.1. Regular Languages and Deterministic Finite Automata.

Let $\mathcal{A}$ denote a finite alphabet. A *word* is a finite sequence of letters from $\mathcal{A}$, also referred to as a *block* or *string*. The set $\mathcal{A}^*$ refers to the set of all finite words over $\mathcal{A}$ and includes the empty string ϵ. For more information on concepts from this subsection, see [**HU79**].

To discuss regular languages, we must first set some notation that is common in regular expressions. Let $w \in \mathcal{A}^*$ be a word and $k \in \mathbb{N}$. Then, w^k represents the word w concatenated with itself k-times; similar notation is used for sets. The Kleene star, $*$, when applied to a word will represent the set containing words resulting from any number of concatenations of that word, including the empty concatenation. We use $|$ to represent the union operation. A *regular expression* is a finite formula comprised of words in $\mathcal{A}^*$ that are combined using concatenation, the Kleene star, and/or unioned. A regular expression is a formula for the words that appear in a regular language.

EXAMPLE 2.1. Let $\mathcal{A} = \{a, b\}$. The regular expression $(a|b)^*b$ represents the language that contains all words that end with b (a word in the language can contain a or b any number of times followed by the letter b).

Regular languages are intimately tied to deterministic finite automata (DFA). To that end, we begin by defining DFA and then provide an example.

DEFINITION 2.2. A *deterministic finite automaton (DFA)* is a 5-tuple $(Q, \mathcal{A}, \delta, q_0, F)$, where Q is the set of states, $\mathcal{A}$ is the alphabet, δ is a partial function from $Q \times \mathcal{A}$ to Q, $q_0 \in Q$ is the *initial state* and $F \subset Q$ is the set of *final states.*

A DFA is essentially a labeled directed graph that includes initial/final state information and is also *deterministic*, i.e. there is at most one outgoing edge from each vertex in the graph with a given label. An *accepting path* in a DFA is a sequence of states $u_0, u_1, \dots u_\ell$ such that $u_0 = q_0$, $u_\ell \in F$, and for each $1 \leq i < \ell$, there exists an $a_i \in \mathcal{A}$ such that $\delta(u_i, a_i) = u_{i+1}$, i.e. a path in the graph that begins at the initial state and terminates at a final state. The language represented by a given DFA is the family of words in $\mathcal{A}^*$ that correspond to accepting paths in that DFA. Note that regular languages can be infinite in size.

DFA are equivalent to regular languages. That is, every regular language can be represented by a DFA and every DFA corresponds to a regular language. Let us now consider an example.

EXAMPLE 2.3. Let $\mathcal{A} = \{a, b\}$ and let L be the regular language where each word in L contains an even number of a's and an even number of b's. We will represent L by a DFA. Consider the DFA in Figure 1. The initial state is denoted by an incoming bold arrow and the final state is denoted by a double circle. In this example, $Q = \{1, 2, 3, 4\}$, $q_0 = 1$, and $F = \{1\}$. The function δ is the transition function that describes the labeled edges in the graph. For example, $\delta(2, b) = 3$ since state 2 transitions to state 3 with the letter b. This DFA "accepts" the word *abab* because you can follow that path in the graph, starting at the initial state and

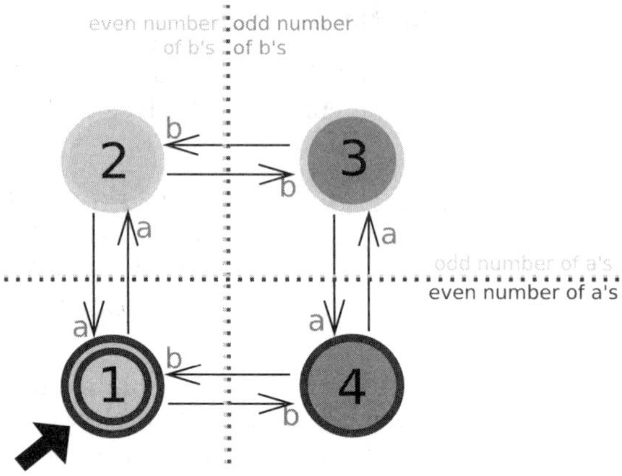

FIGURE 1. Parity automaton for Example 2.3.

ending at a final state. Thus, $abab \in L$. The word abb is not accepted because the path starting at the initial state that follows that word ends at state 2, which is not in F. Thus, $abb \notin L$. Finally, note that $abc \notin L$ because no path exists that follows abc.

DEFINITION 2.4. A *regular language* is a set $L \subset \mathcal{A}^*$ which can be represented by a DFA.

In this paper, our goal is to develop a useful distance function between regular languages. For example, in the introduction we defined the entropy sum distance which computes the entropy of $L_1 \cap \overline{L_2}$ where L_1 and L_2 are two regular languages. Recall that a regular language $\mathcal{L}$ is a subset of $\mathcal{A}^*$ which can be expressed by a regular expression or represented as a DFA. Since regular languages are sets, all set operations are well defined (union, intersection, symmetric difference, and complement). Combining two regular languages using a Boolean operation results in another regular language (see [**HU79**]).

When two regular languages are combined using a Boolean operation, we can view this operation in two ways: (1) two sets are being combined or (2) two DFA are being combined. The first option is natural and straightforward. However, computationally combining two DFA is simpler and more efficient. In what remains of this subsection we describe the construction to combine DFA (see [**HU79**]). With that in mind, we need to review some basic terminology first.

The operation of *trimming* a DFA is to remove any state that is not in some accepting path for that DFA. Observe that trimming a DFA does not change the language represented by the DFA. A state q is removed during trimming if and only if q satisfies at least one of two properties: (1) there is no path from q_0 to q or (2) there is no path from q to any vertex in F. These conditions can be tested using Djikstra's algorithm.

A DFA is *uniform* if δ is a function; in other words, δ is defined over the whole domain of $Q \times \mathcal{A}$. A DFA can be turned into a uniform DFA by adding a state T so that the total states are now $Q' = Q \cup \{T\}$, $\delta(T, a) = T$ for all $a \in \mathcal{A}$ and $\delta(q, a) = T$ whenever $\delta(q, a)$ was not previously defined. This new state T is known as a *trash state*. The existence of a trash state and the DFA being trimmed are

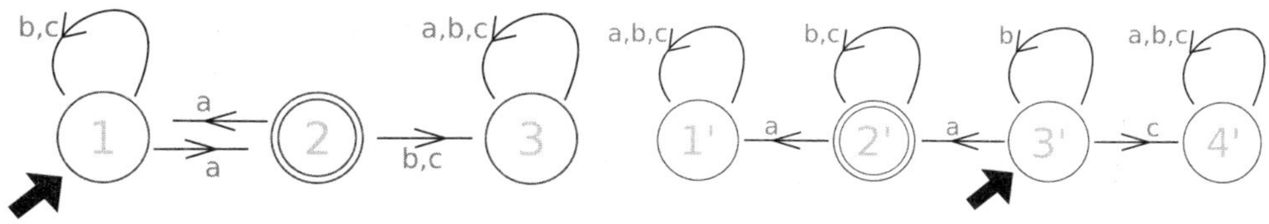

FIGURE 2. Let M_1 be the DFA on the left and M_2 the DFA on the right.

incompatible. Observe that creating a trash state does not change the language represented by the DFA.

To get a feel for how to combine DFA, we begin with a description of a method to compute the union of two DFA, $M_1 = (Q, \mathcal{A}, \delta, q_0, F)$ and $M_2 = (Q', \mathcal{A}, \delta', q_0', F')$. For simplicity, we show the steps performed with an example: let M_1 and M_2 be the two DFA in Figure 2. In this example $\mathcal{A} = \{a, b, c\}$. The states in M_1 are $\{1, 2, 3\}$ where 1 is the initial state and $\{2\}$ is the set of final states. The states in M_2 are $\{1', 2', 3', 4'\}$ where $3'$ is the initial state and $\{2'\}$ is the set of final states. For example, the word *aaba* appears in the language represented by M_1 and the word *bab* appears in the language represented by M_2. Thus, both of these words should be accepted by the DFA $M_1 \cup M_2$. To compute $M_1 \cup M_2$ do the following:

(1) Turn the DFA representing M_1 and M_2 into uniform DFA. (Observe that M_1 and M_2 are already uniform; states 3 and $4'$ are trash states.)

(2) Create a graph with states labeled by ordered pairs (u, v) where $u \in Q = \{1, 2, 3\}$ and $v \in Q' = \{1', 2', 3', 4'\}$.

(3) For every letter $\ell \in \mathcal{A}$ and for every pair of states (u_1, v_1) and (u_2, v_2) place a directed edge from state (u_1, v_1) to state (u_2, v_2) labeled ℓ if there is an edge in M_1 from u_1 to u_2 labeled ℓ and there is an edge in M_2 from v_1 to v_2 labeled ℓ.

(4) The new initial state is $(q_0, q_0') = (1, 3')$ and the new set of final states is given by the formula
$$\{(u, v) : u \in F\} \cup \{(u, v) : v \in F'\}.$$
In this example, the set of final states is
$$\{(2, 1'), (2, 2'), (2, 3'), (2, 4'), (1, 2'), (3, 2')\}.$$

Figure 3 shows the above computations and Figure 4 displays the final DFA for $M_1 \cup M_2$ after it has been trimmed. Note that in Figure 4 both of the words *aaba* and *bab* appear in the DFA representing the union.

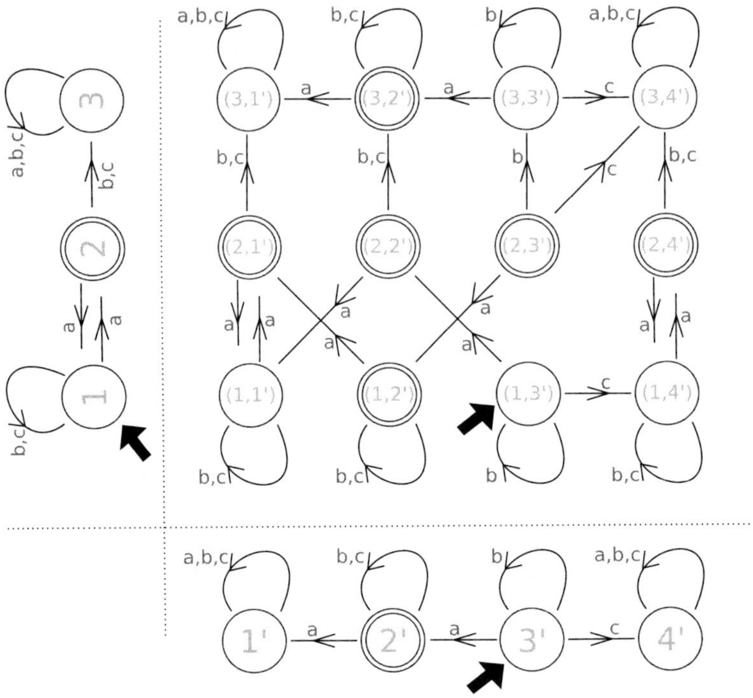

FIGURE 3. Computing $M_1 \cup M_2$.

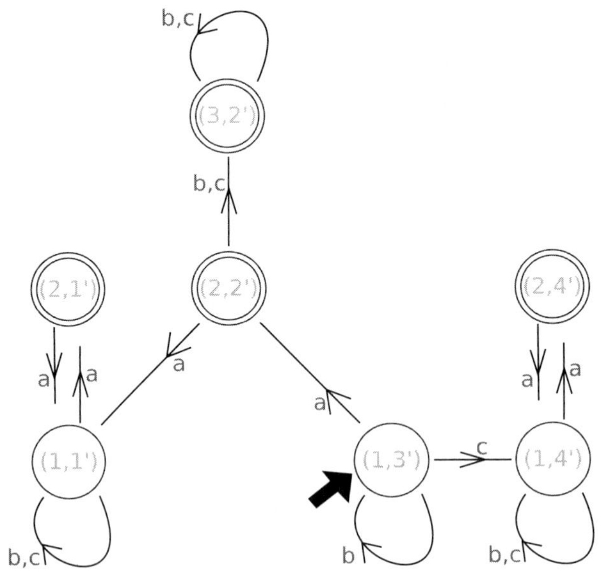

FIGURE 4. The trimmed DFA for $M_1 \cup M_2$.

The procedure for other Boolean operations is similar. Let $\circ$ denote a Boolean operation. Given two DFA, M_1 and M_2, complete steps (1), (2), and (3) above. That is, (1) compute the uniform DFA for M_1 and M_2, (2) create a graph with states labeled by ordered pairs (u, v) where $u \in Q$ and $v \in Q'$, and (3) for every letter $\ell \in \mathcal{A}$ and for every pair of states (u_1, v_1) and (u_2, v_2) place a directed edge from state (u_1, v_1) to state (u_2, v_2) labeled ℓ if there is an edge in M_1 from u_1 to u_2 labeled ℓ and there is an edge in M_2 from v_1 to v_2 labeled ℓ. Now we have a graph representing a general construction and potentially all possible combinations of these regular languages. We are left we identifying the initial and final states; this information will define our new regular language. The initial state for $M_1 \circ M_2$ is the state labeled by the ordered pair composed of original initial states. That is, the initial state is (q_0, q_0'). The set of final states for $M_1 \circ M_2$ is the only part of this construction that depends on the Boolean operation $\circ$. Since the cross product graph constructed thus far with initial state specified potentially represents all Boolean combinations of M_1 and M_2, we need only to narrow down the options by defining which words are accepted by $M_1 \circ M_2$ (e.g. in the regular language represented by $M_1 \circ M_2$). The set of final states is thus defined to be

$$\{(u, v) : u \in F\} \circ \{(u, v) : v \in F'\}.$$

For example, if the operation $\circ$ was symmetric difference and we apply this to the previous example, then the formula for the set of final states is

$$\{(u, v) : u \in F\} \triangle \{(u, v) : v \in F'\}$$

which is equal to

$$\{(2, 1'), (2, 3'), (2, 4'), (1, 2'), (3, 2')\}$$

in this case.

2.2. Sofic Shifts. Let $\mathcal{A}^{\mathbb{Z}}$ denote the *full shift* which is the set of all biinfinite sequences over $\mathcal{A}$, i.e. $\mathcal{A}^{\mathbb{Z}} = \{(x_i)_{i \in \mathbb{Z}} : x_i \in \mathcal{A}\}$. Let $\sigma : \mathcal{A}^{\mathbb{Z}} \to \mathcal{A}^{\mathbb{Z}}$ denote the *shift map* where if $\sigma(x) = y$ then $y_i = x_{i+1}$. A *shift space* X is a subset of the full shift that is closed and shift invariant. This is equivalent to X being such that $X = X_F$ where X_F is a subset of the full shift that does not contain blocks in some set F. Note that if F is finite, then we call this shift space a *shift of finite type*.

Sofic shifts are factors of shifts of finite type and are related to DFA from automata theory. In fact, if the set F above is a regular language, then X_F is a sofic shift [**Wei73**]. We will review the definition of a sofic shift and a basic example.

DEFINITION 2.5. Given a directed labeled graph G, let X_G be the set of labels of biinfinite walks on G. The shift space X is *sofic* if $X = X_G$ for some graph G.

The graph G in the definition of sofic shift is called a *presentation*. Every sofic shift has a presentation that is *right-resolving*, that is, in the graph there is a unique label for each edge leaving a given vertex. Note that this construction is similar to DFA, however there are no initial and final states.

EXAMPLE 2.6 (Golden Mean Shift). Let $\mathcal{A} = \{a, b\}$ and let X be the shift space that is composed of sequences where no two b's are next to each other. This is a sofic shift and a presentation is shown in Figure 5.

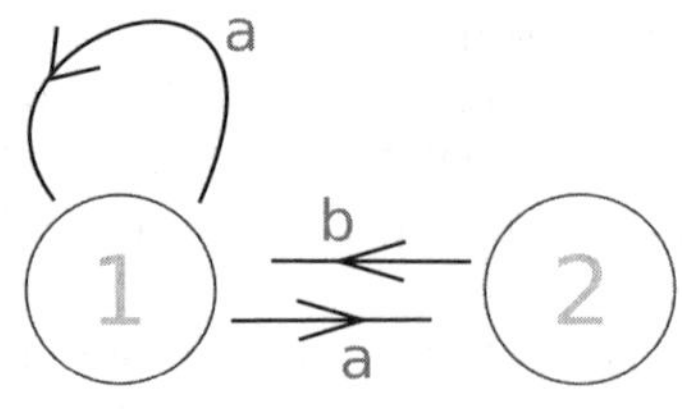

FIGURE 5. The Golden Mean Shift.

2.3. Topological Entropy. Topological entropy is a concept meant to determine the exponential growth of distinguishable orbits of the dynamical system up to arbitrary scale. A positive quantity for topological entropy reflects chaos in the system [**BGKM02**]. This concept was motivated by Kolmogorov and Sinai's theory of measure-theoretic entropy in ergodic theory [**Kol59, Sin59**], which in turn is related to Shannon entropy [**Sha48**].

Let X be a shift space. We call a block *admissible* if there is a point in X where the block appears as a subword. Denote the set of admissible blocks of length n in X by $B_n(X)$.

DEFINITION 2.7. The *topological entropy* of a shift space X is given by

$$h_t(X) = \lim_{n\to\infty} \frac{\log |B_n(X)|}{n}.$$

The topological entropy of sofic shifts is easily computable. Given a directed graph G on m vertices, the corresponding *adjacency matrix* A is an $m \times m$ matrix where the ij-th entry $A_{(ij)}$ is the number of edges from vertex i to vertex j. Using Perron-Frobenius theory it has been proven that the topological entropy of a sofic shift represented by a right-resolving labeled graph G is equal to the log of the spectral radius of the adjacency matrix of G [**LM95**]. That is, the topological entropy is given by the log of the adjacency matrix's largest modulus eigenvalue. Algorithms for computing eigenvalues are well known and run in polynomial time in the width of the matrix [**Fra61**].

EXAMPLE 2.8 (Topological Entropy of Golden Mean Shift). Let X_G be the Golden Mean Shift in Figure 5. The adjacency matrix for G is $\begin{pmatrix} 1 & 1 \\ 1 & 0 \end{pmatrix}$. The eigenvalues of this matrix are $\frac{1\pm\sqrt{5}}{2}$ which means that $h_t(X_G) = log\left(\frac{1+\sqrt{5}}{2}\right)$ (hence the name Golden Mean Shift).

The topological entropy can be computed by analyzing the irreducible components of the graph. An *irreducible component* of a directed graph G is a subgraph H of G such that for every pair of vertices u and v in H, there is a directed path from u to v. Irreducible components can be computed in linear time.

THEOREM 2.9 ([**LM95**]). *Suppose X_G is a sofic shift with presentation G. If $G_1, \ldots, G_k$ are the irreducible components of G and X_{G_i} are the associated sofic shifts, then*

$$h_t(X_G) = \max_{1\le i\le k} h_t(X_{G_i}).$$

2.4. DFA to Sofic Shift. As you can see, sofic shifts are very similar to DFA. The prior method to turn a DFA into a sofic shift used by many other researchers is to simply forget the information regarding initial and final states (for example in [**CSMS03**]). However, information about the resulting sofic shift could be used to understand the original regular language under very limited circumstances. To see why those restrictions are necessary consider any two regular languages L_1, L_2 over the same alphabet, and recall that our construction above for creating (untrimmed) DFA that represent $L_1 \cap L_2$ and $L_1 \triangle L_2$ would lead to the same sofic shift representing two very different languages!

We deal with this issue by describing a new method to transform a DFA into a sofic shift. To explain this method, we will need additional notation. Given a representation G of a sofic shift, the *essential subgraph* is the largest subgraph of G such that each vertex has at least one directed edge entering and at least one directed edge leaving. The essential subgraph of G can be calculated by iteratively deleting vertices that have no directed edge entering that vertex or no directed edge leaving that vertex. Two vertices u, v of directed graph G are *strongly connected* if there exists a path from u to v and a path from v to u. Strongly connected vertices form an equivalence relation, and the irreducible components of directed graph G are the equivalence classes.

DEFINITION 2.10. The *associated sofic shift* of a regular language L is the sofic shift with representation G, where G is constructed from the DFA representing L by (1) trimming the DFA, then (2) ignoring the initial state and final states, and then (3) computing the essential subgraph.

To be clear on this construction, we will walk through an example.

EXAMPLE 2.11. Figure 6 (3) is a representation for the sofic shift that is associated to the regular language defined by the DFA in Figure 6 (1).

Consider the DFA in Figure 6 (1). In this DFA, $\mathcal{A} = \{a, b, c\}$, there are 10 states total, the initial state is 1 and the set of final states is $\{4, 7\}$. Figure 6 (2) shows the trimmed DFA. This operation removed states 8 and 10 since they are not part of any accepting path in the DFA, which can be observed because there is no path starting at 8 or 10 that ends with a final state. Figure 6 (3) shows the essential graph with information about initial/final states suppressed. States 1 and 7 were removed from the trimmed DFA to form the essential graph. The irreducible components are highlighted; there are three. The first irreducible component is composed of states 2 and 3, the second one contains state 4, and the third is composed of states 5, 6, and 9.

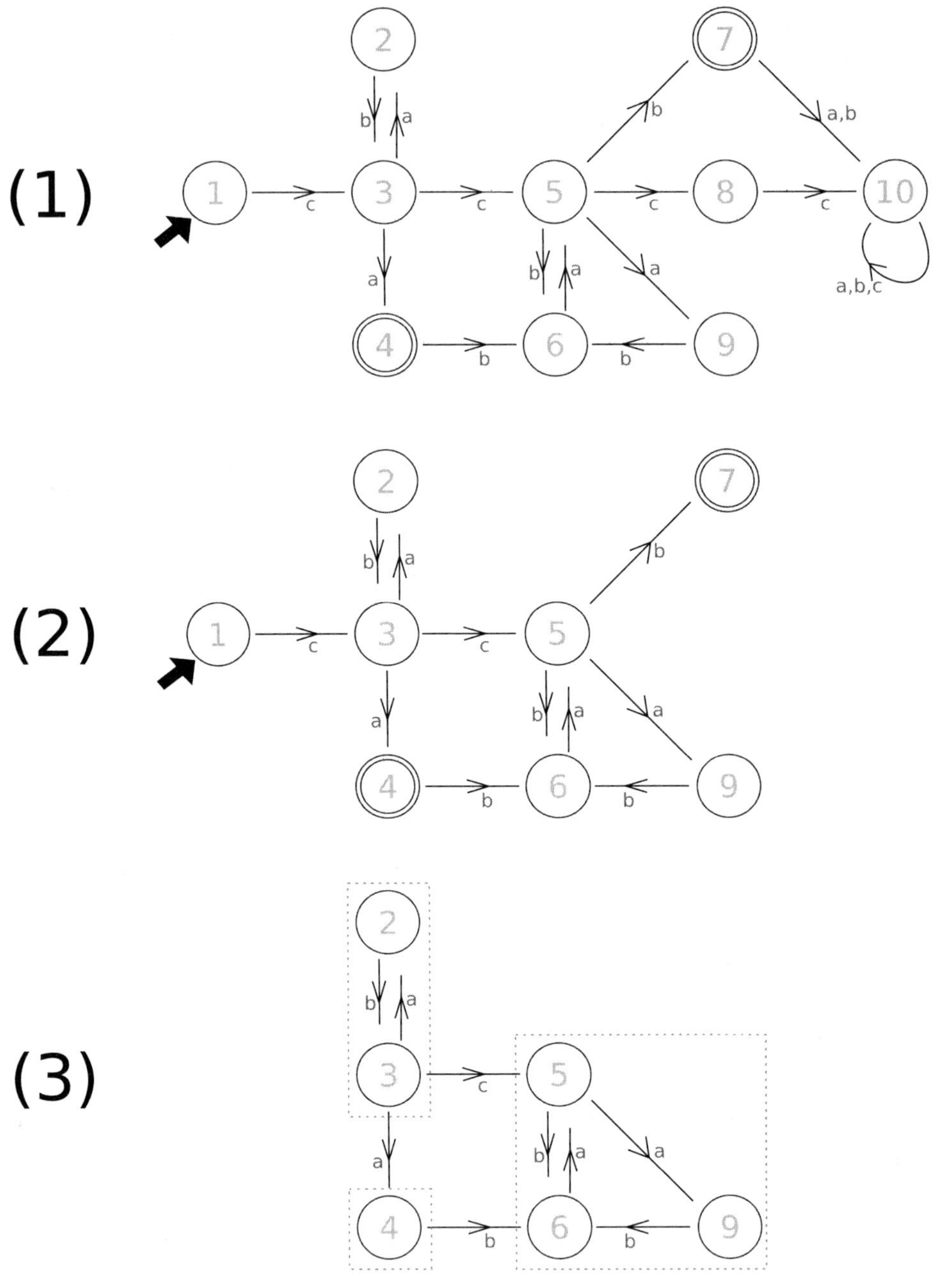

FIGURE 6. The following are displayed: (1) original DFA, (2) trimmed DFA, and (3) essential graph with irreducible components highlighted.

3. Language Entropy

In this section we introduce the language entropy and show that it is the same as the topological entropy of the sofic shift corresponding to the DFA.

Traditionally, the entropy of a regular language L (also called the *channel capacity* [**CM58**] or *information rate* [**CDFI14**]) is defined as $\limsup_{n\to\infty} \frac{\log|W_n(L)|}{n}$. This limit may not exist and so an upper limit is necessary. We will show that this upper limit is realized by the topological entropy of the corresponding sofic shift and define another notion of language entropy, which is preferable since an upper limit is not necessary. We begin by recalling the definition of language entropy.

DEFINITION 3.1. Let L be a non-empty regular language. Define the *language entropy* to be

$$h(L) = \lim_{n\to\infty} \frac{\log|W_{\leq n}(L)|}{n}.$$

We begin by proving a theorem showing the relationship between channel capacity and topological entropy.

THEOREM 3.2. *Let L be a non-empty regular language over the alphabet $\mathcal{A}$, and let X_G be the associated sofic shift. We have that*

$$\limsup_{n\to\infty} \frac{\log|W_n(L)|}{n} = h_t(X_G).$$

Moreover, for a fixed language L there exists a constant c such that there is an increasing sequence of integers n_i satisfying $0 < n_{i+1} - n_i \leq c$ and

$$\lim_{i\to\infty} \frac{\log|W_{n_i}(L)|}{n_i} = h_t(X_G).$$

PROOF. Let λ be the topological entropy of the sofic shift.

Let $(Q, \mathcal{A}, \delta, q_0, F)$ be a DFA for L, and let $w = w_1, \dots, w_n \in L$. For brevity, let $n' = |Q|$. Recall that the Pumping Lemma states that when $n > n'$, there exists a pair i, j such that $1 \leq i < j \leq n'$ and

$$w_1, \dots, w_{i-1} \left(w_i, \dots, w_j\right)^* w_{j+1} \dots, w_n \subseteq L;$$

see [**HU79**] for more information. The proof of this statement uses the fact that if the states of Q seen as w streams by are $q_0, q_1, \dots, q_n$, then there exists a pair i, j as above such that $q_i = q_j$ by the pigeon hole principle.

We claim that $\{q_i, \dots, q_j\}$ are vertices in G (the presentation of X_G that we are working with). For $i < \ell < j$, each vertex q_ℓ has an incoming edge (from vertex $q_{\ell-1}$) and an outgoing edge (to vertex $q_{\ell+1}$). Because $q_i = q_j$, this vertex also has an incoming edge (from vertex q_{j-1}) and an outgoing edge (to vertex q_{i+1}). Therefore, this cycle is part of the essential graph. This proves the claim.

We can iterate this procedure on the word $w_1, \dots, w_{i-1}, w_{j+1} \dots, w_n$ to find another subword that is admissible. We can inductively do this until at most n' characters remain. By construction, if vertices $q_i, q_j \in G$ and $i \leq \ell \leq j$, then $q_\ell \in G$. It follows that there exists an i and a j such that $q_i, \dots, q_j$ is in G and $j - i \geq n - n'$. For large n, most of vertices associated with words of length n will be from the graph G. With that in mind, we just need to count the number of admissible blocks of a certain length in that graph and the number of ways to get into the graph from an initial state and the number of ways to leave the graph to reach a final state. Thus, the number of words of length n in L is

$$|B_n| + |B_{n-1}| \cdot |\mathcal{A}| + \cdots + |B_{n-n'}| \cdot |\mathcal{A}|^{n'} \leq \max(|B_n|, \ldots, |B_{n-n'}|) = |B_n|$$

where the last equality comes from the fact that $\{|B_n|\}$ is a nondecreasing sequence. Hence,

$$\limsup_{n\to\infty} \frac{\log |W_n|}{n} \leq \limsup_{n\to\infty} \frac{\log \left(|B_n| \cdot |\mathcal{A}|^{n'}\right)}{n} = \limsup_{n\to\infty} \frac{\log |B_n|}{n} = \lambda.$$

Let $w = w_1, w_2, \ldots, w_{n-2n'}$ be an admissible block from the sofic shift using vertices $q'_1, q'_2 \ldots, q'_{n-2n'}$. By the definition of a trim graph, there exists paths in our DFA $q_0, q_1, \ldots, q_i$ and $q_j, q_{j+1}, \ldots, q_k$ such that $q_i = q'_1$, $q_j = q'_{n-2n'}$ and $q_k \in F$. We may choose these paths to be minimal, which implies that no state is repeated. Thus, $i \leq n'$ and $k - j \leq n'$. Therefore the path

$$q_0, q_1, \ldots, q_i, q'_2, \ldots, q'_{n-2n'}, q_{j+1}, \ldots, q_k$$

is a valid path in our DFA of length between $n - 2n'$ and n, and corresponds to a word in L that contains w as a subword.

So each admissible block from the sofic shift of length $n - 2n'$ appears in some word of L whose length is between $n - 2n'$ and n. Each word in L of length at most n may contain at most $2n'$ distinct admissible blocks of length $n - 2n'$ (one for each substring starting at offsets 0, 1, 2, ..., $2n'$). Therefore, there exists an m such that $n - 2n' \leq m \leq n$ and $|W_m| \geq \frac{1}{(2n')^2} |B_{n-2n'}(G)|$. Because n' is fixed, this proves the second part of the theorem with $c = 2n'$. It also implies that

$$\limsup_{n\to\infty} \frac{\log |W_n|}{n} \geq \lambda.$$

Which suffices to prove the first part of the theorem. □

Theorem 1.3 is a corollary to the above theorem and is an important statement regarding the connection between topological entropy and language entropy (similar to Shannon's channel capacity). Theorem 1.3 is consistent with remarks made by Chomsky and Miller [**CM58**] that involved undefined assumptions; we show rigorously that this formula is correct for all DFA.

PROOF OF THEOREM 1.3. Let λ be the topological entropy of the sofic shift. To show that $\lim_{n\to\infty} \frac{\log|W_{\leq n}(L)|}{n} = \lambda$ we will show that

$$\lambda \leq \liminf_{n\to\infty} \frac{\log |W_{\leq n}(L)|}{n} \leq \limsup_{n\to\infty} \frac{\log |W_{\leq n}(L)|}{n} \leq \lambda.$$

Let $|W_n(L)| = a_n$. Define a subsequence of $(a_n)_{n=1}^{\infty}$ by $(a_{n_k})_{k=1}^{\infty}$ where $a_{n_k} = \max(a_1, a_2, \ldots, a_k)$. Observe that,

$$\begin{aligned} \limsup_{n\to\infty} \frac{\log |W_{\leq n}(L)|}{n} &= \limsup_{n\to\infty} \frac{\log (a_1 + \cdots + a_n)}{n} \\ &\leq \limsup_{n\to\infty} \frac{\log (n a_{n_k})}{n} \\ &\leq \limsup_{k\to\infty} \frac{\log (a_{n_k})}{n_k} = \lambda \end{aligned}$$

by Theorem 3.2.

For the lower bound we will use the second part of Theorem 3.2. Note that $n - n_k \le c$ where c is given in Theorem 3.2. Thus,

$$\begin{aligned}\liminf_{n\to\infty} \frac{\log |W_{\le n}(L)|}{n} &= \liminf_{n\to\infty} \frac{\log (a_1 + \cdots + a_n)}{n} \\ &\ge \liminf_{n\to\infty} \frac{\log (a_{n_k})}{n_k + (n - n_k)} \\ &> \liminf_{n\to\infty} \frac{\log (a_{n_k})}{n_k + c} = \lambda. \quad \square\end{aligned}$$

Combining the previous two theorems proves Theorem 1.2, which states that language entropy is equal to channel capacity.

Finally, in this section we give some simple properties of language entropy which will be useful later. The first is a simple re-phrasing of Theorem 1.3.

LEMMA 3.3. *For any regular language L and for any $\epsilon > 0$, there exists N such that for all $n \ge N$,*

$$2^{n(h(L)-\epsilon)} \le |W_{\le n}(L)| \le 2^{n(h(L)+\epsilon)}.$$

LEMMA 3.4. *Suppose L_1 and L_2 are regular languages over $\mathcal{A}$. The following hold:*

(1) *If $L_1 \subseteq L_2$, then $h(L_1) \le h(L_2)$.*
(2) $h(L_1 \cup L_2) = \max(h(L_1), h(L_2))$
(3) $\max(h(L_1), h(\overline{L_1})) = \log |\mathcal{A}|$
(4) *If $h(L_1) < h(L_2)$, then $h(L_2 \setminus L_1) = h(L_2)$.*
(5) *If L_1 is finite, then $h(L_1) = 0$.*

PROOF. Each part is proven in turn:

(1) When $L_1 \subseteq L_2$, $W_{\le n}(L_1) \subseteq W_{\le n}(L_2)$. Thus $\lim_{n\to\infty} \frac{\log |W_{\le n}(L_1)|}{n} \le \lim_{n\to\infty} \frac{\log |W_{\le n}(L_2)|}{n}$.

(2) Part 1 implies that $h(L_1 \cup L_2) \ge \max(h(L_1), h(L_2))$. By symmetry, assume that there are infinitely many values of n such that $|W_{\le n}(L_1)| \ge |W_{\le n}(L_2)|$. Because the limits inside the definition of h exist, we have that

$$\begin{aligned}h(L_1 \cup L_2) &\le \liminf_{n\to\infty} \frac{\log (|W_{\le n}(L_1)| + |W_{\le n}(L_2)|)}{n} \\ &\le \liminf_{n\to\infty} \frac{\log |W_{\le n}(L_1)| + \log(2)}{n} \\ &= h(L_1).\end{aligned}$$

(3) Notice that $L_1 \cup \overline{L_1} = \mathcal{A}^*$ and $h(\mathcal{A}^*) = \log |\mathcal{A}|$. The result follows by part 2.

(4) Notice that $L_2 = (L_2 \setminus L_1) \cup (L_1 \cap L_2)$. Since $L_1 \cap L_2 \subset L_1$ we have that $h(L_1 \cap L_2) \le h(L_1) < h(L_2)$ by part 1. Thus $h(L_2) = h((L_2 \setminus L_1) \cup (L_1 \cap L_2)) = \max(h(L_2 \setminus L_1), h(L_1 \cap L_2)) = h(L_2 \setminus L_1)$.

(5) This is trivial. $\square$

4. Entropy Distances

Entropy provides a natural method for dealing with the infinite nature of regular languages. Because it is related to the eigenvalues of the regular language's DFA, it is computable in polynomial time given a DFA for the language. Note that the DFA does not have to be minimal. We can therefore compute the entropy of set-theoretic combinations of regular languages (intersection, disjoint union, etc) as shown in Subsection 2.1 and use those values to determine a distance between the languages.

We will also discuss when certain distance functions are metrics. Recall that a *metric* on the space X is a function $d : X \times X \to \mathbb{R}$ that satisfies

(1) $d(x, y) \geq 0$ with equality if and only if $x = y$ for all $x, y \in X$
(2) $d(x, y) = d(y, x)$ for all $x, y \in X$
(3) $d(x, z) \leq d(x, y) + d(y, z)$ for all $x, y, z \in X$.

An *ultra-metric* is a stronger version of a metric, with the triangle inequality replaced with the ultra-metric inequality:

$$d(x, z) \leq \max\{d(x, y), d(y, z)\}$$

for all $x, y, z \in X$. Also, there exists a weaker version, called a *pseudo-metric*, which allows $d(x, y) = 0$ when $x \neq y$. For any pseudo-metric, the relation $d(x, y) = 0$ is an equivalence relation. Thus, if we mod out by this equivalence relation, the pseudo-metric becomes a metric.

4.1. Entropy Distance.

A natural Jaccard-esque distance function based on entropy is the entropy distance.

DEFINITION 4.1 (Entropy Distance). Suppose L_1 and L_2 are regular languages. Define the *entropy distance* to be $H(L_1, L_2) = \frac{h(L_1 \triangle L_2)}{h(L_1 \cup L_2)}$ if $h(L_1 \cup L_2) > 0$ and $H(L_1, L_2) = 0$ otherwise.

This turns out to be equivalent to a Jaccard limit with added log operations:

COROLLARY 4.2. *Suppose L_1 and L_2 are regular languages. The following relation holds:*

$$\lim_{n\to\infty} \frac{\log |W_{\leq n}(L_1 \triangle L_2)|}{\log |W_{\leq n}(L_1 \cup L_2)|} = H(L_1, L_2).$$

PROOF. Observe the following:

$$\lim_{n\to\infty} \frac{\log |W_{\leq n}(L_1 \triangle L_2)|}{\log |W_{\leq n}(L_1 \cup L_2)|} = \lim_{n\to\infty} \frac{\frac{1}{n}\log |W_{\leq n}(L_1 \triangle L_2)|}{\frac{1}{n}\log |W_{\leq n}(L_1 \cup L_2)|} = \frac{h(L_1 \triangle L_2)}{h(L_1 \cup L_2)} = H(L_1, L_2).$$

Note that we can separate the limits because of Theorem 1.3. □

Note that H is not always a good candidate for a distance function as it only produces non-trivial results for languages that have the same entropy.

PROPOSITION 4.3. *Suppose L_1 and L_2 are regular languages. If $h(L_1) \neq h(L_2)$, then $H(L_1, L_2) = 1$.*

PROOF. WLOG, suppose that $h(L_1) < h(L_2)$. First, $L_1 \cap L_2 \subseteq L_1$ which implies that $h(L_1 \cap L_2) \leq h(L_1)$. Second, $L_2 \subseteq L_1 \cup L_2$, and therefore $h(L_2) \leq h(L_1 \cup L_2)$. All together this gives $h(L_1 \cap L_2) < h(L_1 \cup L_2)$, which implies that $h(L_1 \cap L_2) \neq h(L_1 \cup L_2)$. By Lemma 3.4, $h(L_1 \cup L_2) = \max(h(L_1 \cap L_2), h(L_1 \triangle L_2))$. Thus, $h(L_1 \cup L_2) = h(L_1 \triangle L_2)$. □

As further evidence that H is not a good candidate for a distance function, we show it is an ultra-pseudo-metric. The ultra-metric condition, i.e. $d(x,z) \leq \max(d(x,y), d(y,z))$, is so strong that it can make it difficult for the differences encoded in the metric to be meaningful for practical applications.

THEOREM 4.4. *The function H is an ultra-pseudo-metric.*

PROOF. The first two conditions of an ultra-pseudo-metric are satisfied by the definition of H and from the reflexiveness of $\triangle$ and $\cup$. We now have to verify the ultra-metric inequality.

Suppose L_1, L_2, L_3 are regular languages. We need to show that

$$H(L_1, L_3) \leq \max(H(L_1, L_2), H(L_2, L_3)).$$

Case 1: Suppose $h(L_1) \neq h(L_2)$. By Proposition 4.3, $H(L_1, L_2) = 1$. Since $H(L_1, L_2)$ is a number between 0 and 1, $H(L_1, L_3) \leq \max(H(L_1, L_2), H(L_2, L_3))$.

Case 2: Suppose $h(L_1) = h(L_2)$. If $h(L_3) \neq h(L_2)$, then the above argument holds. Thus, assume that $h(L_1) = h(L_2) = h(L_3)$. By Lemma 3.4, $h(L_1 \cup L_3) = h(L_1 \cup L_2) = h(L_2 \cup L_3)$. Hence, it suffices to show that $h(L_1 \triangle L_3) \leq \max(h(L_1 \triangle L_2), h(L_2 \triangle L_3))$. Using multiple applications of Lemma 3.4 and the fact that larger sets have larger entropy we observe,

$$\begin{aligned}
h\left(L_1 \triangle L_3\right) &= h((L_1 \cap \overline{L_3}) \cup (\overline{L_1} \cap L_3)) \\
&= h((((L_1 \cap \overline{L_3}) \cup (\overline{L_1} \cap L_3)) \cap L_2) \cup (((L_1 \cap \overline{L_3}) \cup (\overline{L_1} \cap L_3)) \cap \overline{L_2})) \\
&= \max(h\left(L_1 \cap \overline{L_3} \cap L_2\right), h\left(L_1 \cap \overline{L_3} \cap \overline{L_2}\right), \\
&\qquad h\left(\overline{L_1} \cap L_3 \cap L_2\right), h\left(\overline{L_1} \cap L_3 \cap \overline{L_2}\right)) \\
&\leq \max(h\left(L_2 \cap \overline{L_3}\right), h\left(L_1 \cap \overline{L_2}\right), h\left(\overline{L_1} \cap L_2\right), h\left(\overline{L_2} \cap L_3\right)) \\
&= \max\left(h\left(L_1 \triangle L_2\right), h\left(L_2 \triangle L_3\right)\right).
\end{aligned}$$

□

4.2. Entropy Sum. In this subsection we define a new (and natural) distance function for infinite regular languages. We call this distance function the *entropy sum distance*. We will prove that not only is this distance function a pseudo-metric, it is also sometimes granular. Granularity lends insight into the quality of a metric. Intuitively, granularity means that for any two points in the space, you can find a point between them. A metric d on the space X is *granular* if for every two points $x, z \in X$, there exists $y \in X$ such that $d(x,y) < d(x,z)$ and $d(y,z) < d(x,z)$, i.e. $d(x,z) > \max(d(x,y), d(y,z))$. Recall the definition of entropy sum distance from the introduction.

DEFINITION 4.5 (Entropy Sum Distance). Suppose L_1 and L_2 are regular languages. Define the *entropy sum distance* to be $H_S(L_1, L_2) = h(L_1 \cap \overline{L_2}) + h(\overline{L_1} \cap L_2)$.

The entropy sum distance was inspired by first considering the entropy of the symmetric difference directly, i.e. $h(L_1 \triangle L_2)$. However, since entropy measures the entropy of the most complex component (Theorem 2.9), more information is gathered by using a sum as above in the definition of entropy sum. We prove this formula to be a pseudo-metric.

THEOREM 4.6. *The function H_S is a pseudo-metric.*

PROOF. The first two conditions of a pseudo-metric are satisfied by the definition of H_S and from the reflexiveness of $\triangle$ and $\cup$. We now have to verify the triangle inequality.

Suppose L_1, L_2, L_3 are regular languages. We need to show that

$$H_S(L_1, L_3) = h\left(L_1 \cap \overline{L_3}\right) + h\left(\overline{L_1} \cap L_3\right) \leq H_S(L_1, L_2) + H_S(L_2, L_3).$$

First observe,

$$\begin{aligned} h\left(L_1 \cap \overline{L_3}\right) &= \max\left(h\left(L_1 \cap \overline{L_3} \cap L_2\right), h\left(L_1 \cap \overline{L_3} \cap \overline{L_2}\right)\right) \\ &\leq \max\left(h\left(L_2 \cap \overline{L_3}\right), h\left(L_1 \cap \overline{L_2}\right)\right). \end{aligned}$$

In a similar fashion, $h\left(\overline{L_1} \cap L_3\right) \leq h(\overline{L_1} \cap L_2) + h(\overline{L_2} \cap L_3)$. Putting these together yields the desired result. □

The next two propositions display when granularity is achieved and when it is not.

PROPOSITION 4.7. *Let L_1 and L_2 be regular languages such that $h(L_1 \cap \overline{L_2}), h(\overline{L_1} \cap L_2) > 0$. Then, there exists two regular languages $R_1 \neq R_2$ such that $H_S(L_1, L_2) > \max(H_S(L_1, R_i), H_S(R_i, L_2))$ for each i.*

PROOF. Let $R_1 = L_1 \cup L_2$ and $R_2 = L_1 \cap L_2$. Notice that $H_S(L_1, R_1) = h(\overline{L_1} \cap L_2)$ and $H_S(R_1, L_2) = h(L_1 \cap \overline{L_2})$. Hence,

$$H_S(L_1, L_2) = h(L_1 \cap \overline{L_2}) + h(\overline{L_1} \cap L_2) > \max(H_S(L_1, R_1), H_S(R_1, L_2)).$$

The statement involving R_2 is analogous. □

PROPOSITION 4.8. *Let L_1 and L_2 be regular languages such that $h(\overline{L_1} \cap L_2) = 0$. For all regular languages L we have that $H_S(L_1, L_2) \leq \max(H_S(L_1, L), H_S(L, L_2))$.*

PROOF. Note that $H_S(L_1, L_2) = h(L_1 \cap \overline{L_2})$. The proof breaks down into two cases:

Case 1: Suppose $h(L_1 \cap \overline{L_2} \cap L) = h(L_1 \cap \overline{L_2})$. Then,

$$\begin{aligned} h\left(L_1 \cap \overline{L_2}\right) &= h\left(L_1 \cap \overline{L_2} \cap L\right) \\ &\leq h\left(L \cap \overline{L_2}\right) \\ &\leq H_S(L, L_2) \\ &\leq \max\left(H_S(L_1, L), H_S(L, L_2)\right). \end{aligned}$$

Case 2: Suppose $h(L_1 \cap \overline{L_2} \cap L) < h(L_1 \cap \overline{L_2})$. Then,

$$\begin{aligned} h\left(L_1 \cap \overline{L_2}\right) &= \max\left(h\left(L_1 \cap \overline{L_2} \cap L\right), h\left(L_1 \cap \overline{L_2} \cap \overline{L}\right)\right) \\ &= h\left(L_1 \cap \overline{L_2} \cap \overline{L}\right) \\ &\leq h\left(L_1 \cap \overline{L}\right) \\ &\leq H_S(L_1, L) \\ &\leq \max\left(H_S(L_1, L), H_S(L, L_2)\right). \end{aligned}$$

□

5. Jaccard Distances

We now turn to investigate several extensions of Jaccard distance to infinite regular languages. We end this section with a discussion about the relationship between Cesáro Jaccard distance and entropy.

5.1. Jaccard Distances using W_n and $W_{\leq n}$. A natural method for applying Jaccard distance to regular languages is to fix n and study words of length n exactly or words of length up to n. Recall the definitions of n Jaccard distance and $\leq n$ Jaccard distance from the introduction.

DEFINITION 5.1 (n Jaccard Distance). Suppose L_1 and L_2 are regular languages. Define the *n Jaccard distance* by

$$J'_n(L_1, L_2) = \frac{|W_n(L_1 \triangle L_2)|}{|W_n(L_1 \cup L_2)|}$$

if $|W_n(L_1 \cup L_2)| > 0$ and $J'_n(L_1, L_2) = 0$ otherwise.

DEFINITION 5.2 ($n_{\leq}$ Jaccard Distance). For regular languages L_1 and L_2, define the *$n_{\leq}$ Jaccard distance* by

$$J_n(L_1, L_2) = \frac{|W_{\leq n}(L_1 \triangle L_2)|}{|W_{\leq n}(L_1 \cup L_2)|}$$

if $|W_{\leq n}(L_1 \cup L_2)| > 0$ and $J_n(L_1, L_2) = 0$ otherwise.

For fixed n, J'_n is a pseudo-metric since it is simply the Jaccard distance among sets containing only length n strings. The following proposition points out one deficiency of J'_n.

PROPOSITION 5.3. *There exists a set $S = \{L_1, L_2, L_3\}$ of infinite unary regular languages with $L_2, L_3 \subset L_1$ such that for all n there exists an $i \neq j$ such that $J'_n(L_i, L_j) = 0$.*

PROOF. Let $L_1 = a^*$, $L_2 = (aa)^*$, and $L_3 = a(aa)^*$. Fix $n \in \mathbb{N}$. If n is even, then $J'_n(L_1, L_2) = 0$, and if n is odd, then $J'_n(L_1, L_3) = 0$. □

The issue with J'_n pointed out by Proposition 5.3 can be proven to not be a problem for J_n: see the first point of Theorem 5.4. On the other hand, the second point of Theorem 5.4 shows that no universal n exists.

THEOREM 5.4. *The function J_n defined above is a pseudo-metric and satisfies the following:*

(1) *Let $S = \{L_1, \ldots, L_k\}$ be a set of regular languages. There exists an n such that J_n is a metric over S. Moreover, we may choose n such that $n \leq \max_{i,j}(s(L_i) + 1)(s(L_j) + 1) - 1$ where $s(L_i)$ represents the number of states in the minimal DFA corresponding to L_i.*
(2) *For any fixed n there exist regular languages L, L' with $L \neq L'$ such that $J_n(L, L') = 0$.*

PROOF. The fact that J_n is a pseudo-metric follows from the fact that the standard Jaccard distance for finite sets is a metric.

(1) Let $S = \{L_1, \ldots, L_k\}$ be a fixed finite set of regular languages. For each $i \neq j$ there exists an $n_{i,j}$ such that $\left|W_{n_{i,j}}(L_i \triangle L_j)\right| \neq 0$ since $L_i \neq L_j$ and only one L_i can be $\emptyset$. Let $n = \max_{i,j} n_{i,j}$. Then J_n is a metric over S. Every regular

language L_i contains a word whose length is at most $s(L_i)$. Now, we simply observe that $s(L_i \triangle L_j) \leq (s(L_i)+1)(s(L_j)+1) - 1$.

(2) Let n be an arbitrary number and let $\mathcal{A}' = \mathcal{A} \cup \{z\}$, where $z \notin \mathcal{A}$. Take an arbitrary regular language L over $\mathcal{A}$. Construct a regular language $L' = L \cup \{z^{n+1}\}$ over $\mathcal{A}'$. L' is the language L with the addition of the element z^{n+1}. When L is considered over alphabet $\mathcal{A}'$, we have: $J_n(L, L') = 0$. □

Due to the fact that one must choose a fixed n, J_n and J'_n cannot account for the infinite nature of regular languages. Limits based on J_n and J'_n are a natural next step. However, the natural limits involving J'_n and J_n do not always exist. An example showing this was given for J'_n in the beginning of the introduction. A similar example applies to J_n.

EXAMPLE 5.5. Consider the languages given by $L_1 = (a|b)^*$ and $L_2 = ((a|b)^2)^*$ where $\mathcal{A} = \{a, b\}$. For these languages, $\lim_{n\to\infty} J_{2n}(L_1, L_2) = 2/3$ and $\lim_{n\to\infty} J_{2n+1}(L_1, L_2) = 1/3$. Hence, $\lim_{n\to\infty} J_n(L_1, L_2)$ does not exist.

The next theorem gives conditions for when the limit of J'_n exists as n goes to infinity. Before the theorem is stated we will need some more terminology.

The *period* of an irreducible graph (or associated adjacency matrix) is the largest integer p such that the vertices can be grouped into classes $Q_0, Q_1, \ldots, Q_{p-1}$ such that if $x \in Q_i$, then all of the out neighbors of x are in Q_j, where $j = i + 1 (mod\ p)$. The period of a reducible graph is the least common multiple of the periods of its irreducible components. See Figure 7 for an example of a regular language whose DFA has period 3. For a more formal definition of periodicity see [**LM95**]. If the graph (or matrix) has period 1 it will be called *aperiodic*. Matrices that are irreducible and aperiodic are called *primitive*. The definition of primitive presented here is equivalent to the condition that there is an n such that all entries of the adjacency matrix A raised to the n-th power (A^n) are positive [**MU16**]. This is illustrated in Figure 7, where the graph is periodic and reducible and all powers of that matrix contain multiple zeroes.

Let A be an adjacency matrix for the DFA M. Then, the ij-th entry in A^n is the number of walks of length n in M of length exactly n. Thus, to study the limit of J'_n we will study the asymptotic growth rate of A^n. We start with an example to show how to compute the words of length n from a DFA.

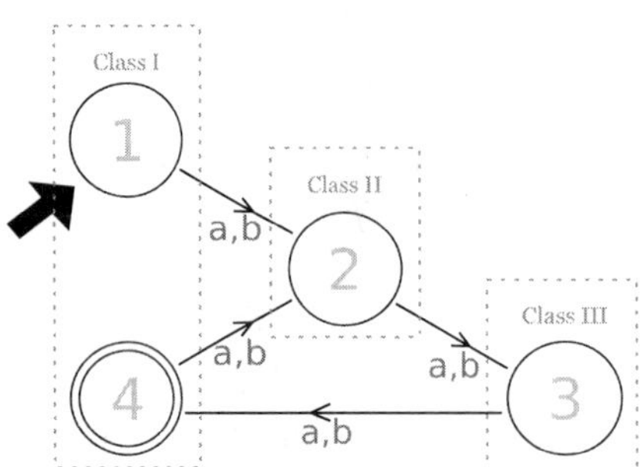

$$A^i = \begin{cases} \begin{pmatrix} 0 & 2^i & 0 & 0 \\ 0 & 0 & 2^i & 0 \\ 0 & 0 & 0 & 2^i \\ 0 & 2^i & 0 & 0 \end{pmatrix} & \text{if } i \equiv 1 \bmod 3 \\ \begin{pmatrix} 0 & 0 & 2^i & 0 \\ 0 & 0 & 0 & 2^i \\ 0 & 2^i & 0 & 0 \\ 0 & 0 & 2^i & 0 \end{pmatrix} & \text{if } i \equiv 2 \bmod 3 \\ \begin{pmatrix} 0 & 0 & 2^i & 0 \\ 0 & 2^i & 0 & 0 \\ 0 & 0 & 2^i & 0 \\ 0 & 0 & 0 & 2^i \end{pmatrix} & \text{if } i \equiv 0 \bmod 3 \end{cases}$$

FIGURE 7. On the left is a DFA of period 3 and on the right is the associated adjacency matrix raised to the ith power.

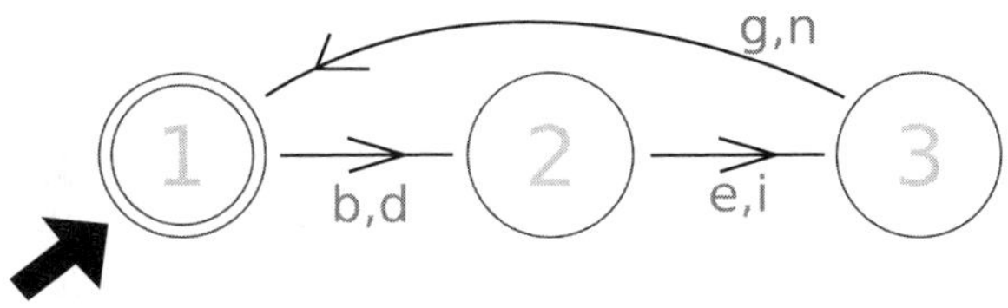

FIGURE 8. DFA corresponding to the regular language given by $[(b|d)(e|i)(g|n)]^*$.

EXAMPLE 5.6. Consider the DFA given in Figure 8. The regular language represented by this DFA is given by the regular expression $[(b|d)(e|i)(g|n)]^*$. Let i_A be the row vector where the j-th entry is 1 if j is the initial state in the DFA and 0 otherwise. Similarly define the column vector f_A for the final states. Thus, $i_A = (1, 0, 0)$ and $f_A^T = (1, 0, 0)$. The number of words of length n in the DFA is given by

$$W_n = \begin{pmatrix} 1 & 0 & 0 \end{pmatrix} \begin{pmatrix} 0 & 2 & 0 \\ 0 & 0 & 2 \\ 2 & 0 & 0 \end{pmatrix}^n \begin{pmatrix} 1 \\ 0 \\ 0 \end{pmatrix} = \begin{cases} 2^n, & n = 3k + 0 \\ 0, & n = 3k + 1 \\ 0, & n = 3k + 2. \end{cases}$$

This DFA is periodic with period 3.

THEOREM 5.7. *Suppose L_1 and L_2 are regular languages. If each irreducible component of the DFA associated to $L_1 \triangle L_2$ and $L_1 \cup L_2$ are aperiodic, then $\lim_{n\to\infty} J'_n(L_1, L_2)$ converges.*

Let us build intuition prior to proving Theorem 5.7, which will also frame the question of convergence in the next subsection. We will first discuss Theorem 5.7 in the case where the DFA associated to regular languages $L_1 \triangle L_2$ and $L_1 \cup L_2$ are primitive. Suppose $A_\triangle$ and $A_\cup$ are the adjacency matrices for $L_1 \triangle L_2$ and $L_1 \cup L_2$ respectively. Perron-Frobenius theory tells us that the eigenvalue of largest modulus of a primitive matrix is real and unique. Let $(v_\triangle, \lambda_\triangle)$ and $(v_\cup, \lambda_\cup)$ be eigenpairs composed of the top eigenvalues for $A_\triangle$ and $A_\cup$ respectively. Notice that $i_\triangle A_\triangle^n f_\triangle$, where $i_\triangle$ is the row vector whose j-th entry is 1 if j is an initial state in $A_\triangle$ and 0 otherwise (a similar definition for final states defining column vector $f_\triangle$ holds), represents words in $L_1 \triangle L_2$ of length n. If we write

$$f_\triangle = c_1 v_\triangle + c_2 w \quad \text{and} \quad f_\cup = d_1 v_\cup + d_2 y$$

then $i_\triangle A_\triangle^n f_\triangle$ converges to $\lambda_\triangle^n c_1 i_\triangle v_\triangle$, and $i_\cup A_\cup^n f_\cup$ converges to $\lambda_\cup^n d_1 i_\cup v_\cup$ as n goes to infinity. This convergence is guaranteed because $\lambda_\cup$ and $\lambda_\triangle$ are unique top eigenvalues. Thus,

$$\lim_{n\to\infty} J'_n(L_1, L_2) = \lim_{n\to\infty} \left(\frac{\lambda_\triangle}{\lambda_\cup}\right)^n \frac{c_1 i_\triangle v_\triangle}{d_1 i_\cup v_\cup}$$

and the limit converges ($\lambda_\triangle \le \lambda_\cup$ because $L_1 \triangle L_2 \subseteq L_1 \cup L_2$).

The general case of Theorem 5.7, which does not assume $L_1 \triangle L_2$ and $L_1 \cup L_2$ have irreducible adjacency matrices, is more complicated. However, the outline of the argument is the same, and we provide it here. The key difference is the use of newer results. An understanding of the asymptotic behavior of A^n for large n was finally beginning to be developed several decades after Chomsky and Miller investigated regular languages. In 1981 Rothblum [**Rot81**] proved the following (a slow treatment of this theory with examples can be found in [**Rot07**]), which we will refer to as *Rothblum's Theorem*:

THEOREM 5.8 (Rothblum's Theorem [**Rot81**]). *Suppose A is a non-negative matrix with largest eigenvalue λ. There exists q and polynomials $S_0(x), S_1(x), \ldots, S_{q-1}(x)$ whose domain is the set of real numbers and whose coefficients are matrices, such that for all whole numbers $0 \le k \le q-1$ we have that*

$$\lim_{n\to\infty} (A/\lambda)^{qn+k} - S_k(n) = 0.$$

Moreover, the number q is the period of the matrix A.

For any given regular language there are infinitely many DFA that represent it. One of the themes of this work is finding a preferred representation and show how that representation can be used to turn intuitive relationships into rigorous ones. We will need a technical lemma that shows how a trim representation of a DFA can prevent a trivial outcome from an application of Rothblum's theorem.

LEMMA 5.9. *Let L be an infinite regular language represented by trimmed DFA $(Q, \mathcal{A}, \delta, q_0, F)$. We use notation i, A, f as before so that $W_m(L) = iA^m f$. Let q and $S_0(x), \ldots, S_{q-1}(x)$ be the integer and polynomials determined by Rothblum's theorem applied to A. Under these conditions, there exists a j such that $iS_j(x)f \neq 0$.*

PROOF. Let $p(x) = \sum_{\ell=0}^{d} \alpha_\ell x^\ell$ be the characteristic polynomial for A. The Cayley-Hamilton theorem states that $p(A) = 0$; therefore $W_m(L) = iA^m f$ follows a linear recursive sequence with characteristic polynomial p. It is well-known (see [**Sta97**]) that $W_m(L) = \sum_t \lambda_t^m \widehat{S}_t(m)$ where $\widehat{S}_t$ is a polynomial and λ_t ranges over the roots of p, which are the eigenvalues of A.

Let ρ denote the spectral radius of A, and fix $0 \le k < q$. Because L is infinite, $\rho \ge 1$. Let $T = \{t : \|\lambda_t\| = \rho\}$ and for each $t \in T$ define $\zeta_t = \lambda_t/\rho$. In reading Rothblum's work [**Rot07**], it is seen that q is chosen such that for each $t \in T$ we have $\zeta_t^q = 1$. Clearly $iS_k(x)f = \sum_{t\in T} \zeta_t^k \widehat{S}_t(x)$.

By way of contradiction, assume that $iS_k(x)f = 0$ for all $0 \le k < q$. This implies that for all m we have that $\sum_{t\in T} \zeta_t^m \widehat{S}_t(m) = 0$. Therefore

$$W_m(L) = \sum_t \lambda_t^m \widehat{S}_t(m) = \sum_{t\notin T} \lambda_t^m \widehat{S}_t(m).$$

Let ρ_* be the magnitude of the largest eigenvalue of A that is not indexed by T. There exist numbers d_* and M_* such that

$$\limsup_{m\to\infty} \frac{W_m(L)}{\rho_*^m m^{d_*}} \le M_*$$

and therefore $\limsup_{m\to\infty} \log(W_m(L))/m \le \rho_* < \rho$.

Recall from Sections 2.3 and 2.4 that the topological entropy of a right-resolving sofic shift is not affected by taking the essential subgraph. Because A comes from a trim DFA, the topological entropy of the associated sofic shift is ρ. However, this contradicts Theorem 3.2. □

PROOF TO THEOREM 5.7. The proof is clear if $L_1 \triangle L_2$ is finite.

Consider the DFA for regular languages $L_1 \cup L_2$ and $L_1 \triangle L_2$. As it costs nothing to trim a DFA, we assume that both DFA are trimmed. We use notation

$A_\triangle, A_\cup, i_\triangle, i_\cup, f_\triangle, f_\cup$ as before. We apply Rothblum's theorem to each of the languages. Observe from [**Rot07**] that the assumptions of the theorem are sufficient for the application of Rothblum's Theorem to produce $q = 1$. So it is well-defined to use as notation for the results from the two instances of applications of Rothblum's Theorem $\rho_\cup, \rho_\triangle, S_\triangle, S_\cup$. By Lemma 5.9 we have that $i_\triangle S_\triangle(x) f_\triangle$ and $i_\cup S_\cup(x) f_\cup$ are each non-zero. As a consequence, there exists integers $d_\triangle$ and $d_\cup$ such that $\lim_{m\to\infty} (i_\triangle A_\triangle f_\triangle/\rho_\triangle)^m m^{-d_\triangle}$ and $\lim_{m\to\infty} (i_\cup A_\cup f_\cup/\rho_\cup)^m m^{-d_\cup}$ are positive reals.

Because $L_1 \triangle L_2 \subseteq L_1 \cup L_2$, we have that $\rho_\cup \geq \rho_\triangle$. Moreover, if $\rho_\cup = \rho_\triangle$, then $d_\cup \geq d_\triangle$. This implies that $\lim_{m\to\infty} (i_\triangle A_\triangle f_\triangle/\rho_\cup)^m m^{-d_\cup}$ converges because it converges to 0 if $\rho_\cup \neq \rho_\triangle$ or $d_\cup \neq d_\triangle$. Because both the top and bottom converge and the bottom does not converge to zero, we see the convergence of

$$\begin{aligned}\lim_{m\to\infty} J'_m(L_1, L_2) &= \lim_{m\to\infty} \frac{W_m(L_1 \triangle L_2)\rho_\cup^{-m} m^{-d_\cup}}{W_m(L_1 \cup L_2)\rho_\cup^{-m} m^{-d_\cup}} \\ &= \frac{\lim_{m\to\infty} W_m(L_1 \triangle L_2)\rho_\cup^{-m} m^{-d_\cup}}{\lim_{m\to\infty} W_m(L_1 \cup L_2)\rho_\cup^{-m} m^{-d_\cup}}.\end{aligned}$$

□

To better illustrate the proof to Theorem 5.7, we will do another example.

EXAMPLE 5.10. Consider the DFA from Figure 9. The regular language represented by this DFA is given by the regular expression $(a|b)^*c(a|b)^*c(a|b)^*c(a|b)^*$. This DFA has four irreducible components, each of period 1. Thus, the DFA and it's adjacency matrix have period 1. The number of words of length n is given by

$$\begin{aligned} W_n &= \begin{pmatrix} 1 & 0 & 0 & 0 \end{pmatrix} \begin{pmatrix} 2 & 1 & 0 & 0 \\ 0 & 2 & 1 & 0 \\ 0 & 0 & 2 & 1 \\ 0 & 0 & 0 & 2 \end{pmatrix}^n \begin{pmatrix} 0 \\ 0 \\ 0 \\ 1 \end{pmatrix} \\ &= \binom{n}{3} 2^{n-3} \\ &= \left(\frac{1}{48} n^3 + \frac{-1}{16} n^2 + \frac{1}{24} n \right) 2^n. \end{aligned}$$

Since $q = 1$ from Rothblum's Theorem, there is one matrix polynomial. Here we have a polynomial of degree 3:

$$\begin{aligned} \begin{pmatrix} 2 & 1 & 0 & 0 \\ 0 & 2 & 1 & 0 \\ 0 & 0 & 2 & 1 \\ 0 & 0 & 0 & 2 \end{pmatrix}^n 2^{-n} &= \begin{pmatrix} 1 & 0 & 0 & 0 \\ 0 & 1 & 0 & 0 \\ 0 & 0 & 1 & 0 \\ 0 & 0 & 0 & 1 \end{pmatrix} + \begin{pmatrix} 0 & 1/2 & -1/8 & 1/24 \\ 0 & 0 & 1/2 & -1/8 \\ 0 & 0 & 0 & 1/2 \\ 0 & 0 & 0 & 0 \end{pmatrix} n \\ &+ \begin{pmatrix} 0 & 0 & 1/8 & -1/16 \\ 0 & 0 & 0 & 1/8 \\ 0 & 0 & 0 & 0 \\ 0 & 0 & 0 & 0 \end{pmatrix} n^2 \\ &+ \begin{pmatrix} 0 & 0 & 0 & 1/48 \\ 0 & 0 & 0 & 0 \\ 0 & 0 & 0 & 0 \\ 0 & 0 & 0 & 0 \end{pmatrix} n^3. \end{aligned}$$

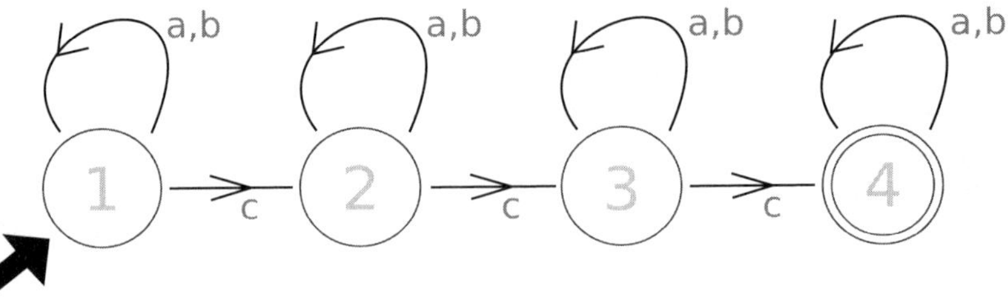

FIGURE 9. DFA corresponding to the regular language given by $(a|b)^*c(a|b)^*c(a|b)^*c(a|b)^*$.

5.2. Cesàro Jaccard. For a sequence of numbers $a_1, a_2, \ldots$, a Cesàro summation is $\lim_{n\to\infty} \frac{1}{n}\sum_{i=1}^{n} a_i$ when the limit exists. The intuition behind a Cesàro summation is that it may give the "average value" of the limit of the sequence, even when the sequence does not converge. For example, the sequence $a_j = e^{\alpha i j}$ (where $i^2 = -1$) has Cesàro summation 0 for all real numbers $\alpha \neq 0$. This follows from the fact that irrational rotations of the circle are uniquely ergodic [**HK03**]. Not all sequences have a Cesàro summation, even when we restrict our attention to sequences whose values lie in $[0, 1]$. For example, the sequence b_i, where $b_i = 1$ when $2^{2n} < i < 2^{2n+1}$ for some $n \in \mathbb{N}$ and $b_i = 0$ otherwise has no Cesàro summation. However, we will be able to show that the Cesàro average of Jaccard distances does exist.

To that end, another limit based distance is the Cesàro average of the J_n or J'_n. Recall the Cesáro Jaccard distance from the introduction.

DEFINITION 5.11 (Cesàro Jaccard Distance). Suppose L_1 and L_2 are regular languages. Define the *Cesàro Jaccard distance* by

$$J_C(L_1, L_2) = \lim_{n\to\infty} \frac{1}{n}\sum_{i=1}^{n} J_i(L_1, L_2).$$

The Cesàro Jaccard distance is theoretically better than the above suggestions in Section 5.1 since we show that it exists. Recall Theorem 1.1 from the introduction, which states that

$$\lim_{n\to\infty} \frac{1}{n}\sum_{i=1}^{n} J_i(L_1, L_2)$$

exists.

We will briefly sketch the proof to Theorem 1.1 with J'_n in the place of J_n. Recall that $|W_n(L_1 \triangle L_2)|$ and $|W_n(L_1 \cup L_2)|$ can be calculated using powers of specific matrices. If we take Q to be the least common multiple of the period from each of the matrices associated with $|W_n(L_1 \triangle L_2)|$ and $|W_n(L_1 \cup L_2)|$, we can immediately see that $\lim_{n\to\infty} J'_{Qn+k}(L_1, L_2)$ exists, via Rothblum's Theorem. Moreover, it will equal zero if they have different values for the largest eigenvalue or the degree of $S_k(x)$. But if they have the same value for the largest eigenvalue and degree of $S_k(x)$, then $\lim_{n\to\infty} J'_{Qn+k}(L_1, L_2)$ will be the ratio of the leading coefficients of the polynomials $S_k(x)$ for the two matrices. The proof finishes by observing that $J'_C(L_1, L_2) = \frac{1}{n}\sum_{i=1}^{n} J'_i(L_1, L_2)$ will be the average of these values.

We will require a new result to show that the more interesting value $J_C(L_1, L_2)$ is well-defined. Note that part (2) of the theorem below is similar to a result in [**Rot81**].

THEOREM 5.12. *Let A be the adjacency matrix for a DFA representing a regular language L, and let λ be the largest eigenvalue of A. Let q and $S_0(x), S_1(x), \ldots, S_{q-1}(x)$ be as in Rothblum's theorem; let d be the largest degree of the polynomials $S_0(x), S_1(x), \ldots, S_{q-1}(x)$. Let $s_\ell = \lim_{n\to\infty} n^{-(d+1)} \sum_{i=1}^n S_\ell(i)$ and $t_\ell = \lim_{n\to\infty} n^{-d} S_\ell(n)$.*

(1) *If $\lambda < 1$, then L is finite.*
(2) *If $\lambda = 1$, then*

$$\lim_{n\to\infty} \frac{1}{n^{d+1}} \sum_{i=1}^{n} A^i = \sum_{i=0}^{q-1} s_\ell.$$

(3) *If $\lambda > 1$, then*

$$\lim_{n\to\infty} \frac{1}{(qn+k)^d} \lambda^{-(qn+k)} \sum_{i=1}^{qn+k} A^i = \frac{1}{1-\lambda^{-q}} \sum_{\ell=k-q+1}^{k} \lambda^{\ell-k} t_\ell$$

where the indices of the t_i are taken modulo q.

PROOF. See [**Rot07**] for a background on linear algebra. The largest eigenvalue of a non-negative matrix is at least the value of the smallest sum of the entries in a row of an irreducible component. Because the adjacency matrix for a DFA has integer entries, this implies that either $\lambda = 0$ or $\lambda \geq 1$. If $\lambda = 0$, then A is nilpotent (in other words, there exists an n' such that $A^{n'} = 0$), which means L is finite. So assume $\lambda \geq 1$.

First consider the case when $\lambda = 1$. For all $p > 0$, we have that $\sum_{i=1}^n x^p - \int_1^n x^p dx \leq O(n^p)$, so s_ℓ is well-defined (it should be clear that s_ℓ is well-defined if $d = 0$). Moreover, it quickly follows from Rothblum's theorem that

$$\lim_{n\to\infty} \frac{1}{n^{d+1}} \sum_{i=1}^{n} A^i = \sum_{i=0}^{q-1} s_\ell.$$

Finally, suppose that $\lambda > 1$. Let $\epsilon > 0$ be an arbitrary number. Let $N = qn + k$, $N_* \approx q(n - \log^2(n))$ such that $N_* \equiv k(mod\ q)$, and $n_* = (N_* - k)/q$. For a matrix M, let $\|M\|_e$ denote the maximum magnitude among the entries of M. Notice that the following terms converge to zero:

- $\|\lambda^{-N} \sum_{i=1}^{N_*} A^i\|_e \leq O\left(n\lambda^{-\log^2(n)}\right) \leq O\left(n^{-1}\right)$.
- For all ℓ and $n' \geq n_*$, we have that $\|S_\ell(n') - S_\ell(n)\|_e / \|S_\ell(n)\|_e \leq O\left(\frac{\log^2(n)}{n}\right)$.
- For all ℓ and $n' \geq n_*$, Rothblum's theorem states that $\|\left(A/\lambda\right)^{qn'+\ell} - S_\ell(n')\|_e$ converges to 0 exponentially.

Let $\delta > 0$ be a number such that $\delta\left(2q\frac{1}{1-\lambda^{-q}} + 1\right) < \epsilon$. Let n be large enough such that each of the following terms is less than δ:

- $\|\lambda^{-N} \sum_{i=1}^{N_*} A^i\|_e$,
- $n^{-d}\|S_\ell(n') - S_\ell(n)\|_e$ for all $0 \leq \ell < q$, $n' > n_*$, and
- $n^{-d}\|\left(A/\lambda\right)^{qn'+\ell} - S_\ell(n')\|_e$ for all $0 \leq \ell < q$, $n' > n_*$.

By the triangle inequality, for all $0 \le \ell < q$, $n' > n_*$, we have that $n^{-d}\| (A/\lambda)^{qn'+\ell} - S_\ell(n)\|_e < 2\delta$. In the following, let all indices of $S_\ell(x)$ be taken modulo q. Because $\sum_{i=N_*+1}^{N} A^i = \sum_{\ell=k-q+1}^{k} \sum_{n'=n_*}^{n} A^{qn'+\ell}$, we have that

$$n^{-d}\left\| \sum_{\ell=k-q+1}^{k} \left(\sum_{n'=n_*}^{n} A^{qn'+\ell} - \lambda^{qn'+\ell} S_\ell(n) \right) \right\| < q \sum_{n'=n_*}^{n} 2\delta\lambda^{q(n'+1)}.$$

Therefore,

$$\begin{aligned}\lambda^{-N} n^{-d}\left\| \sum_{i=N_*+1}^{N} A^i - \sum_{\ell=k-q+1}^{k} \sum_{n'=n_*}^{n} \lambda^{qn'+\ell} S_\ell(n) \right\| &< q \sum_{n'=n_*}^{n} 2\delta\lambda^{q(n'-n)} \\ &\le 2q\delta \frac{1}{1-\lambda^{-q}}.\end{aligned}$$

By the triangle inequality,

$$\begin{aligned}&\left\| \lambda^{-N} n^{-d} \sum_{i=1}^{N} A^i - \sum_{n'=n_*}^{n} \sum_{\ell=k-q+1}^{k} \lambda^{qn'+\ell-N} n^{-d} S_\ell(n) \right\| \\ &\le \lambda^{-N} n^{-d} \left\| \sum_{i=N_*+1}^{N} A^i - \sum_{n'=n_*}^{n} \sum_{\ell=k-q+1}^{k} \lambda^{qn'+\ell} S_\ell(n) \right\| + \left\| n^{-d}\lambda^{-N} \sum_{i=1}^{N_*} A^i \right\|,\end{aligned}$$

which is less than $\delta\left(2q\frac{1}{1-\lambda^{-q}} + 1\right) < \epsilon$. In conclusion,

$$\begin{aligned}\lim_{n\to\infty} n^{-d}\lambda^{-N} \sum_{i=1}^{N} A^i &= \lim_{n\to\infty} \sum_{n'=n_*}^{n} \sum_{\ell=k-q+1}^{k} \lambda^{qn'+\ell-N} n^{-d} S_\ell(n) \\ &= \sum_{\ell=k-q+1}^{k} \lambda^{\ell-k} \lim_{n\to\infty} n^{-d} S_\ell(n) \sum_{n'=n_*}^{n} \lambda^{q(n'-n)} \\ &= \sum_{\ell=k-q+1}^{k} \lambda^{\ell-k} \frac{t_\ell}{1-\lambda^{-q}}.\end{aligned}$$

Note that we can separate the above limits because of Rothblum's Theorem. □

Perhaps the path forward is becoming clear to the reader. Theorem 5.12 is an analogue of Rothblum's theorem for a geometric series of matrices. Repeating the path from before, we must now develop an analogue of Lemma 5.9.

LEMMA 5.13. *Let L be an infinite regular language represented by trimmed DFA $(Q, \mathcal{A}, \delta, q_0, F)$. We use notation i, A, f as before so that $W_m(L) = iA^m f$. Let q and $S_0(x), \ldots, S_{q-1}(x)$ be the integer and polynomials determined by Rothblum's theorem applied to A. Let λ be the largest eigenvalue of A. Let d be the largest degree among the polynomials $iS_j(x)f$ ranging over $j \in [0, q)$. Let $s_\ell = \lim_{n\to\infty} n^{-(d+1)} \sum_{i=1}^{n} S_\ell(i)$ and $t_\ell = \lim_{n\to\infty} n^{-d} S_\ell(n)$.*

(1) *If $\lambda = 1$, then*

$$\lim_{n\to\infty} \frac{1}{n^{d+1}} \sum_{i=1}^{n} W_i(L) = \sum_{i=0}^{q-1} s_\ell > 0.$$

(2) *If* $\lambda > 1$, *then*

$$\lim_{n\to\infty} \frac{1}{(qn+k)^d} \lambda^{-(qn+k)} \sum_{i=1}^{qn+k} W_i(L) = \frac{1}{1-\lambda^{-q}} \sum_{\ell=k-q+1}^{k} \lambda^{\ell-k} t_\ell > 0$$

where the indices of the t_i *are taken modulo* q.

PROOF. The proof is exactly that of Theorem 5.12; we only need to prove that the limits are non-zero. By Lemma 5.9, there exists a j such that $iS_j(x)f$ is a non-zero polynomial. Therefore there exists a j' such that $s_{j'} > 0$ and $t_{j'} > 0$. Because these terms represent the limiting behavior of a counting function, all values are non-negative. And so we see that $\sum_{i=0}^{q-1} s_\ell \geq s_{j'} > 0$ when $\lambda = 1$, and when $\lambda > 1$ we have that for any k the inequality $\sum_{\ell=k-q+1}^{k} \lambda^{\ell-k} t_\ell \geq \lambda^{-k} t_{q'} > 0$. □

We are now prepared to prove Theorem 1.1.

PROOF TO THEOREM 1.1. The proof is clear if $L_1 \triangle L_2$ is finite.

Consider the DFA for regular languages $L_1 \cup L_2$ and $L_1 \triangle L_2$. As it costs nothing to trim a DFA, we assume that both DFA are trimmed. We use notation $A_\triangle, A_\cup, i_\triangle, i_\cup, f_\triangle, f_\cup$ as before. Also, let $\rho_\triangle, \rho_\cup$ be the spectral radius of $A_\triangle, A_\cup$ as before. For simplicity we will assume that $\rho_\triangle, \rho_\cup > 1$ as the other cases follow identically. We apply Lemma 5.13 to $L_1 \triangle L_2$, and let $d_\triangle, q_\triangle, t_{\triangle,0}, \ldots, t_{\triangle,q_\triangle-1}$ be the notation for the resulting values. We also apply Lemma 5.13 to $L_1 \cup L_2$ and use the corresponding notation. Let q be the least common multiple of $q_\triangle$ and $q_\cup$. We can apply the conclusions of Lemma 5.13 about converging limits to each $0 \leq k < q$ by considering $k (\text{mod } q_\triangle)$ or $k (\text{mod } q_\cup)$ as appropriate.

Because $L_1 \triangle L_2 \subseteq L_1 \cup L_2$, we have that $\rho_\cup \geq \rho_\triangle$. Moreover, if $\rho_\cup = \rho_\triangle$, then $d_\cup \geq d_\triangle$. This implies that for each $k \in [0, q)$ we have that

$$\lim_{m\to\infty} \frac{1}{(qm+k)^{d_\cup}} \rho_\cup^{-(qm+k)} \sum_{i=1}^{qm+k} i_\triangle A_\triangle^i f_\triangle$$

converges because it converges to 0 if $\rho_\cup \neq \rho_\triangle$ or $d_\cup \neq d_\triangle$. Because both the top and bottom converge, the limit

$$\lim_{m\to\infty} \frac{(qm+k)^{-d_\cup} \rho_\cup^{-(qm+k)} \sum_{i=1}^{qm+k} i_\triangle A_\triangle^i f_\triangle}{(qm+k)^{-d_\cup} \rho_\cup^{-(qm+k)} \sum_{i=1}^{qm+k} i_\triangle A_\cup^i f_\triangle} = \lim_{m\to\infty} J_{qm+k}(L_1, L_2)$$

is well defined for each k. Notice that the above limit approaches a periodic sequence. Therefore the Cesáro average exists.

□

Note that in $J'_C(L_1, L_2)$ each congruence class k is handled independently and the final answer is the average of such results. On the other hand, in $J_C(L_1, L_2)$ each congruence class k has a limit that is a combination of results from all of the congruence classes. Thus, the total answer is dominated by the overall asymptotic behavior and not just small periodic undercurrents. This point is the key idea in the proof to Lemma 5.13 and the non-zero result for all congruence classes, while Lemma 5.13 only concludes that a single congruence class is non-zero. We illustrate the importance of this distinction via example.

EXAMPLE 5.14. Let $L_1 = ((a|b)^2)^*|c^*$ and $L_2 = ((a|b)^2)^*|d^*$. The languages L_1 and L_2 have $((a|b)^2)^*$ in common and so mutually shared words up to length n grow exponentially. The languages disagree on c^* and d^*, whose words only grow polynomially. Hence, L_1 and L_2 are very similar and should have a small distance. However, J'_C gives equal weight to words of even length and odd length, even though the languages are mostly made up of even-length words.

Rigorously, we have that $\lim_{n\to\infty} J_{2n}(L_1, L_2) = 0$ and $\lim_{n\to\infty} J'_{2n}(L_1, L_2) = 0$. Furthermore, $\lim_{n\to\infty} J_{2n+1}(L_1, L_2) = 0$ and $\lim_{n\to\infty} J'_{2n+1}(L_1, L_2) = 1$. Thus, $J_C(L_1, L_2) = 0$, while $J'_C(L_1, L_2) = \frac{1}{2}$.

We conclude this section with a fact about the Cesáro Jaccard distance.

FACT 5.15. *The Cesàro Jaccard distance inherits the pseudo-metric property from* J_n.

5.3. Relationship between Entropy and Cesáro Jaccard. In Section 5.2 we proved that the Cesàro Jaccard distance is well-defined. As you will see, Cesáro Jaccard and entropy are mostly disjoint in what they measure. Recall Theorem 1.4 from the introduction.

THEOREM 1.4 *Let* L_1, L_2 *be two regular languages.*

(1) *If* $h(L_1 \triangle L_2) \neq h(L_1 \cup L_2)$, *then* $J_C(L_1, L_2) = 0$.
(2) *If* $h(L_1 \cap L_2) \neq h(L_1 \cup L_2)$, *then* $J_C(L_1, L_2) = 1$.
(3) *If* $0 < J_C(L_1, L_2) < 1$, *then the following equal each other:* $h(L_1)$, $h(L_2)$, $h(L_1 \cap L_2)$, $h(L_1 \triangle L_2)$, $h(L_1 \cup L_2)$.

PROOF. Part (1) easily follows from Lemma 3.3.

To see part (2), note that $(L_1 \cup L_2) \cap \overline{(L_1 \cap L_2)} = L_1 \triangle L_2$. It suffices to show that $\lim_{n\to\infty} J_n(L_1, L_2) = 1$. Let $\epsilon < \frac{h(L_1 \cup L_2) - h(L_1 \cap L_2)}{2}$. Then there exists N large enough so that for all $n \geq N$ we have

$$2^{n(h(L)-\epsilon)} \leq |W_{\leq n}(L)| \leq 2^{n(h(L)+\epsilon)}$$

by Lemma 3.3. Let $n \geq N$. Then,

$$\begin{aligned}
\frac{|W_{\leq n}(L_1 \triangle L_2)|}{|W_{\leq n}(L_1 \cup L_2)|} &= \frac{|W_{\leq n}(L_1 \cup L_2)| - |W_{\leq n}(L_1 \cap L_2)|}{|W_{\leq n}(L_1 \cup L_2)|} \\
&= 1 - \frac{|W_{\leq n}(L_1 \cap L_2)|}{|W_{\leq n}(L_1 \cup L_2)|} \\
&\geq 1 - \frac{2^{n(h(L_1 \cap L_2)+\epsilon)}}{2^{n(h(L_1 \cup L_2)-\epsilon)}} \\
&= 1 - 2^{n(h(L_1 \cap L_2) - h(L_1 \cup L_2) + 2\epsilon)}.
\end{aligned}$$

Since $h(L_1 \cap L_2) - h(L_1 \cup L_2) + 2\epsilon < 0$,

$$\lim_{n\to\infty} \frac{|W_{\leq n}(L_1 \triangle L_2)|}{|W_{\leq n}(L_1 \cup L_2)|} \geq 1.$$

As the above limit is always bound above by 1, we have that the limit is equal to 1.

For part (3), the above already implies that if $0 < J_C(L_1, L_2) < 1$, then $h(L_1 \cap L_2)$, $h(L_1 \triangle L_2)$, and $h(L_1 \cup L_2)$ are equal. By symmetry, assume that $h(L_1) \leq h(L_2)$. Because $L_1 \cap L_2 \subseteq L_1$ and $L_2 \subseteq L_1 \cup L_2$, by Lemma 3.4 we have that $h(L_1 \cap L_2) \leq h(L_1) \leq h(L_2) \leq h(L_1 \cup L_2)$. Therefore all five terms are equal. $\square$

To better understand this theorem, consider the following examples corresponding to the three cases of the theorem.

(1): Let $L_1 = ((a|b)^2)^*|c^*$ and $L_2 = ((a|b)^2)^*|d^*$ as in Example 5.14. The union of these two sets is $((a|b)^2)^*|(c^*|d^*)$ and has entropy $\log(2)$, while the symmetric dif ferent of these two sets is $c^*|d^*$ and has entropy 0. This implies that $J_C(L_1, L_2) = 0$.

(2): Let $L_1 = (a|b)^*|c^*$ and $L_2 = (d|e)^*|c^*$. The union of these two sets is $(a|b)^*|(d|e)^*|c^*$ and has entropy $\log(2)$, while the intersection of these two sets is c^* and has entropy 0. This implies that $J_C(L_1, L_2) = 1$.

(3): Let $L_1 = (aa)^*$ and $L_2 = a^*$ as in the introduction. The union of these two sets is a^* and $|W_{\leq n}(L_1 \cup L_2)| = n + 1$, while the symmetric difference of these two sets is $a(aa)^*$ and $|W_{\leq n}(L_1 \triangle L_2)| = \lceil \frac{n}{2} \rceil$. Now notice that

$$\lim_{n\to\infty} J_n(L_1, L_2) = \lim_{n\to\infty} \frac{|W_{\leq n}(L_1 \triangle L_2)|}{|W_{\leq n}(L_1 \cup L_2)|} = \frac{1}{2}.$$

Thus, $J_C(L_1, L_2) = \frac{1}{2}$. Also, note that the entropy of all associated languages is 0.

6. Conclusion

This paper has covered some issues related to the entropy of regular languages and the distance between regular languages.

We considered distances between regular languages based in Jaccard distance and based on entropy. We created new formulations of these distances that resolved issues in prior formulations with non-converging sequences. We also constructed examples showing that our formulations better match an intuitive concept of "closeness" and contain desirable qualities such as granularity (although this only exists under certain conditions). We also compared our distance functions against each other, and we determined that entropy-based distance functions and Jaccard-based distance functions are fundamentally different.

In this paper several formulations of entropy are developed, and it is natural to consider which would be the best to use. In a practical sense it does not matter since all formulations are equivalent (Theorem 3.2) and can be computed using Shannon's determinant-based method [**Sha48**]. However, conceptually, it can be argued that $\lim_{n\to\infty} \frac{log|W_{\leq n}(L)|}{n}$ is the preferable formulation. First, there is a notational argument that prefers using limits that exist. This is a limit that exists (Theorem 1.3), whereas many other limit formulations do not. Second, this limit captures more readily the concept of "number of bits per symbol" that Shannon intended. Because regular languages can have strings with staggered lengths, using W_n forces the consideration of possibly empty sets of strings of a given length. This creates dissonance when the language has non-zero entropy. Instead, the monotonically growing $W_{\leq n}$ more clearly encodes the intuition that the formulation is expressing the number of bits needed to express the next symbol among all words in the language.

Apart from expanding to consider context-free languages and other languages ([**CDFI14**]), one investigation that is absent from this paper is the determination of similarity between languages that are disjoint but obviously similar (i.e. aa^* and ba^*). A framework for addressing such problems is provided in [**CDFI13**], but finding metrics capturing such similarities can be fodder for future efforts.

References

[BGKM02] F. Blanchard, E. Glasner, S. Kolyada, and A. Maass, *On Li-Yorke pairs*, J. Reine Angew. Math. **547** (2002), 51–68, DOI 10.1515/crll.2002.053. MR1900136

[BGvOS04] M. Bodirsky, T. Gärtner, T. von Oertzen, and J. Schwinghammer, *Efficiently computing the density of regular languages*, LATIN 2004: Theoretical informatics, Lecture Notes in Comput. Sci., vol. 2976, Springer, Berlin, 2004, pp. 262–270, DOI 10.1007/978-3-540-24698-5_30. MR2095201

[BP97] M.-P. Béal and D. Perrin, *Symbolic dynamics and finite automata*, Handbook of Formal Languages, Vol. 2, Springer, Berlin, 1997, pp. 463–505. MR1470015

[CDFI13] C. Cui, Z. Dang, T. R. Fischer, and O. H. Ibarra, *Similarity in languages and programs*, Theoret. Comput. Sci. **498** (2013), 58–75, DOI 10.1016/j.tcs.2013.05.040. MR3083514

[CDFI14] C. Cui, Z. Dang, T. R. Fischer, and O. H. Ibarra, *Information rate of some classes of non-regular languages: an automata-theoretic approach (extended abstract)*, Mathematical foundations of computer science 2014. Part I, Lecture Notes in Comput. Sci., vol. 8634, Springer, Heidelberg, 2014, pp. 232–243, DOI 10.1007/978-3-662-44522-8_20. MR3253057

[CGR03] C. Chan, M. Garofalakis, and R. Rastogi, *Re-tree: an efficient index structure for regular expressions*, The VLDB Journal's The International Journal on Very Large Data Bases **12** (2003), no. 2, 102 – 119.

[CM58] N. Chomsky and G. A. Miller, *Finite state languages*, Information and Control **1** (1958), 91–112. MR0108417

[CMR06] C. Cortes, M. Mohri, and A. Rastogi, *On the computation of some standard distances between probabilistic automata*, Implementation and application of automata, Lecture Notes in Comput. Sci., vol. 4094, Springer, Berlin, 2006, pp. 137–149, DOI 10.1007/11812128_14. MR2296453

[CSMS03] T. Ceccherini-Silberstein, A. Machi, and F. Scarabotti, *On the entropy of regular languages*, Theoret. Comput. Sci. **307** (2003), no. 1, 93–102, DOI 10.1016/S0304-3975(03)00094-X. Words. MR2022842

[DMV09] J. Dassow, G. M. Martín, and F. J. Vico, *A similarity measure for cyclic unary regular languages*, Fund. Inform. **96** (2009), no. 1-2, 71–88. MR2588322

[Fra61] J. G. F. Francis, *The QR transformation: a unitary analogue to the LR transformation. I*, Comput. J. **4** (1961/1962), 265–271, DOI 10.1093/comjnl/4.3.265. MR0130111

[HK03] B. Hasselblatt and A. Katok, *A first course in dynamics*, Cambridge University Press, New York, 2003. With a panorama of recent developments. MR1995704

[HU79] J. E. Hopcroft and J. D. Ullman, *Introduction to automata theory, languages, and computation*, Addison-Wesley Series in Computer Science, Addison-Wesley Publishing Co., Reading, Mass., 1979. MR645539

[Kol59] A. N. Kolmogorov, *Entropy per unit time as a metric invariant of automorphisms* (Russian), Dokl. Akad. Nauk SSSR **124** (1959), 754–755. MR0103255

[Koz05] J. Kozik, *Conditional densities of regular languages*, Proceedings of the Second Workshop on Computational Logic and Applications (CLA 2004), Electron. Notes Theor. Comput. Sci., vol. 140, Elsevier Sci. B. V., Amsterdam, 2005, pp. 67–79, DOI 10.1016/j.entcs.2005.06.023. MR2208477

[Kui70] W. Kuich, *On the entropy of context-free languages*, Information and Control **16** (1970), 173–200. MR0269447

[LM95] D. Lind and B. Marcus, *An introduction to symbolic dynamics and coding*, Cambridge University Press, Cambridge, 1995. MR1369092

[MU16] J. Marklof and C. Ulcigrai, *Lecture notes for dynamical systems and ergodic theory*, 2015-2016, `http://www.maths.bris.ac.uk/~majm/DSET/index.html`

[NS08] M.-J. Nederhof and G. Satta, *Computation of distances for regular and context-free probabilistic languages*, Theoret. Comput. Sci. **395** (2008), no. 2-3, 235–254, DOI 10.1016/j.tcs.2008.01.010. MR2424510

[PYY17] A. J. Parker, K. B. Yancey, and M. P. Yancey, *Regular language distance and entropy*, 42nd International Symposium on Mathematical Foundations of Computer Science, LIPIcs. Leibniz Int. Proc. Inform., vol. 83, Schloss Dagstuhl. Leibniz-Zent. Inform., Wadern, 2017, pp. Art. No. 3, 14. MR3755296

[Rot81] U. G. Rothblum, *Expansions of sums of matrix powers*, SIAM Rev. **23** (1981), no. 2, 143–164, DOI 10.1137/1023036. MR618637

[Rot07] U. G. Rothblum, *Chapter 9, nonnegative matrices and stochastic matrices*, Handbook of Linear Algebra, (eds: L. Hogben), Chapman and Hall / CRC, 2007.

[Sha48] C. E. Shannon, *A mathematical theory of communication*, Bell System Tech. J. **27** (1948), 379–423, 623–656, DOI 10.1002/j.1538-7305.1948.tb01338.x. MR0026286

[Sin59] Ja. Sinaĭ, *On the concept of entropy for a dynamic system* (Russian), Dokl. Akad. Nauk SSSR **124** (1959), 768–771. MR0103256

[SS78] A. Salomaa and M. Soittola, *Automata-theoretic aspects of formal power series*, Texts and Monographs in Computer Science, Springer-Verlag, New York-Heidelberg, 1978. MR0483721

[Sta97] R. P. Stanley, *Enumerative combinatorics. Vol. I*, The Wadsworth & Brooks/Cole Mathematics Series, Wadsworth & Brooks/Cole Advanced Books & Software, Monterey, CA, 1986. With a foreword by Gian-Carlo Rota. MR847717

[Wei73] B. Weiss, *Subshifts of finite type and sofic systems*, Monatsh. Math. **77** (1973), 462–474, DOI 10.1007/BF01295322. MR0340556

Institute for Defense Analyses - Center for Computing Sciences, Bowie MD, USA
Email address: `ajpark2@super.org`

Institute for Defense Analyses - Center for Computing Sciences, Bowie MD, USA
Email address: `kbyancey1@gmail.com`

Institute for Defense Analyses - Center for Computing Sciences, Bowie MD, USA
Email address: `mpyancey1@gmail.com`

Contemporary Mathematics
Volume **736**, 2019
https://doi.org/10.1090/conm/736/14832

Good and Bad Functions for Bad Processes

Andrew Parrish and Joseph Rosenblatt

ABSTRACT. Let (X, β, m) be a standard, non-atomic probability space. Consider stochastic processes $(T_n^{\mathcal{R}})$ that structurally depend on $\mathcal{R}$, a representation of a group G as measure-preserving maps in $\mathcal{M}$, the invertible, measure-preserving mappings of (X, β, m). We focus on stochastic processes that are bad in the sense that $T_n^{\mathcal{R}} f$ is generally not convergent in norm and/or a.e. for the typical function $f \in L^r(X)$. The *good functions* $\mathcal{G}^{\mathcal{R}}$ are the functions for which there is convergence. The *bad functions* $\mathcal{B}^{\mathcal{R}}$ are the functions for which there is not convergence. We study the nature and size of both $\mathcal{G}^{\mathcal{R}}$ and $\mathcal{B}^{\mathcal{R}}$, for particular processes and a fixed representation $\mathcal{R}$, and for these classes as $\mathcal{R}$ varies.

1. Introduction

Let (X, β, m) be a standard, non-atomic probability space. We let $\mathcal{M}$ be the group of invertible, measure-preserving mappings τ of (X, β, m). In this article, we focus on the nature of the good and bad functions for bad stochastic processes $T_n^{\mathcal{R}}$ that are defined in terms of a representation in $\mathcal{M}$ of a group G. These will depend on the structure of (T_n) and the nature of the representation $\mathcal{R}$.

While we do give some examples of what we have in mind by considering some classical examples from harmonic analysis, the main focus here will be on two cases. In Section 2, we take the simplest case where $G = \mathbb{Z}$, so we are actually just considering a particular map $\tau = \mathcal{R}(1)$. That is, we are considering the behavior of stochastic process T_n^τ that depend on the choice of an ergodic map $\tau \in \mathcal{M}$. In Section 3, we take more general discrete groups G and focus on examples where the stochastic process is uniformly bounded in norm, but may or may not converge a.e. for the Lebesgue class $L^r(X)$.

1.1. A Classical Example. Here is an example from classical harmonic analysis of the issues that we consider here, although we consider these issues in the context of dynamical systems. Consider the partial sums $S_n f$ of a Fourier series for a function $f \in L^1(0,1)$. The classical norm result is that $S_n f$ converges in L^r-norm for $f \in L^r(0,1)$ with $1 < r < \infty$. There is no such result in $L^1(0,1)$. Indeed, because of the growth of the L^1-norm of the Dirichlet kernel. The Uniform Boundedness Principle says that for a dense G_δ set of functions $f \in L^1(0,1)$, we have $\limsup\limits_{n\to\infty} \|S_n f\|_1 = \infty$. So there are dense sets of good functions and bad functions with respect to the norm behavior in $L^1(0,1)$.

2010 *Mathematics Subject Classification.* Primary 22D40, 37A15, 28D15.

Questions remain nonetheless. For example, what functions $f \in L^1(0,1)$ are good for L^1-norm convergence? Also, what can be said about the functions $f \in L^1(0,1)$ such that $\|S_n f\|_1 \to \infty$ as $n \to \infty$? Indeed, when can we have $\|S_n f\|_1$ increasing to ∞ and $n \to \infty$? Do functions with this property exist? If so, how do we describe them?

On the other hand, while there were many results on the pointwise convergence of the partial sums of the Fourier series, there was no obvious general result even for continuous functions. The conjecture of N. Lusin was that there was at least a.e. convergence in the case of continuous functions. This was settled by the famous theorem of L. Carleson published in 1966: the partial sums $S_n f$ converge a.e. for any $f \in L^2(0,1)$. This was extended by R. Hunt to any function $f \in L^r(0,1)$ with $1 < r < \infty$. However, it was also a classical theorem of A. Kolmogorov that a.e. convergence fails for the general function $f \in L^1(0,1)$. This sets up a classical good function versus bad function scenario with respect to a.e. convergence for the partial sums of the Fourier series.

There are many questions that can be asked about pointwise convergence of the Fourier series/ What functions are good for pointwise a.e. convergence? The good and bad functions for a.e. convergence are certainly both dense in L^1-norm. But the bad functions are a dense G_δ class in both cases. The good and bad functions are certainly Borel sets. However, past this, there does not seem to be any known way to describe the nature of the function $f \in L^1(0,1)$ which guarantees it is good for a.e. convergence of the Fourier series. Also, interestingly enough, while it is a fairly easy classical fact that for $f \in L^1(0,1)$, $|S_n f(x)| = O(\ln n)$ for a.e. x, it is unknown if this is really the best possible pointwise upper bound.

2. Processes Depending on a Map

We are interested here in the case where the representation $\mathcal{R}$ is for the group $G = \mathbb{Z}$. Since $\mathcal{R}$ is completely determined by $\tau = \mathcal{R}(1)$, we frame our results and question in terms of τ itself. As a cornerstone here, we have the norm and pointwise ergodic theorems that describe the behavior of the usual averages $\frac{1}{n}\sum_{k=1}^{n} f \circ \tau^k$. Here the usual restriction for good behavior is $f \in L^1(X)$ when dealing with pointwise a.e. convergence, and $f \in L^r(X), 1 \le r < \infty$, when dealing with L^r-norm convergence.

2.1. Harmonic Series.

2.1. Harmonic Series. Consider the series $\sum_{j=1}^{\infty} \frac{1}{j} f \circ \tau^j$ where $\tau \in \mathcal{M}$ and $f \in L^0(X)$. This series might converge in norm and/or a.e., but that will depend on τ and f in some fashion. So consider the particular stochastic process $(S_n^\tau f)$ given by $S_n^\tau f = \sum_{j=1}^{n} \frac{1}{j} f \circ \tau^j$. We have $S_n^\tau : L^r(X) \to L^r(X)$ for all $0 \le r \le \infty$. Given that τ is ergodic, one can easily use the Rokhlin Lemma to show that the operator norm of S_n^τ on the mean-zero functions in $L^r(X)$ is $\sum_{j=1}^{n} \frac{1}{j}$. So by the Banach-Steinhaus Theorem, $\limsup_{n\to\infty} \|S_n f\|_r = \infty$ for a dense G_δ set of mean-zero functions in $L^r(X)$. This is true for all r, $1 \le r \le \infty$. Hence, for a fixed τ, it is not clear when $(S_n^\tau f)$ converges in $L^r(X)$-norm.

But even more so, it is not clear when it converges a.e. either. Recall this particular instance of Kronecker's Lemma: if $\sum\limits_{j=1}^{\infty} \frac{1}{j} a_j$ converges, then $\frac{1}{n}\sum\limits_{j=1}^{n} a_j \to 0$ as $n \to \infty$. So if f is integrable and mean-zero, and $S_n^\tau f$ converges, then $\frac{1}{n}\sum\limits_{j=1}^{n} f\circ\tau^j$ converges to 0 as $n \to \infty$. This is the conclusion of the Pointwise Ergodic Theorem if τ is ergodic. The naive, but interesting idea, is that using this form of Kronecker's Lemma could be an approach to proving the Ergodic Theorem for all mean-zero $f \in L^1(X)$. However, this does not work. Indeed, for any $\tau \in \mathcal{M}$, there is at least a dense $\mathcal{G}_\delta$ set of functions f, for which one has $\limsup\limits_{n\to\infty} |S_n^\tau f| = \infty$ a.e. Actually, one can prove that this particular stochastic process is bad in norm and a.e. for a dense $\mathcal{G}_\delta$ set of functions in $L^r(X)$ for any $r, 1 \le r \le \infty$. See del Junco and Rosenblatt [**17**].

However, given τ, for some mean-zero functions $f \in L^1(X)$, the sequence $(S_n^\tau f)$ is well-behaved. For example, if τ is a Bernoulli mapping, then there are mean-zero functions $f \in L^1(X)$ such that $(f\circ\tau^j : j \ge 1)$ is an IID sequence. Hence, at least if $f \in L^2(X)$, by Kolmogorov's Three Series Theorem, $(S_n^\tau f)$ converges a.e. and in L^2-norm. Note that by a result due to Sinai (see [**28**] and [**29**]) this example may be extended to any system with positive entropy.

As another example, suppose $f = F - F\circ\tau$ for some $F \in L^1(X)$. We say that f is a τ-coboundary with an integrable transfer function F. There are a number of interesting issues connected with τ-coboundaries that deal with what Lebesgue class f is in and then what Lebesgue class F can be in if we allow τ to vary. See Adams and Rosenblatt [**1, 2**]. In any case, we have this result.

PROPOSITION 2.1. *Given any map τ, consider $f = F - F\circ\tau$ for some $F \in L^1(X)$. Then $S_n^\tau f$ converges a.e.*

PROOF. We compute

$$S_n^\tau f = F\circ\tau - \frac{1}{n}F\circ\tau^{n+1} + \sum_{j=2}^{n}\left(\frac{1}{j} - \frac{1}{j-1}\right)F\circ\tau^j.$$

Hence, because a.e. $\frac{1}{n}F\circ\tau^{n+1} \to 0$ as $n \to \infty$, and because $\sum\limits_{j=1}^{\infty}\frac{1}{j^2}|F\circ\tau^j|$ is integrable, we have $(S_n^\tau f)$ converges a.e. and in L^1-norm for all such τ-coboundaries f. □

These two examples, one specific to the nature of τ and f and the other general (for any ergodic τ), but specific to the nature of f, show that describing the good function $\mathcal{G}^\tau$ may be difficult in general. We do know that at least for ergodic maps τ, the good functions always comprise a dense set in the mean-zero functions, even though it is meager. This is simply because the τ-coboundaries are dense.

But a fundamental question is what happens if we allow τ to vary.

Question: Is every mean-zero function $f \in L^1(X)$ in $\mathcal{G}^\tau$ for some suitable ergodic τ? That is, $\displaystyle\sum_{j=1}^{\infty}\frac{1}{j} f\circ\tau^j$ converges a.e.

One approach to answering this question is to ask that f be a τ-coboundary with the transfer function $F \in L^1(X)$. But there are actually serious limitations to this as described in Adams and Rosenblatt [**1**].

2.2. Coboundary Operator. We consider here a very particular example of a bad process, that behaves well exactly for coboundaries. Let $S_k^\tau f = \sum_{j=0}^{k-1} f \circ \tau^j$, the usual ergodic sum. These sums are L^r-norm bounded if f is a τ-coboundary with a transfer function in $L^r(X)$. The converse of this holds too, but the details get more difficult in $L^1(X)$ and $L^\infty(X)$ where one cannot use the weak compactness of the unit ball. See Lin and Sine [**19**] for background and complete results about these issues.

But also, now consider, as in Lin and Sine [**19**], the *coboundary operators* $T_n^\tau f = \frac{1}{n}\sum_{k=1}^{n} S_k^\tau f$. The basic result is this one.

PROPOSITION 2.2. *Fix $r, 1 \le r < \infty$ and τ is ergodic. For a mean-zero function $f \in L^r(X)$, $\|T_n^\tau\|_r$ is bounded if and only if f is a τ-coboundary with transfer function in $L^r(X)$. Indeed, there is a dichotomy: either*

(1) $\|T_n^\tau f\|_r$ *converges to* ∞*, or*
(2) $f = F - F \circ \tau$ *for a mean-zero* $F \in L^r(X)$*, and* $T_n^\tau f \to F$ *in* $L^r(X)$*-norm.*

PROOF. Because it is so basic to the issues in this article, and for the convenience of the reader, let us prove that $T_n^\tau f$ converges in L^r-norm for some subsequence of values n if and only if the full sequence converges to a mean-zero $F \in L^r(X)$ with $f = F - F \circ \tau$. Indeed, if $f = F - F \circ \tau$, then $T_n^\tau f = \frac{1}{n}\sum_{k=1}^{n} F - F \circ \tau^k$. So if $F \in L^r(X)$ is mean-zero, then $T_n^\tau f$ converges to F in $L^r(X)$ -norm, as $n \to \infty$. On the other hand,

$$\begin{aligned} T_n^\tau f - T_n^\tau f \circ \tau &= \frac{1}{n}\sum_{k=1}^{n}(f - f \circ \tau^k) \\ &= f - \frac{1}{n}\sum_{k=1}^{n} f \circ \tau^k. \end{aligned}$$

But when f is mean-zero, $\frac{1}{n}\sum_{k=1}^{n} f \circ \tau^k$ converges in L^r-norm to zero. Hence, if $T_{n_m}^\tau f$ converges in L^r-norm to F, for some subsequence (n_m), then F is mean-zero and $F - F \circ \tau = f$.

To finish the proof, one needs to argue that the convergence in L^r-norm follows just from some subsequence being bounded in L^r-norm, in the reflexive cases $L^r(X), 1 < r < \infty$, and in the non-reflexive endpoint cases $L^1(X)$. This is of less direct importance in this article, so we refer the reader to Lin and Sine [**19**]. □

This is a very interesting, and structured, example of a bad process. Indeed, the operator norm of T_n^τ on $L^r(X)$ equals the obvious bound $\frac{1}{n}\sum_{k=1}^{n} k = \frac{n+1}{2}$. This is certainly clear if one allows constant functions, but it is even the case using the Rokhlin Lemma for ergodic maps τ if one restricts the operator to the mean-zero functions $f \in L^r(X)$. So by the Uniform Boundedness Principle, for a dense G_δ set of functions $f \in L^r(X)$, we have $\limsup_{n\to\infty} \|T_n^\tau f\|_r = \infty$.

REMARK 2.3. If we make the operator T_n^τ symmetric, then a corresponding result is not at all clear. Indeed, suppose instead of S_k^τ we use $\mathcal{S}_k^\tau f = \sum_{j=-k}^{k} f \circ \tau^j$. Then what is clear is that the corresponding $\mathcal{T}_n^\tau f = \frac{1}{n} \sum_{k=1}^{n} \mathcal{S}_k^\tau f$ is certainly again unbounded in norm. But now if f is a τ-coboundary with transfer function F in $L^r(X)$ for some $r, 1 \le r \le \infty$, then $\mathcal{T}_n^\tau f$ converges in L^r-norm and a.e. to 0. But it is not clear if this is the only way convergence to zero in L^r-norm happens. It is also not by any means clear what property of $\mathcal{T}_n^\tau f$ guarantees that f is a τ-coboundary.

Consider the basic property for unbounded operators: for a dense G_δ set of functions $f \in L^r(X)$, we have $\limsup_{n\to\infty} \|T_n^\tau f\|_r = \infty$. We are interested in the fact that usually one cannot say much more about for which functions f this occurs, or equivalently what we need to know so that this does not happen. The remarkable fact is that in the case of the coboundary operator, one can give a complete analysis.

In addition, for a sequence (T_n) of continuous linear operators on a Banach space, the unboundedness of the operator norms says that for any $f \in L^r(X)$, any K (no matter how large), and any $\epsilon > 0$ (no matter how small), there are functions $g \in L^r(X)$ with $\|f - g\|_r \le \epsilon$, and infinitely many n such that $\|T_n f - T_n g\|_r \le K$. But in the case of coboundary operators, we actually have this phenomenon.

PROPOSITION 2.4. *For the coboundary operator T_n^τ, we have the following specific divergence properties.*
a) For any $f \in L^r(X)$ and for any $\epsilon > 0$, there is $g \in L^r(X)$ with $\|f - g\|_r \le \epsilon$, and still $\|T_n^\tau f - T_n^\tau g\|_r \to \infty$ as $n \to \infty$.

b) For any $\epsilon > 0$ and K, there are functions $f, g \in L^r(X)$ with $\|f - g\|_r \le \epsilon$, such that $T_n^\tau f \to F$ and $T_n g \to G$ in L^r-norm, and $\|F - G\|_r \ge K$.

PROOF. For a), if f is a τ-coboundary, one chooses g that is not, and vice versa. For b), choose F and G to differ by a function D of large L^r-norm, but such that D is τ-almost invariant. Then $f = F - F \circ \tau$ and $g = G - G \circ \tau$ are close together in norm, but F and G are far apart. □

Question: The detailed divergent behavior available in Proposition 2.4 clearly steps far beyond just the lack of a uniform norm bound. But something like this might happen in other cases too, e.g. for partial sums of Fourier series of $L^1(0,1)$ functions. Does it? What types of other information is needed for properties such as a) and b) of Proposition 2.4 to occur?

REMARK 2.5. An additional issue that is unclear for coboundary operators is this: for which f that are not τ-coboundaries with transfer function in $L^r(X)$ do we have actually $\|T_n^\tau f\|_r$ **increasing** to ∞ as $n \to \infty$?

REMARK 2.6. We are constrained in this problem by two category results. Given τ ergodic, the τ-coboundaries a meager set. This is very well known. Also, given $f \in L^r(X)$, there is only a meager set of ergodic mappings τ for which f is a τ-coboundary with a transfer function $F \in L^r(X)$. See Adams and Rosenblatt [**1**].

2.3. Various Examples. We want to give various examples of bad stochastic processes for which the class of good functions is structurally interesting or not interesting, as the case may be. These examples are all of the following type. Take a (probability) finite measure μ on $\mathbb{Z}$. Let $\mu^\tau f = \sum_{k=-\infty}^{\infty} \mu(k) f \circ \tau^k$. Then given a sequence of such probability measures (μ_n), let (μ_n^τ) be for the stochastic process that we consider. Note: these examples are generally uniformly bounded in norm, if not actually $L^1 - L^\infty$ contractions. But they may or may not be norm convergent on various Lebesgue spaces $L^r(X)$.

Of course, this is general enough to include many good stochastic processes and bad stochastic processes. Knowing when a process is good, and for which Lebesgue spaces it is good, is the starting point of our consideration of the dichotomy between good and bad functions.

Example 1: Consider the measures (μ_n) given by $\mu_n = \frac{1}{n}\sum_{k=1}^{n} \delta_k$. Then the stochastic process (μ_n^τ) is the usual one for the Ergodic Theorem. Hence, all functions in $L^r(X), 1 \leq r \leq \infty$, are good in that there is a.e. convergence and, except for $r = \infty$, there is also L^r-norm convergence. However, it is not clear what the good functions are in $L^r(X)$ with $0 \leq r < 1$. For example, given $f \in L^0(X)$, is there always an ergodic mapping τ such that the usual ergodic averages $\frac{1}{n}\sum_{k=1}^{n} f \circ \tau^k$ converges a.e.? See Buczolich [**12**] and Major [**21**] for some results about the behavior of ergodic averages when the function is allowed to be in a larger class than $L^1(X)$.

Example 2: Consider the measures (μ_n) given by $\mu_n = \frac{1}{n}\sum_{k=1}^{n} \delta_{k^2}$. Then the stochastic process (μ_n^τ) is the usual one for the Ergodic Theorem along squares. Bourgain [**9–11**] showed that all functions in $L^r(X), 1 < r \leq \infty$, are good in that there is a.e. convergence and, except for $r = \infty$, there is also L^r-norm convergence. See also Wierdl [**33**]. But later Buczolich and Mauldin [**13**] showed that these averages fail to converge a.e. for a residual class of functions in $L^1(X)$. Indeed, in the L^1-norm topology, there is a dense G_δ set of functions $f \in L^1(X)$ such that $\limsup_{n\to\infty} \frac{1}{n}|\sum_{k=1}^{n} f \circ \tau^{k^2}| = \infty$ a.e.

However, we do not know the answer to the following question.

Question: Is it true that for all $f \in L^1(X)$, there is some ergodic mapping τ for which $\frac{1}{n}\sum_{k=1}^{n} f \circ \tau^{k^2}$ converges a.e.

This type of question is pervasive for processes that depend on a map (or a group of maps). We may have a bad function for a given map that is a good function for another map. The most interesting case is when there are functions that are bad for ALL maps.

It is not clear what property of f and τ would allow this in the manner that we have already observed for the stochastic process $S_n^\tau f$ in Section 2.1. Perhaps the best general question that can be asked here is to give a characterization of the Orlicz class $L^\phi(X)$ between $L^1(X)$ and all the $L^r(X), 1 < r$, for which one has a.e.

convergence i.e. which Orlicz spaces in $L^1(X)$ contain only good functions for the averages along squares.

Example 3: Take for the measures (μ_n) the averages along a lacunary sequence. For example, let $\mu_n = \frac{1}{n}\sum_{k=1}^{n} \delta_{2^k}$. Since lacunary sequences are universally bad, for any ergodic mapping τ, and any $r, 1 \leq r \leq \infty$, there is a residual set of functions $f \in L^r(X)$ such that $\mu_n^\tau f$ fails to converge a.e. So in every $L^r(X), 1 \leq r \leq \infty$, the good functions will be a meager set. However, the good functions in this case are far from empty in general. But there is an ergodic map τ with no good functions (a construction due to T. Adams in a private communication). See Parrish and Rosenblatt [**23**] for relevant background on fully divergent stochastic processes which relates to this. We suspect the following:

Conjecture: For any lacunary sequence (m_k), there is an ergodic mapping τ such that $\frac{1}{n}\sum_{k=1}^{n} f \circ \tau^{m_k}$ fail to converge a.e. for all non-constant functions, and for all $f \in L^1(X)\backslash L^\infty(X)$, we have actually $\limsup_{n\to\infty} \frac{1}{n}|\sum_{k=1}^{n} f \circ \tau^{m_k}| = \infty$ a.e.

Example 4: Bellow [**4**] and Reinhold [**25**] have shown that there are sequences of probability measures (μ_n) which are good for some $L^r(X)$ and not for others. Their examples are actually Cesàro averages along an increasing sequence (m_k) in $\mathbb{Z}^+$; that is, $\mu_n = \frac{1}{n}\sum_{k=1}^{n} \delta_{m_k}$. Combining the results shows these facts:

a) Given $r \geq 1$, there is a sequence (μ_n) such that for all ergodic mappings τ, there is a.e. convergence to (μ_n^τ) on all $L^s(X)$ with $s > r$, and a.e. convergence fails on $L^r(X)$.

b) Given $r > 1$, there is a sequence (μ_n) such that for all ergodic mappings τ, there is a.e. convergence to (μ_n^τ) on all $L^s(X)$ with $s \geq r$, and a.e. convergence fails on all $L^s(X)$ for $s < r$.

c) There is a sequence (μ_n) such that for all ergodic mappings τ, there is a.e. convergence to (μ_n^τ) on all $L^\infty(X)$, but a.e. convergence fails on $L^r(X)$ for any $r < \infty$.

In all of these cases when a.e. convergence fails, there is the usual *generic divergence* i.e. for some dense G_δ set of functions f, we have $\limsup_{n\to\infty} |\mu_n^\tau f| = \infty$ a.e.

Basic Questions: What is the nature of the good functions at the critical index in a), or below the critical index in b) or c)? We also ask if one can get $(\mu_n^\tau f)$ converging a.e. by switching τ to some suitable ergodic mapping, depending on the choice of f?

Example 5: Suppose the sequence (μ_n) consists of probability measures. Can we have the case that for every $f \in L^1(X)$, there exists some ergodic mapping τ such that $(\mu_n^\tau f)$ converges a.e.? Clearly, the answer is negative without some further restrictions. For example, we could have taken $\mu_n = \delta_n$ for all $n \geq 1$. Then if f is not constant and τ is ergodic, $f \circ \tau^n$ fails to converge a.e. and in fact $\limsup_{n\to\infty} |f \circ \tau^n| = \infty$ a.e. whenever $f \notin L^\infty(X)$. What can be said if we assume that (μ_n) is uniformly dissipative i.e. $\sup\limits_{k\in\mathbb{Z}} \mu_n(k) \to 0$ as $n \to \infty$? Is there such a sequence for which there are no non-constant good functions in $L^r(X)$ for any ergodic τ and $r, 0 \le r \le \infty$?

Example 6: We let $\mu_n = \sum\limits_{k=1}^{n} \frac{1}{k}\delta_k$. Now our stochastic process $(\mu_n^\tau f)$ is the partial sums of the harmonic series $\sum\limits_{k=1}^{\infty} \frac{1}{k} f\circ\tau^k$ discussed in Section 1. But we could consider other series examples of this type. Fixing f and τ, is there always a divergent series $\sum\limits_{n=1}^{\infty} a_n$ with $a_n \ge 0$ for all n, such that $\sum\limits_{n=1}^{\infty} a_n f \circ \tau^n$ converges a.e.? Is there always such a divergent series $\sum\limits_{n=1}^{\infty} a_n$ such that $\sum\limits_{n=1}^{\infty} a_n f \circ \tau^n$ diverges a.e. in the sense that $\limsup\limits_{N\to\infty} |\sum\limits_{n=1}^{N} a_n f \circ \tau^n| = \infty$ a.e.?

Example 7: Suppose (μ_n) is some sequence for which a.e. convergence fails for some ergodic τ on some $L^r(X)$. Assume also that actually for some (and hence a generic class of) functions $f \in L^r(X)$, we have $\limsup\limits_{n\to\infty} |\mu_n^\tau f| = \infty$ a.e. Then the same would be true for any other ergodic map σ. One can use the Oxtoby-Ulam Theorem to argue that for the generic function $f \in L^r(X)$, there is a generic class of ergodic maps ω such that $\limsup\limits_{n\to\infty} |\mu_n^\omega f| = \infty$ a.e.. This means that for a generic class of functions $f \in L^r(X)$, we have f not a good function for (μ_n^ω) for a generic set of ergodic maps ω.

Question: Is it the case that for all mean-zero $f \in L^r(X)$, we also have f not a good function for (μ_n^ω) for a generic set of ergodic maps ω? This would mean that for all mean-zero $f \in L^r(X)$, f can only be a good function for some map ω in a meager set of options.

Example 8: When are there *sentinel functions*? This means that we are considering some class $\mathcal{S}$ of stochastic processes (μ_n) and some set $\mathcal{T}$ of ergodic maps τ. The function $f \in L^r(X)$ is a *sentinel function* for $\mathcal{S}$ and $\mathcal{T}$ when, given $(\mu_n^\tau f)$ converges a.e. for some $(\mu_n) \in \mathcal{S}$ and some $\tau \in \mathcal{T}$, we actually have $(\mu_n^\tau g)$ converges a.e. for all $g \in L^r(X)$. Sentinel functions are a different type of "good" function: one that forces good behavior for an entire Lebesgue space $L^r(X)$. Examples of this phenomenon are in Adams and Rosenblatt [**1**]. This is not a very likely event. It needs at least some maximal inequality for the stochastic processes in $\mathcal{S}$ and the maps in $\mathcal{T}$. So given this, a sentinel function is simply one that forces there to be additionally a dense class of good functions.

Example 9: We consider a sequence (t_n) of non-zero real numbers converging to 0. Let $T_n f(x) = f(t_n + x)$ for $f \in L^1(\mathbb{R})$. These operators converge in L^r-norm for all $f \in L^r(X)$ with $1 \le r < \infty$. However, they generally fail to converge a.e. Indeed, even their averages fail to converge a.e. See Bellow [**6**] and Bourgain [**8**]. If f is bounded, and bounded support, and Riemann integrable, then it is a good function because it is continuous a.e. But if f is not continuous a.e., then it is not immediately clear if it is a bad function for some (t_n). As it happens, if f is the characteristic function of a set of positive measure without interior, one may always construct a sequence (t_n) for which $T_n f(x)$ diverges a.e. We conjecture that the good functions in this instance consist of those functions that are equal a.e. to a Riemann integrable function.

2.4. Pointwise Bounds. As part of trying to understand the nature and extent of bad functions for bad processes, one can ask how badly does an average such as $\frac{1}{n}\sum_{k=1}^{n} f\circ\tau^{m_k}$ diverge a.e., when it typically does not converge a.e.? This question was addressed in Akcoglu, Jones, and Rosenblatt [**3**] where the conjecture was that if $\frac{1}{w_n}\sum_{k=1}^{n} f\circ\tau^{m_k}$ always converges to 0 a.e. for all $f\in L^1(X)$ and all ergodic maps, then it must be the case that $\sum_{n=1}^{\infty}\frac{1}{w_n}<\infty$. But in Quas and Wierdl [**24**], this was shown not to be the case by producing a controlling rate (w_n) for which $\sum_{n=1}^{\infty}\frac{1}{w_n}=\infty$. With the appropriate understanding of what the *iterated logarithm rate* means, they showed this:

THEOREM 2.7. *If $w_n = n\ln(n)\ln\ln(n)\cdots$, then $f\,\frac{1}{w_n}\sum_{k=1}^{n} f\circ\tau^{m_k}$ converges to 0 a.e. for all $f\in L^1(X)$, all measure-preserving maps, and all increasing sequences m_k.*

REMARK 2.8. They also showed that this rate is optimal in that for an ergodic map, if the weights w_n are slower than the iterated logarithm rate, then for some $f\in L^1(X)$ one actually has $\limsup_{n\to\infty}\frac{1}{w_n}|\sum_{k=1}^{n} f\circ\tau^{2^k}|=\infty$ a.e. Quas and Wierdl [**24**] also show that this remarkable result has a version when one considers only functions in $L^r(X), 1<r<\infty$. But in this case the rate is actually much simpler: now $w_n = n\ln^{1/r} n$. One can think of this as the type of result mentioned in the introduction for Fourier series, where the worst growth of partial sums for $f\in L^1(0,1)$ is $O(\ln(n))$.

REMARK 2.9. Since the other extreme is the Pointwise Ergodic Theorem itself, it is natural to ask for rate results where one has control on the powers (m_k) only in that the gap $m_{k+1}-m_k$ is bounded by some growth rate $\rho(n)$. It is interesting that as soon as a bounded gap is allowed, one can have sequences of powers that make the ergodic average a bad process. So even in this case, and more generally, it would be very interesting if there were a controlling rate $\omega=(\omega_n)$ corresponding to ρ such that $\frac{1}{\omega_n}\sum_{k=1}^{n} f\circ\tau^{m_k}\to 0$ as $n\to\infty$, for all ergodic maps τ and all $f\in L^1(X)$ as long as the gap in (m_k) is controlled by ρ. Even better would be an optimal rate. It would be very surprising if the iterated logarithm rate is actually needed as soon as one allows for a finite bounded gap in the sequence of powers.

3. Processes Depending on a Group of Maps

As observed in some of the examples in Section 2.3, a general class of processes in which there are challenges to describing the good and bad functions are ones that are uniformly bounded, but nonetheless may not converge in norm generally. There is also a strong difference here between conditions that guarantee norm convergence and ones that guarantee a.e. convergence. Since this is such a pervasive phenomenon, we will focus on the particular case of ergodic theorems for averaging operators of discrete group actions given by a representation $\mathcal{R}$. Much of what we cite, describe, and prove have adaptations for non-discrete groups too. Sometimes this change is more than a technical one. For brevity's sake, we do not consider this class of operators explicitly.

There are two general ways to consider the problem of averaging for group actions. One of these is to take a sequence of finite sets (F_n) and use these to define the operators. We denote the cardinality of a finite set F by $|F|$. We let our operators $T_n^{\mathcal{R}} f$ be defined to be $\frac{1}{|F_n|} \sum_{g \in F_n} f \circ \mathcal{R}(g)$ for $f \in L^r(X)$. This is the simplest way to average over sets. But sometimes we are forced to weight these averages to get positive results. In this case, we would have a sequence of probability measures μ_n supported on F_n, and we let $T_n f = \sum_{g \in F_n} \mu_n(g) f \circ \mathcal{R}(g)$ for all $f \in L^r(X)$.

The other basic method attempts to list the group elements in a sequential order. Now we have some fixed sequence (g_k) in G instead of a sequence of finite sets F_n. We take $T_n f = \frac{1}{n} \sum_{k=1}^{n} f \circ \mathcal{R}(g_k)$ for all $f \in L^r(X)$. Of course, we will want the sequence to (essentially) exhaust the group. This is like the first method only the sets $F_n = \{g_1, \dots, g_n\}$ increase slowly in size. This again is the simplest sequence version of averaging processes, but just as for averaging on finite sets sometimes it is necessary to weight the averages. Then we would have probability measures μ_n defined on $(g_k : 1 \leq k \leq n)$, and take $T_n f = \sum_{k=1}^{n} \mu_n(g_k) f \circ \mathcal{R}(g_k)$ for all $f \in L^r(X)$.

The first of these methods easily generalizes to continuous groups where we would replace the finite sets F_n by sets of finite measure K_n, and the summation operation over F_n by integration over K_n. There is a vast literature of these types of averaging methods, but in order to make our case for the value of studying good and bad functions we will stick to processes defined for discrete group actions.

3.1. Examples for discrete group actions. It is important to understand that having bad processes for which good and bad functions create an interesting dichotomy requires some effort. It is far from automatic even in general groups. For example, see the basic theorem in Rosenblatt [**26**]. This result uses weights on the maps in a critical fashion.

THEOREM 3.1. *In any discrete group, there are always weighted averaging methods that are good for the entire Lebesgue space $L^1(X)$ both for pointwise a.e. averaging and norm averaging.*

REMARK 3.2. The results in Rosenblatt [**26**] have easy adaptations in the non-discrete group case.

This basic result already opens many questions that we cannot answer at this time. The most basic are structural: can the sequence of weighted measures be replaced by averages of the form $\frac{1}{|F_n|} \sum_{g \in F_n} f \circ \mathcal{R}(g)$? How about replacing the sequence of weighted averages by Cesàro averages along a sequence (g_n) that enumerates G? Indeed, can this be done with one enumeration for a given $\mathcal{R}_1$, but it requires a different enumeration for a different representation $\mathcal{R}_2$? Note: the strength of the classical ergodic theorem for representations of $\mathbb{Z}$, is that only one enumeration is needed!

Beyond these basic questions are ones that focus more on good and bad functions, in the style of the theorems of Bellow [**4**] and Reinhold [**25**]. Take an explicit representation $\mathcal{R}$ of G. Can we always construct averaging methods that are good for one Lebesgue class (or set of classes) and not for another (or any other) larger one? When this is possible, the structure of the intermediate classes of good functions is not clear even in the results in [**4, 25**]. For example, one could ask if there is an averaging method for ergodic maps along a sequence of powers such that the good functions are exactly the union of all $L^r(X), 1 < r \leq \infty$ inside $L^1(X)$. The answer is negative: there is always a good function outside the union of these Lebesgue spaces. But it is not clear if there is always an entire Orlicz class of good functions that is larger than $\bigcup\limits_{r>1} L^r(X)$.

REMARK 3.3. Can we extend the results of Quas and Wierdl [**24**] discussed in Remark 2.4? It is not so clear how to phrase rate results for group actions, but here is one possible approach. Suppose we have a sequence of finite sets $F = (F_n)$ in a discrete group G. Consider the ergodic sums $S_F^{\mathcal{R}} f = \sum\limits_{g \in F_n} f \circ \mathcal{R}(g)$. What is the optimal rate w_n such that for all increasing sequences of finite sets and all representations $\mathcal{R}$, we have $\frac{1}{w_n} S_{F_n} f \to 0$ as $n \to \infty$, for any $f \in L^1(X)$? How does this rate depend on the group? What happens if we replace $L^1(X)$ by $L^r(X)$ for a fixed $r, 1 < r < \infty$?

REMARK 3.4. In line with the above general questions about group actions, here is one of the most interesting and difficult examples. There are very strong results for averaging of actions of the free group on two generators $\mathbf{F}_2$. The finite sets used F_n are the spherical elements in terms of the word metric i.e. the words of length n with respect to the standard symmetric generating set. The results of Nevo and Stein [**22**] and Bufetov [**14**] proved a.e. convergence for actions $\mathcal{R}$ on $L^r(X)$ with $1 < r < \infty$, and even on $L \log L$. There are also convergence results that are better in case one uses F_{2n} for the averaging set. But then resolving the open case, Tao [**30**] proved these averages fail to converge a.e. for some representation $\mathcal{R}$. See [**30**] for references. The obvious question is what are the good functions in general for representations of $\mathbf{F}_2$ if they are not all of $L^1(X)$? Note: because the group is not amenable, issues of transference are more difficult, if not impossible.

3.2. General Group Examples. There are many deep results about the behavior of averaging operators based on the nature of the representation $\mathcal{R}$, or perhaps even the nature of any representation $\mathcal{R}$ of G. In some ways, this can be seen from studying averaging methods that are divergent as in Parrish and Rosenblatt [**23**]. These are constructed knowing that the map τ is rigid. The constructions do not work generally if the map is strong mixing.

In the same fashion, if the action $\mathcal{R}$ has rapidly vanishing correlation functions, then the behavior of averaging operators is dramatically impacted. For example, see the ergodic theorems of A. Tempelman. A good general reference is his text [**31**]. The representation theoretic properties are what give the strong spectral estimates that are behind the convergence results in Chou, Lau, and Rosenblatt [**16**].

A common thread in all of this work is the unexpected good behavior of what is sometimes called unaveraged convergence, even though the operators are given by a probability measure and so there is inherently averaging occurring. In any case, in these examples there is a completely unresolved issue of which functions are the good ones. For example, consider a probability measure μ on a discrete group G. We let $T_\mu f = \sum_{g\in G} \mu(g) f \circ \mathcal{R}(g)$ for $f \in L^1(X)$. Now take the convolution powers μ^n of the probability measures μ to form our averaging operators $T_{\mu^n} f = \sum_{g\in G} \mu^n(g) f \circ \mathcal{R}(g)$. Here $T_{\mu^n} = T_\mu^n$, the nth composition of $T_\mu f$ itself. In the strong spectral case, the operator norm of T when restricted to the mean-zero functions in $L^2(X)$ is strictly less than one. This can usually be extended to all $L^r(X), 1 < r < \infty$. But there is no result describing what the good functions in $L^1(X)$ will be. For details of how this occurs see Chou, Lau, and Rosenblatt [**16**] and Rosenblatt [**27**].

Question: Suppose $T_\mu^n f$ is L^r-norm and a.e. convergent for all $r, 1 < r < \infty$. Then for what functions $f \in L^1(X)$, do the operator powers $T_\mu^n f$ converge in L^1-norm and/or a.e.?

3.3. The Bellow-Losert Construction for Amenable Group Actions. As a counterpoint to the issues of identifying good and bad functions as was discussed in the sections above, we want to demonstrate how one can generally get classical type averages for finite sets or sequences, at least for amenable groups, that have good behavior on all of $L^1(X)$.

3.3.1. *Preliminaries and statement of result.* Our aim here is to prove the existence of a zero-density sequence (of sets) that remains pointwise L^1-good for an action of a finitely-generated amenable group G. The new setting necessitates a more particular definition of density. Define *density with respect to a Følner sequence* as follows.

DEFINITIONS 3.5. *A set S has a* density of α with respect to a Følner sequence *$\{F_n\}$ if*

$$\lim_{n\to\infty} \frac{|S \cap F_n|}{|F_n|} = \alpha.$$

Replacing the limit with a limit superior or inferior results in a upper or lower density with respect to $\{F_n\}$, respectively.

We should note at this stage that this definition, when viewed in the setting of $\mathbb{Z}$ or $\mathbb{Z}^d$ actions, is more nuanced than the traditional definition of density in $\mathbb{Z}$. For example, the sparse block sequences of Bellow and Losert [**7**] and their multidimensional analogues in [**18**], would each have zero density with respect to some Følner sequences and a density of one with respect to others.

In the setting of amenable groups, the crucial property for convergence of ergodic averages taken over a Følner sequence is that of being tempered.

DEFINITIONS 3.6. *A* tempered *Følner sequence is one for which, for some $C > 0$,*

$$\left| \bigcup_{k\le n} F_k^{-1} F_{n+1} \right| \le C\,|F_{n+1}|$$

for all $n \in \mathbb{N}$.

THEOREM 3.7. *Let G be an amenable group. Then there is a tempered Følner sequence, $\{G_n\}$, and a set S, of zero density with respect to $\{G_n\}$, such that for all representations $\mathcal{R}$,*

$$\lim_{n\to\infty} \frac{1}{|S\cap G_n|} \sum_{s\in S\cap G_n} f(\mathcal{R}(s)x) = \int_X f\, d\mu$$

for a.e. $x \in X$.

This theorem is a corollary of two theorems due to Lindenstrauss.

THEOREM 3.8 (Pointwise Ergodic Theorem for Amenable Groups, [**20**]). *Let G be an amenable group acting via an ergodic representation $\mathcal{R}$ on a measure space (X,μ). If $\{F_n\}$ is a tempered Følner sequence then for any $f \in L^1(X)$*

$$\lim_{n\to\infty} \frac{1}{|F_n|} \sum_{g\in F_n} f(\mathcal{R}(g)x) = \int_X f\, d\mu$$

for a.e. $x \in X$.

THEOREM 3.9 (Proposition 1.5, [**20**]). *Every Følner sequence has a tempered subsequence.*

3.3.2. *Proof of Theorem* 3.7. The goal of the proof will be to construct S and $\{G_n\}$ in such a way as to insure that $\{S \cap G_n\}$ is, itself, a Følner sequence. The conclusion then follows from Theorem 3.9 and the pointwise theorem for amenable groups. In spirit, the construction is a successor to that of Below and Losert in [**7**].

Let G be an amenable group and let $\{F_n\}$ be a tempered Følner sequence in G. We construct a nested tempered Følner sequence with a specific rate of growth as follows. Let $u(x)$ be a continuous, increasing, and convex function on $\mathbb{R}$ and define

$$B_k = \bigcup_{n\le k} F_n,$$

where F_k is the first element of $\{F_n\}$ such that $|F_k| > u(k)|B_{k-1}|$. So we have that

$$|B_k| > u(k)|B_{k-1}|,$$

and $u(x)$ sets a lower bound on the rate of growth for the B_k.

$\{B_k\}$ is nested and inherits the Følner property from $\{F_n\}$. We now wish to show that it is a tempered sequence. Since the sequence is nested, we have that

$$\bigcup_{j\le k} B_j^{-1} B_{k+1} = B_k^{-1} B_{k+1}.$$

By our selection of F_k in our construction of B_k and the temper of $\{F_n\}$, we have

$$\begin{aligned}
\left|B_k^{-1}B_{k+1}\right| &= \left|B_k^{-1}B_k \cup B_k^{-1}\left(B_{k+1}\setminus B_k\right)\right| \\
&= \left|B_k^{-1}B_k \cup \left(\left(\bigcup_{j\le k} F_k\right)^{-1} F_{k+1}\right)\right| \\
&\le \left|B_k^{-1}B_k\right| + C\left|F_{k+1}\right|.
\end{aligned}$$

We now place a final requirement on our function $u(x)$:

$$u(k) \ge c|F_k|\frac{|B_{k-1}|}{|B_k|}$$

for some constant c.

With this requirement, we have that

$$\begin{aligned}\left|B_k^{-1}B_k\right| &\le |B_k|^2\\ &\le |B_k|\frac{|F_{k+1}|}{u(k+1)}\\ &\le \frac{|B_{k+1}|}{c}.\end{aligned} \tag{3.1}$$

Combining the two inequalities, we have

$$\left|\Big(\bigcup_{j\le k} B_j\Big)^{-1} B_{k+1}\right| \le \frac{|B_{k+1}|}{c} + C\,|F_{k+1}| \le C'\,|B_{k+1}|\,.$$

We will now construct our S and $\{G_n\}$ from our nested, tempered Følner sequence, $\{B_k\}$, making use of the lower bounds on the growth of its members.

We first select a subsequence of $\{B_k\}$ with the property that

$$\lim_{j\to\infty}\frac{\sum_{m=1}^{j-1}|B_{k_m}|}{\left|B_{k_j}\right|}\to 0.$$

Let $g_j \in G$ be such that $B_{k_j+1}\cap g_jB_{k_j}=\emptyset$. Define

$$S=\bigcup_{j=1}^{\infty} g_jB_{k_j},\ \text{ and } G_n = B_{k_n+1}\cup g_nB_{k_n}.$$

The sequence $\{G_n\}$ remains nested and Følner. To show that it is tempered, observe that

$$\begin{aligned}\left|G_n^{-1}G_{n+1}\right| &= \left|\Big(B_{k_n+1}\cup g_nB_{k_n}\Big)^{-1}\Big(B_{k_{n+1}+1}\cup g_{n+1}B_{k_{n+1}}\Big)\right|\\ &\le \left|B_{k_n+1}^{-1}B_{k_{n+1}+1}\right| + \left|B_{k_n+1}^{-1}g_{n+1}B_{k_{n+1}}\right| + \left|\Big(g_nB_{k_n}\Big)^{-1}B_{k_{n+1}+1}\right|\\ &\qquad + \left|g_nB_{k_n}^{-1}g_{n+1}B_{k_{n+1}}\right|\\ &\le c_1\left|B_{k_{n+1}+1}\right| + c_2\left|B_{k_{n+1}}\right| + c_3\left|B_{k_{n+1}+1}\right| + c_4\left|B_{k_{n+1}}\right|\\ &\le C\,|G_{n+1}|\,,\end{aligned}$$

using calculations similar to those employed in 3.1.

By our choice of a subsequence, then, and the inequality 3.1, we have

$$\begin{aligned}\frac{|S\cap G_n|}{|G_n|} &= \frac{|S|}{|G_n|}\\ &\le \frac{2c\left|B_{k_j}\right|}{\left|B_{k_j}\right|^2}\end{aligned}$$

which tends to 0 as $j\to\infty$.

To conclude, note that

$$S_n=\bigcup_{j=1}^{n} g_jB_{k_j}$$

is nested and, due to the growth rate of the subsequence B_{k_j}, inherits the Følner property from $\{B_k\}$. Applying Theorem 3.9, we pass to a tempered subsequence of $\{S_n\}=\{G_n\cap S\}$ and Theorem 3.8 completes the proof.

NOTE 3.10. In the conclusion of the proof, we might wish to show that

$$S_n = \bigcup_{j=1}^{n} g_j B_{k_j}$$

is, itself, a tempered Følner sequence. This is made difficult by our lack of a parallel to the Cone Condition of [**5**].

REMARK 3.11. We might wish to find a set S for which the intersections with any tempered Følner sequence would produce a.e. pointwise convergence for L^1 functions. However, even in the case of certain block constructions in $\mathbb{Z}^2$, we find that the sequence of balls centered at the origin produce averages that diverge when they are allowed to intersect only a small portion of the next block [**18**]. The solution there was to make geometric requirements on the sequences which positioned each block so that the Følner sequence intersected with enough of the block and the imbalance did not occur. In the proof above, we construct the Følner sequence in such a way as to ensure that we only intersect whole blocks.

References

[1] T. Adams and J. Rosenblatt, *Coboundaries and Moving Averages*, 29 pages, preprint.

[2] T. Adams and J. Rosenblatt, *Joint coboundaries*, Dynamical systems, ergodic theory, and probability: in memory of Kolya Chernov, Contemp. Math., vol. 698, Amer. Math. Soc., Providence, RI, 2017, pp. 5–33, DOI 10.1090/conm/698/14034. MR3716084

[3] M. Akcoglu, R. L. Jones, and J. M. Rosenblatt, *The worst sums in ergodic theory*, Michigan Math. J. **47** (2000), no. 2, 265–285, DOI 10.1307/mmj/1030132533. MR1793624

[4] A. Bellow, *Perturbation of a sequence*, Adv. Math. **78** (1989), no. 2, 131–139, DOI 10.1016/0001-8708(89)90030-3. MR1029097

[5] A. Bellow, R. Jones, and J. Rosenblatt, *Convergence for moving averages*, Ergodic Theory Dynam. Systems **10** (1990), no. 1, 43–62, DOI 10.1017/S0143385700005381. MR1053798

[6] A. Bellow, Two problems, Proceedings Oberwolfach Conference on Measure Theory (June 1987), Springer Lecture Notes in Math 945, 1987.

[7] A. Bellow and V. Losert, *On sequences of density zero in ergodic theory*, Conference in modern analysis and probability (New Haven, Conn., 1982), Contemp. Math., vol. 26, Amer. Math. Soc., Providence, RI, 1984, pp. 49–60, DOI 10.1090/conm/026/737387. MR737387

[8] J. Bourgain, *Almost sure convergence and bounded entropy*, Israel J. Math. **63** (1988), no. 1, 79–97, DOI 10.1007/BF02765022. MR959049

[9] J. Bourgain, *On the maximal ergodic theorem for certain subsets of the integers*, Israel J. Math. **61** (1988), no. 1, 39–72, DOI 10.1007/BF02776301. MR937581

[10] J. Bourgain, *On the pointwise ergodic theorem on L^p for arithmetic sets*, Israel J. Math. **61** (1988), no. 1, 73–84, DOI 10.1007/BF02776302. MR937582

[11] J. Bourgain, *An approach to pointwise ergodic theorems*, Geometric aspects of functional analysis (1986/87), Lecture Notes in Math., vol. 1317, Springer, Berlin, 1988, pp. 204–223, DOI 10.1007/BFb0081742. MR950982

[12] Z. Buczolich, *Non-L^1 functions with rotation sets of Hausdorff dimension one*, Acta Math. Hungar. **126** (2010), no. 1-2, 23–50, DOI 10.1007/s10474-009-8204-0. MR2593316

[13] Z. Buczolich and R. D. Mauldin, *Divergent square averages*, Ann. of Math. (2) **171** (2010), no. 3, 1479–1530, DOI 10.4007/annals.2010.171.1479. MR2680392

[14] A. I. Bufetov, *Convergence of spherical averages for actions of free groups*, Ann. of Math. (2) **155** (2002), no. 3, 929–944, DOI 10.2307/3062137. MR1923970

[15] J.-P. Conze, *Convergence des moyennes ergodiques pour des sous-suites* (French), Contributions au calcul des probabilités, Soc. Math. France, Paris, 1973, pp. 7–15. Bull. Soc. Math. France, Mém. No. 35, DOI 10.24033/msmf.113. MR0453975

[16] C. Chou, A. T. M. Lau, and J. Rosenblatt, *Approximation of compact operators by sums of translations*, Illinois J. Math. **29** (1985), no. 2, 340–350. MR784527

[17] A. del Junco and J. Rosenblatt, *Counterexamples in ergodic theory and number theory*, Math. Ann. **245** (1979), no. 3, 185–197, DOI 10.1007/BF01673506. MR553340

[18] P. LaVictoire, A. Parrish, and J. Rosenblatt, *Multivariable averaging on sparse sets*, Trans. Amer. Math. Soc. **366** (2014), no. 6, 2975–3025, DOI 10.1090/S0002-9947-2014-06084-4. MR3180737

[19] M. Lin and R. Sine, *Ergodic theory and the functional equation* $(I - T)x = y$, J. Operator Theory **10** (1983), no. 1, 153–166. MR715565

[20] E. Lindenstrauss, *Pointwise theorems for amenable groups*, Invent. Math. **146** (2001), no. 2, 259–295, DOI 10.1007/s002220100162. MR1865397

[21] P. Major, *A counterexample in ergodic theory*, Acta Sci. Math. (Szeged) **62** (1996), no. 1-2, 247–258. MR1412932

[22] A. Nevo and E. M. Stein, *A generalization of Birkhoff's pointwise ergodic theorem*, Acta Math. **173** (1994), no. 1, 135–154, DOI 10.1007/BF02392571. MR1294672

[23] A. Parrish and J. Rosenblatt, *Full divergence and maximal functions with cancellation*, Colloq. Math. **152** (2018), no. 1, 97–121, DOI 10.4064/cm7230-8-2017. MR3778899

[24] A. Quas and M. Wierdl, *Rates of divergence of non-conventional ergodic averages*, Ergodic Theory Dynam. Systems **30** (2010), no. 1, 233–262, DOI 10.1017/S0143385709000054. MR2586353

[25] K. Reinhold-Larsson, *Discrepancy of behavior of perturbed sequences in L^p spaces*, Proc. Amer. Math. Soc. **120** (1994), no. 3, 865–874, DOI 10.2307/2160481. MR1169889

[26] J. Rosenblatt, *Ergodic group actions*, Arch. Math. (Basel) **47** (1986), no. 3, 263–269, DOI 10.1007/BF01192003. MR861875

[27] J. Rosenblatt, *Translation-invariant linear forms on $L_p(G)$*, Proc. Amer. Math. Soc. **94** (1985), no. 2, 226–228, DOI 10.2307/2045380. MR784168

[28] Ja. G. Sinaĭ, *A weak isomorphism of transformations with invariant measure* (Russian), Dokl. Akad. Nauk SSSR **147** (1962), 797–800. MR0161960

[29] Ja. G. Sinaĭ, *On a weak isomorphism of transformations with invariant measure* (Russian), Mat. Sb. (N.S.) **63 (105)** (1964), 23–42. MR0161961

[30] T. Tao, *Failure of the L^1 pointwise and maximal ergodic theorems for the free group*, Forum Math. Sigma **3** (2015), e27, 19 pp., DOI 10.1017/fms.2015.28. MR3482275

[31] A. Tempelman, *Ergodic theorems for group actions: Informational and thermodynamical aspects*, Mathematics and its Applications, vol. 78, Kluwer Academic Publishers Group, Dordrecht, 1992. Translated and revised from the 1986 Russian original. MR1172319

[32] M. Wierdl, *Perturbation of plane curves and sequences of integers*, Illinois J. Math. **42** (1998), no. 1, 139–153. MR1492044

[33] M. Wierdl, *Pointwise ergodic theorem along the prime numbers*, Israel J. Math. **64** (1988), no. 3, 315–336 (1989), DOI 10.1007/BF02882425. MR995574

Department of Mathematics and Computer Science, Eastern Illinois State University, 600 Lincoln Avenue, Charleston, IL 61920-3099

Email address: `ajparrish@eiu.edu`

Department of Mathematical Sciences, Indiana University-Purdue University Indianapolis, 402 N. Blackford Street, Indianapolis, IN 46202-3217

Email address: `rosnbltt@illinois.edu`

Contemporary Mathematics
Volume **736**, 2019
https://doi.org/10.1090/conm/736/14836

Isomorphisms of cubic rational maps that preserve an infinite measure

Rachel L. Rossetti

Abstract. We study rational functions of degree 3 that preserve Lebesgue measure on the real line. These maps are known as cubic generalized Boole transformations. We calculate the Krengel entropy and describe its role as an invariant for c-isomorphisms between certain classes of these maps. We also characterize the 1-isomorphisms for a specific subset of these maps.

1. Introduction

In this paper, we study rational maps of degree 3 that are ergodic, exact and preserve Lebesgue measure on the real line. These maps are generalized Boole transformations and related to classical inner functions. We provide an analysis of the Krengel entropy and c-isomorphism classes of these maps.

Generalizations of the classical Boole function ([**9**], [**3**]) provide some natural examples of infinite-measure-preserving systems with chaotic dynamics. These maps preserve Lebesgue measure on $\mathbb{R}$ and extend to inner functions on $\mathbb{C}$. Their dynamics have been studied by many ([**1**], [**2**], [**6**], [**7**], [**10**], [**13**], and [**14**]).

A negative generalized Boole transformation has the form

$$S(x) = -x - \beta - \sum_{k=1}^{N} \frac{p_k}{t_k - x}, \quad \beta, p_k, t_k \in \mathbb{R}, p_k > 0, \tag{1.1}$$

with all t_k distinct.

Maps of the form (1.1) were studied in [**5**] and [**6**] and shown to be conservative, exact and ergodic with respect to Lebesgue measure on $\mathbb{R}$. It was also proved in [**6**] that the Krengel entropy, an entropy defined for transformations preserving an infinite measure, can be computed using Rohlin's formula [**17**]. Furthermore, it was shown that Boole functions are quasi-finite, so the three most common definitions of entropy for transformations preserving an infinite measure (Krengel [**12**], Parry [**15**], and Poisson [**18**]) are all equivalent.

In [**7**], it was proved that quadratic Boole functions admit no periodic orbits of period 2 and thus are conformally conjugate to maps in the unique parameter space of rational maps with this property ([**8**], [**11**]). This conformally conjugate form was used to provide a complete characterization of the c-isomorphism classes of quadratic Boole functions [**7**].

2010 *Mathematics Subject Classification.* Primary 37A40, 37F10, 26A18 .

We show that cubic Boole functions behave somewhat differently to their quadratic relatives. Although there exist rational maps of degree 3 that admit no nontrivial period 2 orbits ([**8**], [**11**]), the cubic Boole functions do not arise as part of that family. Furthermore, given the additional parameters in degree 3, the c-isomorphism classes are more difficult to characterize than in degree 2. In this paper, we focus mainly on cubic Boole functions with symmetry properties. The main result of this paper gives a precise characterization of the 1-isomorphism classes of cubic Boole maps that are symmetric.

2. Background and Notation

By $(X, \mathcal{B}, \mu, T)$ we denote a nonsingular measurable dynamical system. Throughout this paper, we assume that $T : (X, \mathcal{B}, \mu) \to (X, \mathcal{B}, \mu)$, where X is a topological space, $\mathcal{B}$ is the σ-algebra of Borel sets, and μ is a σ-finite measure on X. In the definitions that follow, all sets A are measurable and each statement holds up to sets of μ measure 0. We assume $T : X \to X$ is *nonsingular*, so $\mu(A) = 0$ if and only if $\mu(T^{-1}A) = 0$. We say T is *conservative* if there exists some $n \in \mathbb{N}$ such that $\mu(T^{-n}A \cap A) > 0$ for every set A with $\mu(A) > 0$. We say T is *ergodic* if $T^{-1}(A) = A$ implies $\mu(A) = 0$ or $\mu(X \setminus A) = 0$. We say T is *exact* if $X = \bigcup_{n \geq 1} T^{-n} \circ T^n(A)$ for every A with $\mu(A) > 0$.

The map T is *n-to-1* if for almost every $x \in X$, the set $T^{-1}(x)$ contains precisely n distinct points. Given a nonsingular n-to-1 transformation, we define $\mathcal{P} = \{P_i\}_{i=1}^n$ to be a *Rohlin partition* of X if $T : P_i \to X$ is one-to-one and onto for each $i = 1, ..., n$ and $\bigcup_{i=1}^n P_i = X \bmod \mu$. Given $(X, \mathcal{B}, \mu, T)$ with Rohlin partition $\mathcal{P} = \{P_i\}_{i=1}^n$, we denote each branch $T|_{P_i}$ by T_i. The *Jacobian* of T is defined by $J_T(x) = \sum_{i=1}^n \mathbb{1}_{P_i}(x) \frac{d\mu T_i}{d\mu}(x)$, where $\mathbb{1}_A$ denotes the characteristic function on the set A. Clearly if $X = \mathbb{R}$, $\mu = \lambda$, and T is piecewise C^1, then $J_T(x) = |T'(x)|$.

A set $A \in \mathcal{B}$ is called a *sweep-out set* for T if $\bigcup_{n=0}^\infty T^{-n}A = X \mod \mu$. For $x \in A$ we let $\phi_A(x)$ be the *first-return-time* of x to A, so $\phi_A(x) = \min\{n : T^n(x) \in A\}$. The *induced transformation*, $T_A : A \to A$, is defined by

$$T_A(x) = T^{\phi_A(x)}(x) \text{ for } x \in A.$$

If $(X, \mathcal{B}, \mu, T)$ is a measure-preserving system and A is a sweep-out set for T, then T_A is a measure-preserving transformation of $(A, \mathcal{B}|_A, \mu|_A)$, where $\mathcal{B}|_A = \{B \cap A : B \in \mathcal{B}\}$ and $\mu|_A(B) = \mu(A \cap B)$.

In [**12**] Krengel gave the following definition of entropy for infinite measure-preserving transformations. Let $(X, \mathcal{B}, \mu, T)$ be a conservative σ-finite measure-preserving system. If $A \in \mathcal{B}$ is such that $0 < \mu(A) < \infty$, and A is a sweep-out set for T, then $h_{\text{Kr}}(T) = h(T_A, \mu|_A)$ (i.e. the traditional Kolmogorov-Sinai entropy of the induced system).

In 1934, Szegó proved that if $G : (\mathbb{R}, \mathcal{B}, \lambda) \to (\mathbb{R}, \mathcal{B}, \lambda)$ is a rational function that preserves Lebesgue measure, λ, on $\mathbb{R}$, then G is of the form $G = \pm S$ for some S of the form (1.1) (Szegö's solution [**19**] was in response to a problem posed by Pólya in 1931 [**16**] and was later extended by Letac [**13**]). In [**6**], it was shown that $(\mathbb{R}, \mathcal{B}, \lambda, S)$, where S is a Boole function of the form (1.1), is conservative, ergodic, and exact. A formula for the Krengel entropy of $(\mathbb{R}, \mathcal{B}, \lambda, S)$ was also given in [**6**].

THEOREM 2.1 ([**6**]). *Any rational function $G : (\mathbb{R}, \mathcal{B}, \lambda) \to (\mathbb{R}, \mathcal{B}, \lambda)$ which is λ-preserving and conservative has Krengel entropy given by*

$$h_{\mathrm{Kr}}(G) = \int_{\mathbb{R}} \log |G'(x)| d\lambda(x). \tag{2.1}$$

The following proposition proves that $\log |G'(x)|$ is integrable and (2.1) is finite.

PROPOSITION 2.2. *If $G : (\mathbb{R}, \mathcal{B}, \lambda) \to (\mathbb{R}, \mathcal{B}, \lambda)$ is λ-preserving and conservative, then $\log |G'|$ is Riemann (and therefore Lebesgue) integrable.*

PROOF. By the discussion above we know $G = \pm S$, where S is of the form (1.1). We have $|G'(x)| = 1 + \sum_{k=1}^{N} \frac{p_k}{(t_k - x)^2}$. Assume the poles are in increasing order, so $t_i < t_{i+1}$. Note that $\log |G'|$ is smooth on $\mathbb{R} \setminus \{t_1, t_2, ..., t_N\}$.

On the compact set $[t_i + \varepsilon, t_{i+1} - \varepsilon]$ (between the poles) we have that $\log |G'|$ is continuous and therefore bounded and integrable.

Now we look near the poles. If $x \in [t_i - \varepsilon, t_i + \varepsilon]$, then we write

$$|G'(x)| = 1 + \sum_{k=1}^{i-1} \frac{p_k}{(t_k - x)^2} + \frac{p_i}{(t_i - x)^2} + \sum_{k=i+1}^{N} \frac{p_k}{(t_k - x)^2}. \tag{2.2}$$

When $k \neq i$, each term $\frac{p_k}{(t_k - x)^2}$ is bounded, so we need only show

$$\int_{[t_i - \varepsilon, t_i + \varepsilon]} \log \left(C \cdot \frac{p_i}{(t_i - x)^2} \right) d\lambda(x) < \infty, \tag{2.3}$$

where C is a positive constant that bounds $1 + \sum_{k \neq i} \frac{p_k}{(t_k - x)^2}$. We use the change of variables $y = (t_i - x)$ combined with the observation that the constant $C \cdot p_i$ does not affect the integrability to simplify the left-hand-side of (2.3) and obtain $\int_{[-\varepsilon, \varepsilon]} \log (1/y^2) \, d\lambda(y)$. Note that $\log (1/y^2)$ is symmetric about the y-axis, so we have

$$2 \int_{[0, \varepsilon]} \log \left(\frac{1}{y^2} \right) d\lambda(y) = 2 \lim_{\tau \to 0} 2y - y \log (y^2) \Big|_{\tau}^{\varepsilon} = 2(2\varepsilon + \varepsilon \log(\varepsilon^2)), \tag{2.4}$$

which is finite for all ε and equals 0 as $\varepsilon \to 0$.

Finally, we consider large x and prove $\int_{[t_N + \varepsilon, \infty)} \log |G'(x)| d\lambda(x) < \infty$, and note that the proof of $\int_{(-\infty, t_1 - \varepsilon]} \log |G'(x)| d\lambda(x) < \infty$ is similar. Choose M large enough such that for all $x \in (M, \infty)$ we have $x - t_N \geq \frac{x}{2}$ and $|G'(x)| = 1 + \sum_{k=1}^{N} \frac{p_k}{(t_k - x)^2} \leq 1 + \frac{C}{(t_N - x)^2}$, where $C = N \cdot \max\{p_k : k = 1..N\}$. By our choice of M we have, $1/(t_N - x)^2 \leq 4/x^2$ and

$$\begin{aligned} \int_{(M,\infty)} \log |G'(x)| d\lambda(x) &< \int_{(M,\infty)} \log \left(1 + \frac{C}{(t_N - x)^2} \right) d\lambda(x) \\ &< \int_{(M,\infty)} \log \left(1 + \frac{4C}{x^2} \right) d\lambda(x) < \int_{(M,\infty)} \frac{4C}{x^2} d\lambda(x) < \infty, \end{aligned}$$

where the second to last inequality comes from the fact that $\log(1 + x) < x$ for all $x \in \mathbb{R}$.

□

We are interested in the relationship between Krengel entropy and isomorphisms of Boole transformations on $\mathbb{R}$. For transformations that preserve a probability measure, it is well-known that entropy is an isomorphism invariant. In the infinite setting, we must account for a less restrictive type of isomorphism called a c-isomorphism.

DEFINITION 2.3. Let $(X_1, \mathcal{B}_1, m_1, T_1)$ and $(X_2, \mathcal{B}_2, m_2, T_2)$ be two infinite-measure preserving systems. Suppose there are two sets $M_1 \in \mathcal{B}_1$ and $M_2 \in \mathcal{B}_2$ with $m_1(X_1 \setminus M_1) = 0$ and $m_2(X_2 \setminus M_2) = 0$ such that $T_1(M_1) \subseteq M_1$ and $T_2(M_2) \subseteq M_2$. For $c \in (0, \infty]$ we say $(X_1, \mathcal{B}_1, m_1, T_1)$ is *c-isomorphic* to $(X_2, \mathcal{B}_2, m_2, T_2)$ if there exists an invertible map $\phi : M_1 \to M_2$ such that for all $A \in \mathcal{B}_2|_{M_2}$,

(1) $\phi^{-1}(A) \in \mathcal{B}_1|_{M_1}$,
(2) $m_1(\phi^{-1}(A)) = c \cdot m_2(A)$, and
(3) $(\phi \circ T_1)(x) = (T_2 \circ \phi)(x)$ for all $x \in M_1$.

If (1)-(3) hold, we write $\phi : T_1 \to^c T_2$, and call ϕ a *c-isomorphism*.

The following proposition and corollaries can be found in [**7**].

PROPOSITION 2.4. *If S_1 and S_2 are two Boole transformations of the form (1.1), and $\phi : S_1 \to^c S_2$ is a c-isomorphism, then*

$$h_{\mathrm{Kr}}(S_1) = c \cdot h_{\mathrm{Kr}}(S_2). \tag{2.5}$$

COROLLARY 2.5. *Krengel entropy is a 1-isomorphism invariant for Boole transformations. That is, if S_1 and S_2 are 1-isomorphic, then $h_{\mathrm{Kr}}(S_1) = h_{\mathrm{Kr}}(S_2)$.*

COROLLARY 2.6. *If S_1 and S_2 are two Boole transformations, then there is at most one $c \in (0, \infty]$ such that $\phi : S_1 \to^c S_2$ is a c-isomorphism.*

Thus, Krengel entropy is a 1-isomorphism invariant for Boole transformations. That is, if $h_{\mathrm{Kr}}(S_1) \neq h_{\mathrm{Kr}}(S_2)$, then S_1 and S_2 are not 1-isomorphic. They could, however, be c-isomorphic for some $c \neq 1$, and in this case there is at most one $c \in (0, \infty]$ such that S_1 is c-isomorphic to S_2.

3. Degree 3 Boole Transformations

We now turn our focus to Boole transformations of degree 3. From now on

$$S_{(\beta, p_1, p_2, t_1, t_2)}(x) = -x - \beta - \frac{p_1}{t_1 - x} - \frac{p_2}{t_2 - x}, \tag{3.1}$$

where $\beta, p_1, p_2, t_1, t_2 \in \mathbb{R}$ and $p_1, p_2 > 0$.

Let S be of the form (3.1) and relabel the set of poles $\{t_1, t_2\}$ to be in increasing order, so $t_1 < t_2$. Let $\mathcal{W} = \{W_1, W_2, W_3\}$, where $W_1 = (\infty, t_1)$, $W_2 = (t_1, t_2)$, and $W_3 = (t_2, \infty)$.

LEMMA 3.1. *For each S of the form (3.1), S maps each interval in $\mathcal{W}$ diffeomorphically onto $\mathbb{R}$, and every point in $\mathbb{R}$ has precisely 3 distinct preimages in $\mathbb{R}$.*

PROOF. Note that S is smooth and uniformly decreasing on W_i for $i = 1, 2, 3$. Furthermore,

$$\lim_{x \to t_1^+} S(x) = \infty \qquad \lim_{x \to t_2^+} S(x) = \infty \qquad \lim_{x \to -\infty} S(x) = \infty$$
$$\lim_{x \to t_1^-} Sx) = -\infty \qquad \lim_{x \to t_2^-} S(x) = -\infty \qquad \lim_{x \to \infty} S(x) = -\infty.$$

Therefore, $\mathcal{W}$ is a Rohlin partition for S on $\mathbb{R}$, and S maps each interval in $\mathcal{W}$ diffeomorphically onto $\mathbb{R}$. There are 3 intervals in $\mathcal{W}$, and this is exactly $\deg(S)$. (See Figure 1 for the general shape of S.) □

We can extend S to a map on the Riemann sphere $\widehat{\mathbb{C}}$, where $\widehat{\mathbb{R}}$ denotes the great circle on $\widehat{\mathbb{C}}$ whose image is the real line under stereographic projection. Proposition 1.5 in [**7**] shows that the Julia set of S is $\widehat{\mathbb{R}}$.

LEMMA 3.2. *A cubic Boole function, S, of the form (3.1), has a neutral fixed point at ∞ with multiplier -1 and three real fixed points that are repelling.*

PROOF. It is clear that S fixes ∞. By Lemma 3.1, S has three real fixed points, $x_i \in W_i$ for $i = 1, 2, 3$. Taking the derivative yields, $S'(x) = -1 - \frac{p_1}{(t_1-x)^2} - \frac{p_2}{(t_2-x)^2}$ for all $x \in \mathbb{R} \setminus \{t_1, t_2\}$. Therefore, $S'(x) < -1$ for all $x \in \mathbb{R}$, so the three real fixed points are repelling. As $x \to \infty$, we have that $1/S'(x) \to -1$, so the fixed point at ∞ is neutral (parabolic) with multiplier -1. □

In [**7**], the authors proved that quadratic Boole functions admit no periodic orbits of period 2 and thus are conformally conjugate (and therefore c-isomorphic) to maps in the unique parameter space of rational maps with this property ([**8**], [**11**]). Rational maps lacking periodic orbits of a particular period are extremely rare. If a rational map of degree $d \geq 2$ lacks period n orbits, then (d, n) is one of the pairs $(2, 2)$, $(2, 3)$, $(3, 2)$, or $(4, 2)$ [**4**]. In [**11**], Hagihara completely characterized the $(3, 2)$ case with the following families of rational maps:

$$\begin{aligned}
(3.2)\qquad R_1(z) &= \frac{z^3 - 2az^2 - z}{-z^2 - \frac{2}{a}z + 1}, \quad \text{where } a \in \mathbb{C} \setminus \{0, \pm i\} \\
R_2(z) &= \frac{z^3 - z}{-z^2 + az + 1}, \quad \text{where } a \in \mathbb{C} \setminus \{0\} \\
R_3(z) &= \frac{z^3 + az^2 - z}{(a^2 - 1)z^2 - 2az + 1}, \quad \text{where } a \in \mathbb{C} \setminus \{0\}.
\end{aligned}$$

We prove the following proposition which shows that unlike the quadratic case, cubic Boole functions must contain period 2 orbits.

PROPOSITION 3.3. *A cubic Boole function, S, of the form (3.1), is not conformally conjugate to a member of the families R_1, R_2, or R_3. Therefore, S contains nontrivial period 2 orbits.*

PROOF. By Lemma 3.2, S has a neutral fixed point at ∞ and three repelling fixed points in $\mathbb{R}$. By [**11**], every map in the family R_1 has three neutral fixed points and one repelling fixed point. Also, every map of the form R_2 has two neutral fixed points, one repelling fixed point, and the last fixed point may become attracting depending on the parameter. Therefore, S cannot be conformally conjugate to a member of R_1 or R_2. Finally, R_3 has a neutral fixed point with three distinct immediate basins, each with 2 disjoint Fatou components, forming a period 2 cycle (i.e. 6 petals). Theorem 3.3 in [**7**] shows that S has only one immediate basin at ∞ consisting of two distinct Fatou components (i.e. 2 petals). Therefore, S cannot be conformally conjugate to a member of R_3. □

It is convenient to write S in a normalized form to simplify some of the calculations in the upcoming sections. In the notation of (3.1), we say a Boole function is in *normal form* if it has the form $S_{(\beta,1,p,0,t)}$, where $t > 0$. In other words, one of the poles is 0 with numerator 1, and the other pole is positive with numerator p. Throughout this paper, we use $N_{(\beta,p,t)}$ to denote a normalized degree 3 Boole function, so

$$N_{(\beta,p,t)}(x) = -x - \beta - \frac{1}{-x} - \frac{p}{t-x}, \quad \text{where } p, t > 0 \text{ and } \beta \in \mathbb{R}. \tag{3.3}$$

LEMMA 3.4. *A Boole transformation, $S_{(\beta,p_1,p_2,t_1,t_2)}$, of the form (3.1) is $\sqrt{p_1}$-isomorphic to a normalized Boole transformation, $N_{(\beta',p',t')}$, where $\beta' = \frac{2t_1+\beta}{\sqrt{p_1}}$, $p' = \frac{p_2}{p_1}$, and $t' = \frac{t_2-t_1}{\sqrt{p_1}} > 0$.*

PROOF. We first move the smaller of the poles to 0 via the conjugating maps

$$\psi_{(t_1)}(x) = x - t_1 \quad \text{and} \quad \psi^{-1}_{(t_1)}(x) = x + t_1. \tag{3.4}$$

We have

$$\begin{aligned}\tilde{S}(x) &= (\psi_{(t_1)} \circ S_{(\beta,p_1,p_2,t_1,t_2)} \circ \psi^{-1}_{(t_1)})(x) \\ &= -(x+t_1) - \beta - \frac{p_1}{t_1-(x+t_1)} - \frac{p_2}{t_2-(x+t_1)} - t_1 \\ &= -x - (\beta + 2t_1) - \frac{p_1}{-x} - \frac{p_2}{(t_2-t_1)-x}.\end{aligned}$$

Now, we move the numerator to 1 via the conjugating maps,

$$\phi_{(\sqrt{p_1})}(x) = \frac{x}{\sqrt{p_1}} \quad \text{and} \quad \phi^{-1}_{(\sqrt{p_1})}(x) = \sqrt{p_1}x. \tag{3.5}$$

We have

$$(\phi_{(\sqrt{p_1})} \circ \tilde{S} \circ \phi^{-1}_{(\sqrt{p_1})})(x) = \frac{1}{\sqrt{p_1}}\left(-\sqrt{p_1}x - (\beta+2t_1) - \frac{p_1}{-\sqrt{p_1}x} - \frac{p_2}{(t_2-t_1)-\sqrt{p_1}x}\right)$$

$$= -x - \frac{\beta+2t_1}{\sqrt{p_1}} - \frac{1}{-x} - \frac{p_2/p_1}{((t_2-t_1)/\sqrt{p_1}) - x}. \tag{3.6}$$

Since $t_2 - t_1 > 0$, then (3.6) is a normalized Boole transformation, $N_{(\beta',p',t')}$, where $\beta' = \frac{2t_1+\beta}{\sqrt{p_1}}$, $p' = \frac{p_2}{p_1}$, and $t' = \frac{t_2-t_1}{\sqrt{p_1}} > 0$. □

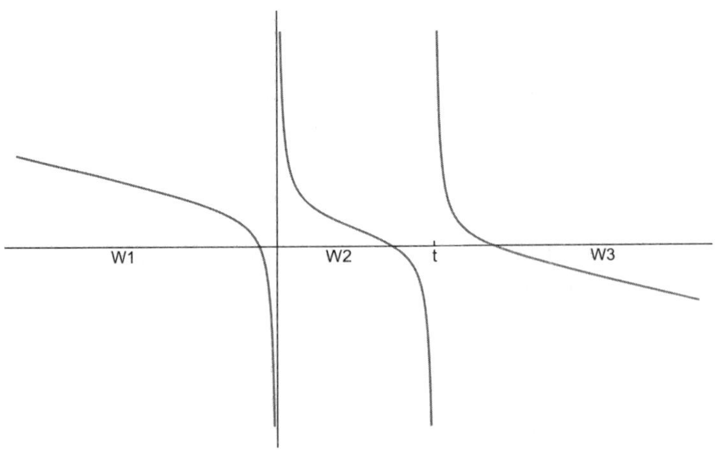

FIGURE 1. A normalized degree 3 Boole transformation, $N(\beta, p, t)$, with Rohlin partition $\mathcal{W} = \{W_1, W_2, W_3\}$.

4. Symmetric Degree 3 Boole Transformations

A degree three Boole function is *symmetric* if the poles are $-t$ and t with matching numerators p. That is,

$$S_{\mathrm{sym}(\beta,\mathrm{p},\mathrm{t})}(x) = -x - \beta - \frac{p}{-t-x} - \frac{p}{t-x}, \tag{4.1}$$

where $p > 0$ and $t, \beta \in \mathbb{R}$. Note that $S_{\mathrm{sym}(\beta,\mathrm{p},\mathrm{t})} = S_{\mathrm{sym}(\beta,\mathrm{p},-\mathrm{t})}$. For consistency we assume throughout that $t > 0$ and write $S_{\mathrm{sym}(\beta,\mathrm{p},\mathrm{t})}$.

We begin the study of symmetric Boole transformations by explicitly computing the Krengel entropy.

THEOREM 4.1. *If $S_{\mathrm{sym}(\beta,\mathrm{p},\mathrm{t})}(x)$ is of the form (4.1), then*

$$h_{\mathrm{Kr}}(S_{\mathrm{sym}(\beta,\mathrm{p},\mathrm{t})}) = 2\pi\left(\sqrt{p-t^2-\sqrt{p^2-4pt^2}} + \sqrt{p-t^2+\sqrt{p^2-4pt^2}}\right). \tag{4.2}$$

Before proving Theorem 4.1, we prove a helpful lemma.

LEMMA 4.2. *On any interval in $\mathbb{R}\setminus\{-t,t\}$, the antiderivative of $f(x) = \log|S'_{\mathrm{sym}(\beta,\mathrm{p},\mathrm{t})}(x)|$ is given by*

$$\begin{aligned} F(x) = {}& x\log\left(1+\frac{p}{(t-x)^2}+\frac{p}{(-t-x)^2}\right) + 2t\log|t-x| - 2t\log|t+x| \\ & + 2\sqrt{p-t^2-\sqrt{p^2-4pt^2}}\arctan\left(\frac{x}{\sqrt{p-t^2-\sqrt{p^2-4pt^2}}}\right) \\ & + 2\sqrt{p-t^2+\sqrt{p^2-4pt^2}}\arctan\left(\frac{x}{\sqrt{p-t^2+\sqrt{p^2-4pt^2}}}\right). \end{aligned} \tag{4.3}$$

PROOF. We have $f(x) = \log|S'_{\mathrm{sym}(\beta,\mathrm{p},\mathrm{t})}(x)| = \log\left(1+\frac{p}{(t-x)^2}+\frac{p}{(-t-x)^2}\right)$. Integration by parts yields

$$\begin{aligned} F(x) = {}& x\log\left(1+\frac{p}{(t-x)^2}+\frac{p}{(-t-x)^2}\right) \\ & + \int -x\left(\frac{4px(3t^2+x^2)}{(t-x)(t+x)((t^2-x^2)^2+2p(t^2+x^2))}\right)dx. \end{aligned} \tag{4.4}$$

We consider the remaining integral in (4.4) and use partial fractions to obtain

$$\int -\frac{2t}{t-x} - \frac{2t}{t+x} + \frac{2\left(p-t^2-\sqrt{p^2-4pt^2}\right)}{p-t^2-\sqrt{p^2-4pt^2}+x^2} + \frac{2\left(p-t^2+\sqrt{p^2-4pt^2}\right)}{p-t^2+\sqrt{p^2-4pt^2}+x^2}dx.$$

Finding the antiderivative for the integrand above yields the result. □

We are now ready to prove Theorem 4.1.

PROOF OF THEOREM 4.1. By Theorem 2.1 we have

$$h_{\mathrm{Kr}}(S_{\mathrm{sym}(\beta,\mathrm{p},\mathrm{t})}) = \int_{\mathbb{R}} \log\left(1+\frac{p}{(t-x)^2}+\frac{p}{(-t-x)^2}\right)d\lambda(x). \tag{4.5}$$

By Proposition 2.2 we know $\log|S'_{\mathrm{sym}(\beta,\mathrm{p},\mathrm{t})}|$ is integrable and (4.5) is finite. Let $f(x) = \log|S'_{\mathrm{sym}(\beta,\mathrm{p},\mathrm{t})}(x)|$, and let $N > t$. We have

$$\int_{-N}^{N} f(x)d\lambda(x) = \lim_{\varepsilon\to 0}\int_{-N}^{-t-\varepsilon} f(x)d\lambda(x) + \int_{-t+\varepsilon}^{t-\varepsilon} f(x)d\lambda(x) + \int_{t+\varepsilon}^{N} f(x)d\lambda(x). \tag{4.6}$$

By Lemma 4.2, the antiderivative of f on any interval not containing a pole, is given by F of the form (4.3), so by the fundamental theorem of calculus (4.6) is equal to

$$\lim_{\varepsilon\to 0}[F(-t-\varepsilon) - F(-t+\varepsilon)] + [F(t-\varepsilon) - F(t+\varepsilon)] + [F(N) - F(-N)]. \tag{4.7}$$

Since f is integrable on $\mathbb{R}$, we know its antiderivative, F, is continuous, so the first two terms of (4.7) go to 0 zero as $\varepsilon \to 0$. Therefore, $\int_{-N}^{N} f(x)d\lambda(x) = F(N) - F(-N)$, and evaluating using (4.3) yields

$$2N\log\left(1 + \frac{p}{(t-N)^2} + \frac{p}{(-t-N)^2}\right) + 4t\log|t-N| - 4t\log|t+N|$$
$$+ 2a\left(\arctan\left(\frac{N}{a}\right) - \arctan\left(\frac{-N}{a}\right)\right) + 2b\left(\arctan\left(\frac{N}{b}\right) - \arctan\left(\frac{-N}{b}\right)\right),$$

where $a = \sqrt{p - t^2 - \sqrt{p^2 - 4pt^2}}$ and $b = \sqrt{p - t^2 + \sqrt{p^2 - 4pt^2}}$. Taking the limit as $N \to \infty$ yields the result. □

We leverage the entropy formula for symmetric Boole transformations to prove the following proposition that gives a necessary criterion for the existence of a 1-isomorphism between $S_{\mathrm{sym}(\beta,1,\mathrm{t})}$ and $S_{\mathrm{sym}(\gamma,1,\mathrm{r})}$.

PROPOSITION 4.3. *If $S_{\mathrm{sym}(\beta,1,\mathrm{t})}$ and $S_{\mathrm{sym}(\gamma,1,\mathrm{r})}$ are of the form (4.1), and $S_{\mathrm{sym}(\beta,1,\mathrm{t})}$ is 1-isomorphic to $S_{\mathrm{sym}(\gamma,1,\mathrm{r})}$, then $t = r$.*

PROOF. By Proposition 2.4, we know $h_{\mathrm{Kr}}(S_{\mathrm{sym}(\beta,1,\mathrm{t})}) = h_{\mathrm{Kr}}(S_{\mathrm{sym}(\gamma,1,\mathrm{r})})$. By Theorem 4.1 we have $h_{\mathrm{Kr}}(S_{\mathrm{sym}(\beta,1,\mathrm{t})}) = 2\pi\left(\sqrt{1-t^2-\sqrt{1-4t^2}} + \sqrt{1-t^2+\sqrt{1-4t^2}}\right)$ and $h_{\mathrm{Kr}}(S_{\mathrm{sym}(\gamma,1,\mathrm{r})}) = 2\pi\left(\sqrt{1-r^2-\sqrt{1-4r^2}} + \sqrt{1-r^2+\sqrt{1-4r^2}}\right)$. Setting the entropy equations equal to each other and squaring both sides yields

$$2 - 2r^2 + 2\sqrt{2r^2+r^4} = 2 - 2t^2 + 2\sqrt{2t^2+t^4}. \tag{4.8}$$

Moving $2 - 2r^2$ to the other side, then squaring both sides again yields

$$\begin{aligned} 8r^2 + 4r^4 &= \left(2r^2 - 2t^2 + 2\sqrt{2t^2+t^4}\right)^2 \\ &= 8t^4 + 8t^2 + 4r^4 - 8r^2t^2 + 8r^2\sqrt{2t^2+t^4} - 8t^2\sqrt{2t^2+t^4}. \end{aligned} \tag{4.9}$$

Collecting everything on one side yields

$$-8t^4 - 8t^2 + 8t^2\sqrt{2t^2+t^4} + r^2(8 + 8t^2 - 8\sqrt{2t^2+t^4}) = 0. \tag{4.10}$$

Factoring the left-hand-side gives

$$8(r-t)(r+t)(1 + t^2 - \sqrt{2t^2+t^4}) = 0. \tag{4.11}$$

Note that there does not exist a t such that $1 + t^2 = \sqrt{2t^2+t^4}$, since squaring both sides yields $1 + 2t^2 + t^4 = 2t^2 + t^4$. Therefore, $t = r$ or $t = -r$. We assume both $r, t > 0$, since they are coming from a normalized Boole transformation, so we have $t = r$. □

REMARK 4.4. By Lemma 3.4, a symmetric degree 3 Boole transformation, $S_{\mathrm{sym}(\beta,\mathrm{p},\mathrm{t})}$, of the form (4.1), is $\sqrt{p}$-isomorphic to a normalized Boole transformation, $N_{(\frac{-2t+\beta}{\sqrt{p}},1,\frac{2t}{\sqrt{p}})}$, of the form (3.3) (by letting $t_1 = -t$ and $t_2 = t$). If we define $\eta(x) = \frac{t-x}{\sqrt{p}}$, then $\eta^{-1}(x) = t - \sqrt{p}x$, so we also have

$$\begin{aligned}(\eta \circ S_{\mathrm{sym}(\beta,\mathrm{p},\mathrm{t})} \circ \eta^{-1})(x) &= -x - \frac{-2t-\beta}{\sqrt{p}} - \frac{1}{x} - \frac{1}{\frac{2t}{\sqrt{p}} - x} \\ &= N_{(\frac{-2t-\beta}{\sqrt{p}},1,\frac{2t}{\sqrt{p}})}. \end{aligned} \tag{4.12}$$

Therefore, $S_{\mathrm{sym}(\beta,\mathrm{p},\mathrm{t})}$ is $\sqrt{p}$-isomorphic to $N_{(\frac{-2t+\beta}{\sqrt{p}},1,\frac{2t}{\sqrt{p}})}$ and $N_{(\frac{-2t-\beta}{\sqrt{p}},1,\frac{2t}{\sqrt{p}})}$, so the normal form of $S_{\mathrm{sym}(\beta,\mathrm{p},\mathrm{t})}$ is not unique. Note that this in contrast to quadratic Boole functions, as it was shown in [**7**] that the analogous normal form of a Boole function of degree 2 is unique. The main result of this section (Theorem 4.5 along with Corollary 4.8) shows that there are precisely two normal forms for $S_{\mathrm{sym}(\beta,1,\mathrm{t})}$.

We now investigate the 1-isomorphism classes of normalized Boole transformations of the form (3.3) where $p = 1$ (i.e. $N_{(\beta,1,t)}$ is a normal form of a symmetric Boole map). We prove the following theorem.

THEOREM 4.5. *Two normalized cubic Boole transformations, $N_{(\beta,1,t)}$ and $N_{(\gamma,1,r)}$, are 1-isomorphic if and only if one of the following holds:*

(1) *$r = t$ and $\gamma = \beta$, or*
(2) *$r = t$ and $\gamma = -(2t + \beta)$.*

In order to prove Theorem 4.5 we need the following proposition and auxiliary lemma.

PROPOSITION 4.6. *If $N_{(\beta,1,t)}$ is a normalized cubic Boole transformation, then $N'_{(\beta,1,t)}(x) < -1$ for all $x \in \mathbb{R} \setminus \{0,1\}$, both $N'_{(\beta,1,t)}$ and $N''_{(\beta,1,t)}$ have poles at $x = 0$ and $x = t$, and $\max_{x\in W_2}\{N'_{(\beta,1,t)}(x)\}$ exists and is equal to $-1-8/t^2$. Furthermore, for a fixed $x \in \mathbb{R}$, if $Y = \{y \in \mathbb{R} : N'_{(\beta,1,t)}(x) = N'_{(\beta,1,t)}(y)\}$, then $|Y| \leq 4$.*

PROOF. Define a Rohlin partition, $\mathcal{W}$, for $N_{(\beta,1,t)}$ as in Lemma 3.1, so $W_1 = (-\infty, 0), W_2 = (0,t)$, and $W_3 = (t, \infty)$. By Lemma 3.1, $N_{(\beta,1,t)}$ is smooth on W_i for $i = 1,2,3$. We differentiate $N_{(\beta,1,t)}$ and obtain

$$N'_{(\beta,1,t)}(x) = -1 - \frac{1}{x^2} - \frac{1}{(t-x)^2} \quad \text{and} \quad N''_{(\beta,1,t)}(x) = \frac{2}{x^3} - \frac{2}{(t-x)^3}. \tag{4.13}$$

By (4.13), it is clear that both $N'_{(\beta,1,t)}$ and $N''_{(\beta,1,t)}$ have poles at $x = 0$ and $x = t$, and $N'_{(\beta,1,t)}(x) < -1$ for all $x \in \mathbb{R} \setminus \{0,t\}$. Also, $N_{(\beta,1,t)}$ is concave down on W_1, so there exists at most one $y_1 \in W_1$ such that $N'_{(\beta,1,t)}(x) = N'_{(\beta,1,t)}(y)$. Similarly, $N_{(\beta,1,t)}$ is concave up on W_3, so there exists at most one such $y_3 \in W_3$. On W_2, $N_{(\beta,1,t)}$ has one inflection point where the concavity changes from positive to negative. Let $d \in W_2$ be the inflection point, then $N''_{(\beta,1,t)}(d) = 0$, and by (4.13) we see that $d = t/2$. Since $\lim_{x\to 0^+} N_{(\beta,1,t)} = \infty$ and $\lim_{x\to t^-} N_{(\beta,1,t)} = -\infty$, we know that the absolute maximum of $N'_{(\beta,1,t)}$ on W_2 exists and must occur at d. Therefore,

$$\max_{x\in W_2}\{N'_{(\beta,1,t)}(x)\} = N'_{(\beta,1,t)}(t/2) = -1 - \frac{1}{(t/2)^2} - \frac{1}{(t-(t/2))^2} = -1 - \frac{8}{t^2}.$$

Thus, there exist at most two $y_{21}, y_{22} \in W_2$ such that $N'_{(\beta,1,t)}(x) = N'_{(\beta,1,t)}(y)$. Therefore, $Y \subseteq \{y_1, y_{21}, y_{22}, y_3\}$. □

LEMMA 4.7. *If $z \in W_3$ and $z > \frac{1}{2}(t + t\sqrt{3})$, then $N'_{(\beta,1,t)}(z) > \max_{x \in W_2}\{N'_{(\beta,1,t)}(x)\}$.*

PROOF. Let $Y = \{y \in \mathbb{R} : N'_{(\beta,1,t)}(t/2) = N'_{(\beta,1,t)}(y)\}$. By Proposition 4.6, we know that $N'_{(\beta,1,t)}(t/2) = -1 - 8/t^2$ is the absolute maximum of $N'_{(\beta,1,t)}(x)$ on W_2, so $y_{21} = y_{22} = t/2$ (i.e. the only $y \in Y \cap W_2$ must be $t/2$ itself). Therefore, there are exactly three points, $y \in \mathbb{R}$, such that $N'_{(\beta,1,t)}(t/2) = N'_{(\beta,1,t)}(y)$. Computing the derivative and using Proposition 4.6 yields $-1 - \frac{8}{t^2} = -1 - \frac{1}{(t-y)^2} - \frac{1}{y^2}$. Solving for y yields $y_1 = \frac{1}{2}(t - \sqrt{3}t)$, $y_{21} = y_{22} = t/2$, and $y_3 = \frac{1}{2}(t + \sqrt{3}t)$. Recall that $N'_{(\beta,1,t)}(x) \to -1$ as $x \to \infty$, so for $z > y_3$ we have that $N'_{(\beta,1,t)}(z) > N'_{(\beta,1,t)}(t/2) = \max_{x \in W_2}\{N'_{(\beta,1,t)}(x)\}$.

□

We are now ready to prove Theorem 4.5

PROOF OF THEOREM 4.5. ($\Rightarrow$) Let $\phi : N_{(\beta,1,t)} \to N_{(\gamma,1,r)}$ be a 1-isomorphism. Let $\psi_{(t/2)}(x) = x - t/2$. Then $(\psi_{(t/2)} \circ N_{(\beta,1,t)} \circ \psi^{-1}_{(t/2)}) = S_{\text{sym}(\beta+\text{t},1,\text{t}/2)}$, so $N_{(\beta,1,t)}$ is 1-isomorphic to $S_{\text{sym}(\beta+\text{t},1,\text{t}/2)}$. Similarly, $N_{(\gamma,1,r)}$ is 1-isomorphic to $S_{\text{sym}(\gamma+\text{r},1,\text{r}/2)}$ via the conjugating map $\psi_{(r/2)}(x) = x - r/2$. Therefore, $(\psi_{(r/2)} \circ \phi \circ \psi^{-1}_{(t/2)}) : S_{\text{sym}(\beta+\text{t},1,\text{t}/2)} \to S_{\text{sym}(\gamma+\text{r},1,\text{r}/2)}$ is a 1-isomorphism. By Proposition 4.3 we have $\frac{r}{2} = \frac{t}{2}$, therefore $r = t$.

By the Definition 2.3, we have

$$\phi \circ N_{(\beta,1,t)} = N_{(\gamma,1,t)} \circ \phi, \text{ for almost every } x \in \mathbb{R}. \tag{4.14}$$

Let $\phi_*\lambda$ denote the pushforward measure; i.e. $\phi_*\lambda(A) = \lambda(\phi^{-1}A)$ for all measurable A. By Definition 2.3 and the fact that ϕ is a 1-isomorphism, we have $\frac{d\phi_*^{-1}\lambda}{d\lambda} = 1$. By the chain rule, taking the Jacobian of both sides yields

$$|N'_{(\beta,1,t)}(x)| = |N'_{(\gamma,1,t)}(\phi(x))|, \text{ for almost every } x \in \mathbb{R}. \tag{4.15}$$

Let $\phi(x) = y$. Recall that $N_{(\beta,1,t)}$ and $N_{(\gamma,1,t)}$ are uniformly decreasing on $\mathbb{R}\setminus\{0, t\}$, so (4.15) becomes

$$\frac{1}{(t-x)^2} + \frac{1}{x^2} = \frac{1}{(t-y)^2} + \frac{1}{y^2}. \tag{4.16}$$

Combining the fractions on each side yields

$$\frac{t^2 - 2ty + 2y^2}{y^2(t-y)^2} = \frac{t^2 - 2tx + 2x^2}{x^2(t-x)^2}. \tag{4.17}$$

Multiplying by the denominator of the left-hand-side and collecting everything on one side yields

$$t^2 - 2ty + 2y^2 - y^4\left(\frac{t^2 - 2tx + 2x^2}{x^2(t-x)^2}\right) + y^3 2t\left(\frac{t^2 - 2tx + 2x^2}{x^2(t-x)^2}\right) - y^2t^2\left(\frac{t^2 - 2tx + 2x^2}{x^2(t-x)^2}\right) = 0.$$

Combining the fractions yields

$$\frac{-(t-x-y)(y-x)(t^3x - t^2x^2 + t^3y - 2t^2xy + 2tx^2y - t^2y^2 + 2txy^2 - 2x^2y^2)}{x^2(t-x)^2} = 0.$$

Setting the numerator equal to 0 yields the following four candidates for $y = \phi(x)$,

$$(1)\quad y = x$$

$$(2)\quad y = -x + t$$

$$(3)\quad y = \frac{t^3 - 2t^2x + 2tx^2 - t\sqrt{t^4 - 4t^2x^2 + 8tx^3 - 4x^4}}{2(t^2 - 2tx + 2x^2)}$$

$$(4)\quad y = \frac{t^3 - 2t^2x + 2tx^2 + t\sqrt{t^4 - 4t^2x^2 + 8tx^3 - 4x^4}}{2(t^2 - 2tx + 2x^2)}.$$

Let $d = \max\{\frac{1}{2}(t + \sqrt{3}t), N^{-1}_{(\beta,1,t)}(0) \cap W_3, N^{-1}_{(\gamma,1,t)}(0) \cap W_3\}$, and suppose $x > d$ (i.e. x is large as in Lemma 4.7 and larger than the right-most root of both $N_{(\beta,1,t)}$ and $N_{(\gamma,1,t)}$). By Lemma 4.7, $Y = \{y \in \mathbb{R} : N'_{(\beta,1,t)}(x) = N'_{(\gamma,1,r)}(y)\} = \{y_1, y_3\}$, where $y_1 \in W_1$ and $y_3 \in W_3$, and we see that case (3) and (4) above are not real. Therefore, $\phi(x) = x$ or $\phi(x) = -x + t$. Almost every x in the interval $I = (d, \infty)$ satisfies the above requirements. Therefore, for almost every $x \in I$, we have $\phi(x) = x$ or $\phi(x) = -x + k$. Let $A = \{x \in I : \phi(x) = x\}$ and $B = \{x \in I : \phi(x) = -x + t\}$.

<u>Claim 1</u>: If $\lambda(A) > 0$, then $\gamma = \beta$.

By assumption (4.14) holds for almost every $x \in \mathbb{R}$. Pick $x \in A$ such that (4.14) holds for x and $N_{(\beta,1,t)}(x)$. Therefore, by (4.14) for x, we have

$$(4.18)\qquad (\phi \circ N_{(\beta,1,t)})(x) = (N_{(\gamma,1,t)} \circ \phi)(x) = N_{(\gamma,1,t)}(x),$$

and by (4.15) for $N_{(\beta,1,t)}(x)$

$$(4.19)\qquad N'_{(\beta,1,t)}(N_{(\beta,1,t)}(x)) = N'_{(\gamma,1,t)}(\phi(N_{(\beta,1,t)}(x))).$$

Note that $N_{(\beta,1,t)}(z) - N_{(\gamma,1,t)}(z) = \beta - \gamma$ (i.e. $N_{(\beta,1,t)}$ and $N_{(\beta,1,t)}$ differ by a constant). Therefore, $N'_{(\beta,1,t)}(z) = N'_{(\gamma,1,t)}(z)$ for all $z \in \mathbb{R}$, and in particular

$$(4.20)\qquad N'_{(\beta,1,t)}(N_{(\beta,1,t)}(x)) = N'_{(\gamma,1,t)}(N_{(\beta,1,t)}(x)).$$

Combining (4.19) and (4.20) yields

$$(4.21)\qquad N'_{(\gamma,1,t)}(\phi(N_{(\beta,1,t)}(x))) = N'_{(\gamma,1,t)}(N_{(\beta,1,t)}(x)).$$

By Proposition 4.6, if $\phi(N_{(\beta,1,t)}(x)) = y$, then y is one of four possible points $y_1 \in W_1$, $y_{21}, y_{22} \in W_2$, or $y_3 \in W_3$. Since $x > d$ both $N_{(\beta,1,t)}(x) < 0$ and $N_{(\gamma,1,t)}(x) < 0$. Therefore, $\phi(N_{(\beta,1,t)}(x)) = y_1 = N_{(\beta,1,t)}(x)$. Furthermore, by (4.18) we have $N_{(\beta,1,t)}(x) = N_{(\gamma,1,t)}(x)$, so $\gamma = \beta$. This proves Claim 1. In fact, we have also shown that if $\lambda(A) > 0$, then $\lambda(I \setminus A) = 0$.

<u>Claim 2</u>: If $\lambda(B) > 0$, then $\gamma = -(2t + \beta)$.

By the same argument as above pick $x \in B$ such that

$$(4.22)\qquad (\phi \circ N_{(\beta,1,t)})(x) = (N_{(\gamma,1,t)} \circ \phi)(x) = N_{(\gamma,1,t)}(-x + t).$$

and

$$(4.23)\qquad N'_{(\gamma,1,t)}(\phi(N_{(\beta,1,t)}(x))) = N'_{(\gamma,1,t)}(N_{(\beta,1,t)}(x)).$$

Assume also that x is large enough such that

$$(4.24)\qquad N_{(\gamma,1,t)}(-x + t) > t.$$

Our assumption that $\lambda(B) > 0$ implies $\lambda(I \setminus A) \neq 0$, so by the note at the end of Claim 1, we have $\lambda(A) = 0$. Therefore, $\lambda(I \setminus B) = 0$, and we know a large enough x exists in B to satisfy (4.24). Now, given (4.23), by Proposition 4.6 if $\phi(N_{(\beta,1,t)}(x)) = y$, then y is one of four possible points $y_1 \in W_1$, $y_{21}, y_{22} \in W_2$, and $y_3 \in W_3$. A calculation shows, $N'_{(\gamma,1,t)}(-N_{(\beta,1,t)}(x) + t) = N'_{(\gamma,1,t)}(N_{(\beta,1,t)}(x))$, so $-N_{(\beta,1,t)}(x)+t$ is a candidate for one of the y's. Since $x > d$, we know $N_{(\beta,1,t)}(x) < 0$, so $-N_{(\beta,1,t)}(x) + t > t$. Therefore, $-N_{(\beta,1,t)}(x) + t = y_3$. Also, x satisfies (4.24), so by (4.22), we have $N_{(\gamma,1,t)}(-x + t) = y_3 = \phi(N_{(\beta,1,t)}(x))$.Therefore, $N_{(\gamma,1,t)}(-x + t) = -N_{(\beta,1,t)}(x) + t$. Writing out both sides of this equation yields

$$-\left(-x-\beta-\frac{1}{-x}-\frac{1}{t-x}\right)+t=-(-x+t)-\gamma-\frac{1}{-(-x+t)}-\frac{1}{t-(-x+t)},$$

and simplifying gives $\gamma = -2t - \beta$.

($\Leftarrow$) If $r = t$ and $\gamma = \beta$, then let $\phi(x) = x$. If $r = t$ and $\gamma = -(2t + \beta)$, then let $\phi(x) = -x + t$. In either case, we have $\phi : N_{(\beta,1,t)} \to N_{(\gamma,1,r)}$, is a 1-isomorphism. □

COROLLARY 4.8. *A normalized Boole transformation, $N_{(\widehat{\beta},1,\widehat{t})}$, is obtained via 1-isomorphism from a symmetric Boole transformation, $S_{\text{sym}(\beta,1,\text{t})}$ as in (4.1), if and only if $t = \frac{\widehat{t}}{2}$ and $\beta = \widehat{t} + \widehat{\beta}$ or $\beta = -\widehat{t} - \widehat{\beta}$.*

PROOF. ($\Leftarrow$) As in Lemma 3.4, let $t_1 = -t$ and $t_2 = t$, then conjugating $S_{\text{sym}(\beta,1,\text{t})}$ by $\psi_{(-t)} = x - (-t)$ yields $\widehat{t} = 2t$ and $\widehat{\beta} = -2t + \beta$. Simultaneously solving these equations for t and β yield $t = \frac{\widehat{t}}{2}$ and $\beta = \widehat{t} + \widehat{\beta}$. Similarly, let $\eta_{(t)}(x) = t - x$ (as in Remark 4.4), then conjugating $S_{\text{sym}(\beta,1,\text{t})}$ by $\eta_{(t)}$ yields $\widehat{t} = 2t$ and $\widehat{\beta} = -2t - \beta$. Simultaneously solving these equations yields $t = \widehat{t}/2$ and $\beta = -\widehat{t} - \widehat{\beta}$.

($\Rightarrow$) Assume $\phi : S_{\text{sym}(\beta,1,\text{t})} \to^1 N_{(\widehat{\beta},1,\widehat{t})}$ is a 1-isomorphism. Also, assume that $\psi : S_{\text{sym}(\beta,1,\text{t})} \to^1 N_{(\beta',1,t')}$ is a 1-isomorphism. Then $\phi \circ \psi^{-1} : N_{(\beta',1,t')} \to N_{(\widehat{\beta},1,\widehat{t})}$ is a 1-isomorphism. By Theorem 4.5, we have $\widehat{t} = t'$ and $\widehat{\beta} = \beta'$ or $\widehat{\beta} = -2t' - \beta'$. Let $\psi = \psi_{(-t)}$ be defined as in the proof of the previous direction, then $t' = \widehat{t} = 2t$ and $\beta' = \beta - 2t$. If $\widehat{\beta} = \beta'$, then $\beta = \widehat{t} + \widehat{\beta}$. If $\widehat{\beta} = -2t' - \beta'$, then $\beta = -\widehat{t} - \widehat{\beta}$. □

COROLLARY 4.9. *If $S_{\text{sym}(\beta,1,\text{t})}$ and $S_{\text{sym}(\gamma,1,\text{r})}$ are two symmetric generalized Boole transformations as in (4.1), then $S_{\text{sym}(\beta,1,\text{t})}$ is 1-isomorphic to $S_{\text{sym}(\gamma,1,\text{r})}$ if and only if $r = t$ and $\gamma = \pm\beta$.*

PROOF. ($\Rightarrow$) Let $S_{\text{sym}(\beta,1,\text{t})}$ be 1-isomorphic to $S_{\text{sym}(\gamma,1,\text{r})}$. The normal forms of $S_{\text{sym}(\beta,1,\text{t})}$ are $N_{(-2t+\beta,1,2t)}$ and $N_{(-2t-\beta,1,2t)}$. Similarly, the normal forms of $S_{\text{sym}(\gamma,1,\text{r})}$ are $N_{(-2r+\gamma,1,2r)}$ and $N_{(-2r-\gamma,1,2r)}$. Then by Theorem 4.5, we have $2t = 2r$, so $t = r$. Also, $-2t + \beta = -2t + \gamma$ or $-2t + \beta = -2t - \gamma$, so $\beta = \pm\gamma$.
($\Leftarrow$) This direction is clear using the conjugating maps $\phi(x) = \pm Id$. □

Corollary 4.8 gives us a mechanism for reverse engineering symmetric Boole functions that are c-isomorphic to a normalized Boole function.

EXAMPLE 4.10. Consider the following normalized Boole function

$$N_{(3,1,4)} = -x - 3 - \frac{1}{-x} - \frac{1}{4-x}. \tag{4.25}$$

By Corollary 4.8, we know $N_{(3,1,4)}$ is the normalized form of a symmetric Boole transformation where $t = \widehat{t}/2 = 4/2 = 2$ and $\beta = \widehat{t} + \widehat{\beta} = 4 + 3$ or $\beta = -\widehat{t} - \widehat{\beta} = -4 - 3$. Therefore, $N_{(3,1,4)}$ came from one of the following two symmetric Boole transformations via a 1-isomorphism,

$$S_{\mathrm{sym}(7,1,2)}(x) = -x - 7 - \frac{1}{-2-x} - \frac{1}{2-x}, \text{ or}$$
$$S_{\mathrm{sym}(-7,1,2)}(x) = -x + 7 - \frac{1}{-2-x} - \frac{1}{2-x}.$$

We can push this a step further and recover some symmetric Boole maps that are c-isomorphic to $N_{(3,1,4)}$. Using the conjugating maps in Lemma 3.4 (letting $t_1 = -t$ and $t_2 = t$), we have $S_{\mathrm{sym}(\beta,\mathrm{p},\mathrm{t})}$ is $\sqrt{p}$-isomorphic to $N_{(\widehat{\beta},1,\widehat{t})}$, where $t = \sqrt{p}\cdot\widehat{t}/2$ and $\beta = \sqrt{p}(\widehat{\beta} + \widehat{t})$ or $\beta = \sqrt{p}(-\widehat{\beta} - \widehat{t})$. In other words, $N_{(3,1,4)}$ came from one of the following two symmetric Boole transformations via a $\sqrt{p}$-isomorphism,

$$S_{\mathrm{sym}(7\sqrt{\mathrm{p}},\mathrm{p},2\sqrt{\mathrm{p}})} = -x - 7\sqrt{p} - \frac{p}{-2\sqrt{p}-x} - \frac{p}{2\sqrt{p}-x}, \text{ or}$$
$$S_{\mathrm{sym}(-7\sqrt{\mathrm{p}},\mathrm{p},2\sqrt{\mathrm{p}})} = -x + 7\sqrt{p} - \frac{p}{-2\sqrt{p}-x} - \frac{p}{2\sqrt{p}-x}.$$

References

[1] J. Aaronson, *Ergodic theory for inner functions of the upper half plane* (English, with French summary), Ann. Inst. H. Poincaré Sect. B (N.S.) **14** (1978), no. 3, 233–253. MR508928

[2] J. Aaronson and K. K. Park, *Predictability, entropy and information of infinite transformations*, Fund. Math. **206** (2009), 1–21, DOI 10.4064/fm206-0-1. MR2576257

[3] R. L. Adler and B. Weiss, *The ergodic infinite measure preserving transformation of Boole*, Israel J. Math. **16** (1973), 263–278, DOI 10.1007/BF02756706. MR0335751

[4] I. N. Baker, *Fixpoints of polynomials and rational functions*, J. London Math. Soc. **39** (1964), 615–622, DOI 10.1112/jlms/s1-39.1.615. MR0169989

[5] R. L. Bayless (2013) Entropy of Infinite Measure-Preserving Transformations, PhD thesis, University of North Carolina at Chapel Hill.

[6] R. L. Bayless, *Ergodic properties of rational functions that preserve Lebesgue measure on* $\mathbb{R}$, Real Anal. Exchange **43** (2018), no. 1, 137–153, DOI 10.14321/realanalexch.43.1.0137. MR3816436

[7] R.L. Bayless-Rossetti and J. Hawkins (2018) A special class of infinite measure-preserving quadratic rational maps. Dynamical Systems. 1-16.

[8] A. F. Beardon, *Iteration of rational functions: Complex analytic dynamical systems*, Graduate Texts in Mathematics, vol. 132, Springer-Verlag, New York, 1991. Complex analytic dynamical systems. MR1128089

[9] G. Boole (1857) On the comparison of transcendence with certain applications to the theory of definite integrals. Philos. Trans. Roy. Soc. London. Vol. 147 Part III, 745–803

[10] Y.-Y. Chen, *Generalized Boole transformations with infinitely many singularities*, ProQuest LLC, Ann Arbor, MI, 2016. Thesis (Ph.D.)–Indiana University. MR3593172

[11] R. Hagihara, *Quadratic rational maps lacking period 2 orbits*, Proc. Amer. Math. Soc. **137** (2009), no. 9, 3077–3090, DOI 10.1090/S0002-9939-09-09852-9. MR2506466

[12] U. Krengel, *Entropy of conservative transformations*, Z. Wahrscheinlichkeitstheorie und Verw. Gebiete **7** (1967), 161–181, DOI 10.1007/BF00532635. MR0218522

[13] G. Letac, *Which functions preserve Cauchy laws?*, Proc. Amer. Math. Soc. **67** (1977), no. 2, 277–286, DOI 10.2307/2041287. MR0584393

[14] T. Y. Li and F. Schweiger, *The generalized Boole's transformation is ergodic*, Manuscripta Math. **25** (1978), no. 2, 161–167, DOI 10.1007/BF01168607. MR0499081

[15] W. Parry, *Entropy and generators in ergodic theory*, W. A. Benjamin, Inc., New York-Amsterdam, 1969. MR0262464

[16] G. Polya (1931) Problem. Jber. deutsch. Math. Verin. Vol. 40, 2. Abt. p.81.

[17] V.A. Rohlin (1963) Exact endomorphisms of a Lebesgue space, AMS Transl. Ser. 2(39), 1–36.
[18] E. Roy, *Poisson suspensions and infinite ergodic theory*, Ergodic Theory Dynam. Systems **29** (2009), no. 2, 667–683, DOI 10.1017/S0143385708080279. MR2486789
[19] G. Szegő (1934) Problem Solution. Jber. deutsh. Math. Verin. Vol. 43, 2. Abt. 17–20.

Department of Mathematics, Agnes Scott College, 141 E. College Ave. Box 1093 Decatur, Georgia 30030

Email address: rrossetti@agnesscott.edu

Contemporary Mathematics
Volume **736**, 2019
https://doi.org/10.1090/conm/736/14847

Orbit classification and asymptotic constants for d-symmetric covers

Martin Schmoll

ABSTRACT. We show that the parameter space of cyclic degree d covers of a marked Riemann surface (X, x_0) fully branched over two distinguished points is $\mathcal{C}_d(X) = (X_{ab} - H_1(X;\mathbb{Z})\widetilde{x}_0)/dH_1(X;\mathbb{Z})$, where $\pi_{ab} : X_{ab} \to X$ is the universal abelian cover. Pullback of a holomorphic 1-form ω on X gives 1-forms and translation structures on all covers of X. We call degree d cyclic covers with the pull back 1-form d-symmetric covers and use the induced flat geometry on $\mathcal{C}_d(X, x_0, \omega) = (\mathcal{C}_d(X), \pi_{ab}^*\omega)$ to study the geometry of individual d-symmetric covers. The geometry on $\mathcal{C}_d(\mathbb{C}/\mathbb{Z}[i])$ allows a straightforward $\mathrm{SL}_2\mathbb{Z}$ orbit classifiaction for d symmetric torus covers and en route a classification of their Teichmüller curves. For general d-symmetric covers we present formulas for asymptotic quadratic growth rates (Siegel-Veech constants) of geodesic loops and other geodesic segments in terms of the parameter space geometry. Combining the orbit classification and formulas for Siegel-Veech constants, we carry out nearly complete calculations for d-symmetric covers.

1. Introduction and Results

This paper is concerned with geometric problems and classification problems for cyclic covers and studies those from a global and geometric viewpoint. Instead of looking at a particular cover over a fixed base surface, or the (finite) set of, say cyclic torus covers of fixed degree and branching, we view a cover as point in a parameter surface, obtained by varying the relative branching loci. Under our assumptions the parameter surface will itself be a cover of the base surface. A flat metric, induced by a 1-form on the base surface, pulls back to any cover, in particular to the parameter curve. Once equipped with a pull back 1-form we call a cyclic cover of degree d a d-symmetric differential. Using the parameter curves, it is almost elementary to classify Teichmüller curves defined by d-symmetric differentials that cover tori $\mathbb{C}/\Lambda$ equipped with the standard 1-form dz inherited from $\mathbb{C}$. The covering and translation structure of the parameter space has been used in **[S2, EMS]** and more recently by Duryev **[D18]** to study Teichmüller curves for genus 2 torus covers.

In general both discovery and classification of Teichmüller curves is challenging. As for the classification part, this is even more true for torus covers since those are abundant. In fact, covers of $\mathbb{T} := \mathbb{C}/\mathbb{Z}[i]$ branched only over rational

2010 *Mathematics Subject Classification.* 14H15, 14H52, 30F30, 30F60, 37C85, 58D15, 58D27.

The author was partially supported by Simons Collaboration Grant #318898.

points give Teichmüller curves. Despite recent progress by Eduard Duryev [**D18**] the Teichmüller curve classification is still open in genus 2. Duryev uses the flat geometry of the moduli curve paired with some topological properties. Our initial motivation was to give a class of examples, where the asymptotic growth rate can be can be determined for all points of the parameter curve and not only for generic covers , i.e. those that are not Teichmüller curves, as in [**EMS**]. There is a good amount of literature devoted to Teichmüller curves, such as the Teichmüller curves stemming from the billiard in a regular polygon found by Veech [**V1**], and some induced by genus two surfaces described by Calta [**C**] and independently McMullen [**McM1**],...,[**McM5**]. More examples were found by Bouw and Möller [**BM**] and Hooper [**H**]. Cyclic covers on the other hand can be constructed easily and have been used to built examples with interesting properties and as a periscope to study flat surfaces in higher genus. For results in this direction see [**EKZ11**], [**FMZ1**] and for an application [**FS**].

1.1. Results. Let X be a compact Riemann surface. A cyclic cover $Y \to X$ of degree d is called *d-symmetric*, if it is fully branched over two *distinct* points $x_0 \neq x_1$ in X. A cover is *fully branched*, if all branch points have maximal order d. Let us call two d-symmetric covers $Y_1 \to X$ and $Y_2 \to X$ *isomorphic*, if there exists an orientation preserving homeomorphism $Y_1 \to Y_2$ that is $\mathbb{Z}_d$ equivariant. In addition we identify $\mathbb{Z}_d$ covers that agree up to a change of the $\mathbb{Z}_d$ action by a $\mathbb{Z}_d$ homomorphism, given by multiplication with a unit $\mathbb{Z}_d^*$. Let $\pi_{ab} : (X_{ab}, x_{ab}) \to (X, x_0)$ be the universal abelian cover of X marked in $x_{ab} \in \pi_{ab}^{-1}(x_0)$. The universal abelian cover is the regular cover with deck group $H_1(X, \mathbb{Z}) \cong \mathbb{Z}^{2g}$, g the genus of X. For an integer $d > 1$ we denote the $\mathbb{Z}$-submodule $dH_1(X, \mathbb{Z}) \subset H_1(X, \mathbb{Z})$ by dH_1 and $H_1(X, \mathbb{Z})$ by H_1. Let $\pi_d : X_{ab}/dH_1 \to X$ be the quotient cover and $x_d = x_{ab} + dH_1$. We show:

THEOREM 1. *Fix $x_0 \in X$, then isomorphy classes of d-symmetric covers $\pi : Y \to X$ branched over x_0 and $x \in X \backslash \{x_0\}$ are parameterized by $((X_{ab} - H_1 \cdot x_{ab})/dH_1, x_d)$. If the class of the cover π is given by the point $z_\pi \in X_{ab}/dH_1$, then $\pi_d(z_\pi) = x$.*

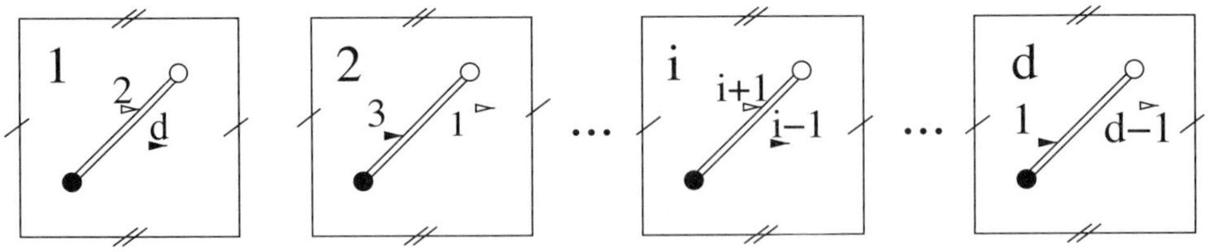

FIGURE 1. A d-symmetric cover of the standard torus.

Torus covers. If X is a complex torus, we may identify it with its Jacobian $Jac(X) \cong \mathbb{C}/\Lambda$, where Λ is the lattice $\{\int_\gamma \omega : \gamma \in H_1(X; \mathbb{Z})\}$ generated by a nontrivial holomorphic one form $\omega \in \Omega(X)$. In particular $H_1(X; \mathbb{Z}) \cong \Lambda$. Put $\mathbb{T}_\Lambda := \mathbb{C}/\Lambda$, and recall the subset of points $\mathbb{T}_\Lambda[n] \subset \mathbb{T}_\Lambda$ that become $[0]$ when multiplied with $n \in \mathbb{N}$ are the *n-torsion points* of $\mathbb{T}_\Lambda$. Denote the n-torsion points, that are not already m-torsion points for some $m|n$ by $\mathbb{T}_\Lambda(n)$ and call them the *primitive n-torsion* points. If $\pi : Y \to \mathbb{C}/\Lambda$ is a cover branched over two distinct points $[z_1]$ and $[z_0]$ define $b_\pi = [z_1 - z_0] \in \mathbb{T}_\Lambda$.

COROLLARY 1 (Torus covers). *Let $\Lambda \subset \mathbb{C}$ be a lattice, then any d-symmetric torus cover $\pi : Y \to \mathbb{C}/\Lambda$ is represented by a point $z_\pi \in (\mathbb{C} - \Lambda)/d\Lambda$, so that $\pi_d(z_\pi) = b_\pi$.*

Tori, that are a quotient of $\mathbb{C}$ by a lattice inherit a natural euclidean metric from $\mathbb{C}$. Since the metric tensor $|dz|^2 = dz \otimes d\bar{z}$ depends only on the differential dz, we obtain a metric on Y via the pullback $\pi^* dz = \omega$. In fact any non trivial holomorphic differential $\omega \in \Omega(X)$ on a Riemann surface X determines a flat metric away from the zeros of X, this pair is denoted by $(X, |\omega|)$. The Riemannian metric given by a 1-form is an euclidean metric with cone point type singularities at the zeros of ω, for more see **[Zo]**. Returning to tori, let $\mathrm{SL}(\Lambda) \subset \mathrm{SL}_2\mathbb{R}$ be the group of orientation preserving real linear maps of $\mathbb{C}$ that stabilize the lattice Λ. These maps descend to $\mathbb{C}/\Lambda$ and define orientation preserving affine homeomorphisms of $\mathbb{C}/\Lambda$ fixing $[0] \in \mathbb{C}/\Lambda$. The group $\mathrm{SL}(\Lambda)$ acts on branched covers of $\mathbb{C}/\Lambda$ by post-composition and acts on the torus $(\mathbb{C} - \Lambda)/d\Lambda$, since the set of removed points is $\mathrm{SL}(\Lambda)$ invariant. The $\mathrm{SL}(\Lambda)$ action on covers and on their parameter space are compatible in the sense:

$$A \cdot \#^d_z(\mathbb{C}/\Lambda, dz) = \#^d_{Az}(\mathbb{C}/\Lambda, dz). \tag{1}$$

Here $\#^d_z(\mathbb{C}/\Lambda, dz)$ denotes a d-symmetric torus cover in the isomorphy class given by $z \in (\mathbb{C} - \Lambda)/d\Lambda$. The dotted $\mathrm{SL}(\Lambda)$-action is the one on covers and Az denotes the action on $\mathbb{C}/\,\mathrm{d}\,\Lambda$ induced from the real linear action of $A \in \mathrm{SL}(\Lambda)$ on $\mathbb{C}$. The following is well known:

PROPOSITION 1. *For any $n \in \mathbb{N}$ the primitive n-torsion points lie on a single* $\mathrm{SL}(\Lambda)$ *orbit. Those are all finite* $\mathrm{SL}(\Lambda)$ *orbits. All other* $\mathrm{SL}(\Lambda)$ *orbits are dense in* $\mathbb{T}_\Lambda$.

One can show this statement about finite orbits by successively applying parabolic elements from $\mathrm{SL}(\Lambda)$ by moving a general primitive n-torsion point into a particular primitive n-torsion point. One can interpret this fact in terms of flat geometry: The complex linear bijective map *multiplication by $n \in \mathbb{N}$*, that is $z \mapsto nz$, maps the integer lattice $\Lambda = \mathbb{Z}[i]$ to the lattice $n\mathbb{Z}[i]$, is $\mathrm{SL}_2\mathbb{Z}$ equivariant and hence induces an $\mathrm{SL}_2\mathbb{Z}$-equivariant bijective map $(\mathbb{C}/\mathbb{Z}[i], dz) \to (\mathbb{C}/n\mathbb{Z}[i], dz)$. Since there is a map in $\mathrm{SL}_2\mathbb{Z}$ mapping $z \in \mathbb{Z}[i]$ to 1, if and only if $\gcd(\mathrm{Re}\, z, \mathrm{Im}\, z) = 1$ the orbit classification is equivalent to the existence of a line segment from 0 to some $z + n\mathbb{Z}[i]$ that does not contain a point of $n\mathbb{Z}[i]$ in its interior. Thus on the torus, there is a line segment from $[0]$ to $[z]$, that is, besides its endpoints, completely contained in $(\mathbb{C} - n^{-1}\mathbb{Z}[i])/\mathbb{Z}[i]$, or in other words: $[z]$ can be *illuminated* from $[0]$ on $(\mathbb{C} - n^{-1}\mathbb{Z}[i])/\mathbb{Z}[i]$. For the integer lattice $\mathbb{Z}[i]$ we note:

COROLLARY 2. *For $d \in \mathbb{N}$ every finite* $\mathrm{SL}_2\mathbb{Z}$ *orbit on $\mathbb{C}/d\mathbb{Z}[i]$ is a set of primitive n torsion points for some $n \in \mathbb{N}$. The stabilizer of the point $[d/n]$ is the congruence group $\Gamma_1(n) \subset \mathrm{SL}_2\mathbb{Z}$.*

These rather elementary observations on the $\mathrm{SL}_2\mathbb{Z}$ action on tori become powerful statements when we interpret those tori as parameter spaces of d-symmetric covers. Using Corollary 2 for instance, one easily obtains the $\mathrm{SL}_2\mathbb{Z}$ orbit decomposition for d-symmetric covers branched over a given rational point. For $n \in \mathbb{Z}$, let

$D(n)$ denote the *number of positive divisors of* n. If further $d \in \mathbb{N}$ let d_n be the maximal divisor of d that is coprime to n, i.e. $d_n | d$ is maximal with $\gcd(d_n, n) = 1$.

THEOREM 2. *[Covers with torsion branching] The set of d-symmetric covers π with $b_\pi \in \mathbb{T}_\Lambda(n)$ consists of d^2 covers that lie on $D(d_n)$ different $\mathrm{SL}(\Lambda)$ orbits.*

This statement is shown as part of Proposition 6 on page 211 and Corollary 5 on page 211. The following corollary generalizes the appearance of either two or one $\mathrm{SL}(\Lambda)$-orbits for torsion torus covers in genus 2, see [**Ka1, HL**] and [**McM2**], for d-symmetric torus covers:

COROLLARY 3. *If d is prime and $n \in \mathbb{N}$, then the d-symmetric covers with relative branching $b_\pi \in \mathbb{T}_\Lambda(n)$ are contained in one $\mathrm{SL}(\Lambda)$-orbit if $d|n$, otherwise* $\gcd(d, n) = 1$ *and the covers lie on two $\mathrm{SL}(\Lambda)$-orbits.*

In particular the four genus 2-symmetric covers with relative branching $b_\pi \in \mathbb{T}_\Lambda(n)$ are on one $\mathrm{SL}(\Lambda)$ orbit, if n is even and on two $\mathrm{SL}(\Lambda)$ orbits, if n odd.

Each finite $\mathrm{SL}(\Lambda)$ orbit of d-symmetric covers is (roughly speaking) the intersection of a single Teichmüller curve with the covers of a fixed torus. For more on Teichmüller curves see section 6. Reformulating Corollary 3 and Theorem 3 in terms of Teichmüller curves gives:

COROLLARY 4. *For any $d, n \in \mathbb{N}$ the Teichmüller curve determined by any primitive n-torsion point of $(\mathbb{C} - \Lambda)/d\Lambda$ is isometric to $\mathbb{H}/\Gamma_1(n)$. If d is prime and n a multiple of d, then every d-symmetric cover with branching $b_\pi \in \mathbb{T}[n]$ is on one Teichmüller curve. If d is prime and n is not a multiple of d, then there are* 2 *Teichmüller curves containing the covers with branching $b_\pi \in \mathbb{T}(n)$.*

In summary, for d-symmetric torus covers the $\mathrm{SL}(\Lambda)$ orbit classification is equivalent to an, in this case elementary, illumination principle. Consider a set $A \subset X$ of a translation surface (X, ω). We say that a point $p \in \overline{A}$ is *visible* or *illuminated* from another point in $q \in \overline{A}$, if there is a regular line segment, i.e. one that does not contain zeros of ω, from p to q that is contained in A. In words, we require the segment away from its endpoints to be in A. The following states that some torsion points in the space of d-symmetric covers cannot be illuminated.

PROPOSITION 2. *If $d \in \mathbb{N}$, the only points on $(\mathbb{C} - \Lambda)/d\Lambda$ that are not illuminated from* $[0]$ *are those primitive n torsion points with $n < d$ so, that n does not divide d.*

To explain the relevance of illumination, we look at the above spaces as parameter spaces of certain covers of a surface, say X. In this case a given point z in parameter space represents a cover $X_z \to X$ and a line illuminating z projects to a line segment $s \subset X$ in the complement of the zero set of $\omega \in \Omega(X)$. This regular line segment allows us to construct the translation cover X_z via a cutting, copy and paste construction along s. The existence of such an illuminating segment allows us to apply maps from $\mathrm{SL}(X) \subset \mathrm{SL}_2\mathbb{R}$, if non trivial, to move the cover around in parameter space and shorten, in the finite orbit case even minimize, the length of the line segment.

1.2. Counting line segments on d-symmetric torus covers. The standard foliation of $\mathbb{R}^2$ by oriented parallel lines tangent to $\theta \in S^1$ descend to line foliations $\mathcal{F}_\theta(\mathbb{C}/\Lambda)$ on the translation torus $\mathbb{C}/\Lambda$. By pullback we obtain direction foliations on every (translation) cover of a translation torus. Direction foliations can be defined on any translation surface, see section 3. We are mainly interested in two types of leaves of direction foliations, the *compact leaves* and the *saddle connections*. A saddle connection is a leaf that does not contain any zero of ω, or marked point, but is bounded at both ends by those. Compact leaves on the other hand appear in families of parallel, isotopic loops that cover a maximal open cylinder bounded by saddle connections. The *length* of a compact leaf equals the circumference, or *width* of the cylinder the leaf determines with respect to the induced euclidean metric.

Below we restrict our presentation to d-symmetric covers $\#_z^d(\mathbb{T}, dz)$ of $\mathbb{T}$. Let us consider the counting functions:

$$N_C(\#_z^d(\mathbb{T}, dz), T) := \#\{\text{iso. classes of cpt. leaves on } \#_z^d(\mathbb{T}, dz) \text{ of length} \leq T\}$$

and its equivalent $N_{SC}(\#_z^d(\mathbb{T}, dz), T)$ for saddle connections. For the torus $\mathbb{T}$ itself one has the elementary $\lim_{T\to\infty} \frac{N_C(\mathbb{T},T)}{T^2} = 6/\pi$ obtained by counting integer lattice points visible from the origin. This quadratic asymptotic constant exists for d-symmetric covers as well, and can be expressed using flat geometric data of the parameter space equipped with the pullback flat structure from $\mathbb{T}$. The general formalism is presented and applied to genus 2 torus covers in **[EMS]**. Below we calculate the normalized quadratic constants $\mathbf{c}_C(S) := \frac{\pi}{6} \lim_{T\to\infty} \frac{N_C(S,T)}{T^2}$ for d-symmetric torus covers. Those constants are generally known as Siegel-Veech constants, see **[V3, EMZ]** and **[EMS]**. Foundational work on the asymptotic growth rates, such as quadratic estimates, has been done by Howard Masur **[Ma1, Ma2, Ma3]** and William Veech **[V1, V2, V3]**.

Since the Siegel-Veech constants are independent of the (unimodular) lattice, we restrict some of the following statements to the integer lattice. If $\Lambda = \mathbb{Z}[i]$ the horizontal direction on $\mathbb{T}$ is periodic and so the horizontal direction of $(\mathbb{C} - \mathbb{Z}[i])/d\mathbb{Z}[i]$ decomposes into d maximal cylinders foliated by horizontal leaves. More precisely, denote by $\mathscr{C}_k \subset (\mathbb{C} - \mathbb{Z}[i])/d\mathbb{Z}[i]$ the maximal open cylinder containing the image of the point $(k - 1/2)i \subset \mathbb{C}$ under the covering map $\mathbb{C} - \mathbb{Z}[i] \to (\mathbb{C} - \mathbb{Z}[i])/d\mathbb{Z}[i]$. Further denote the upper boundary of the cylinder $\mathscr{C}_k$ by $\partial^{top}\mathscr{C}_k$. Each boundary contains d saddle connections. Attached to each cylinder and boundary is a datum, say $(w_{k1}, ..., w_{kn_k}) \in \mathbb{N}^{n_k}$ for $\mathscr{C}_k$ and $(b_{k1}, ..., b_{km_k}) \in \mathbb{N}^{m_k}$ for $\partial\mathscr{C}_k$, that records the length of every horizontal cylinder on each d-symmetric differential $\#_z^d(\mathbb{T}, dz)$ whenever $z \in \mathscr{C}_k$ and $z \in \partial\mathscr{C}_k$ respectively. This definition uses, that the circumference of horizontal cylinders on $\#_z^d(\mathbb{T}, dz)$ only depend on the cylinder $\mathscr{C}_k$, or saddle connection $\partial\mathscr{C}_k$ that contains z. Any periodic direction on $(\mathbb{C}-\mathbb{Z}[i])/d\mathbb{Z}[i]$ has a rational slope, so that there is an element of $\mathrm{SL}_2\mathbb{Z}$ mapping it to the horizontal direction.

THEOREM 3. **[Cylinders]** *Under the previous assumptions and conventions, if* $z \in (\mathbb{C} - \mathbb{Z}[i])/d\mathbb{Z}[i]$ *has infinite* $\mathrm{SL}_2\mathbb{Z}$ *orbit, the Siegel-Veech constant for maximal cylinders for* $\#_z^d(\mathbb{T}, dz)$ *is:*

$$\mathbf{c}_C(\#_z^d(\mathbb{T}, dz)) = \frac{1}{d} \sum_{k=1}^{d} \sum_{j=1}^{n_k} \frac{1}{(w_{jk})^2}. \tag{2}$$

If $z \in (\mathbb{C} - \mathbb{Z}[i])/d\mathbb{Z}[i]$ *has finite* $\mathrm{SL}_2\mathbb{Z}$ *orbit, say* $\mathscr{O}_z := \mathrm{SL}_2\mathbb{Z} \cdot z$*, then:*

$$\mathbf{c}_C(\#_z^d(\mathbb{T}, dz)) = \frac{1}{|\mathscr{O}_z|} \sum_{k=1}^{d} \left(\sum_{j=1}^{n_k} \frac{|\mathscr{O}_z \cap \mathscr{C}_k|}{(w_{kj})^2} + \sum_{j=1}^{m_k} \frac{|\mathscr{O}_z \cap \partial^{top}\mathscr{C}_k|}{(b_{kj})^2} \right). \tag{3}$$

We show these formulas for general d-symmetric differentials in section 5.

To calculate the Siegel-Veech constants in Theorem 3 for d-symmetric torus covers we need the circumferences and multiplicities of horizontal cylinders for each surface in a cylinder $\mathscr{C}_k$.

PROPOSITION 3. *The d-symmetric surface* $\#_z^d(\mathbb{T}, dz)$ *has the following horizontal cylinder decomposition:*

$$\left.\begin{array}{l} \gcd(k,d) \textit{ cylinders of width } \frac{d}{\gcd(k,d)} \textit{ and} \\ \gcd(k+1,d) \textit{ cylinders of width } \frac{d}{\gcd(k+1,d)} \end{array}\right\} \quad \textit{if } z \in \mathscr{C}_k,$$

and

$$\gcd(k,d) \textit{ cylinders of width } \frac{d}{\gcd(k,d)}, \qquad \textit{if } z \in \partial^{top}\mathscr{C}_k$$

It remains to determine how many points of each finite $\mathrm{SL}_2\mathbb{Z}$ orbit happen to be in a particular horizontal cylinder, or on a particular saddle connection in $(\mathbb{C}/d\mathbb{Z}[i], d\mathbb{Z}[i])$. This question is considered in section 10.

Weighted counting formulas. In [**EKZ11**, **EKZ14**] it was shown, that *Lyapunov exponents* of the Teichmüller geodesic flow and Siegel-Veech constants are related. For this relation one needs to count cylinders weighted by their area:

$$N_{\mathrm{C},\alpha}(S,T) := \sum_{\substack{C \in \mathrm{C}(S) \\ |C| < T}} \mathrm{area}^\alpha(C),$$

where $\mathrm{C}(S)$ denotes the set of cylinders on S. Note, that for $\alpha = 0$ this is the counting problem that Siegel-Veech constant given by formula 2 in Theorem 3. The normalized Siegel-Veech constant is again defined as:

$$\mathbf{c}_{\mathrm{C},\alpha}(S) = \frac{\pi}{6} \lim_{T \to \infty} \frac{N_{\mathrm{C},\alpha}(S,T)}{T^2}.$$

Below we use Euler's totient: $\varphi(p) = \#\{j \in \mathbb{Z}_d : \gcd(j,p) = 1\}$.

THEOREM 4. *The Siegel-Veech constants for the weighted asymptotic quadratic growth rate of cylinders on* $\#_z^d(\mathbb{T}_\Lambda, dz)$ *with* $z \in \mathbb{C}/d\Lambda$ *of infinite* $\mathrm{SL}(\Lambda)$*-orbit is:*

$$\mathbf{c}_{C,\alpha}(\#_z^d(\mathbb{T}, dz)) = \frac{2}{1+\alpha} \sum_{p|d} \frac{\varphi(p)}{p^{3-\alpha}}. \tag{4}$$

Note, that $\alpha \mapsto \mathbf{c}_{C,\alpha}$ is continuous. For $\alpha = 0$, suppressing its dependence on d, we obtain the standard cylinder Siegel-Veech constant:

$$\mathbf{c}_C = \mathbf{c}_{C,0} = 2 \sum_{p|d} \frac{\varphi(p)}{p^3}. \tag{5}$$

If on the other hand $\alpha = 1$, that is weighing with the cylinder area, we obtain the sum of Lyapunov exponents [**EKZ11**]:

$$\sum_{i=1}^{d} \lambda_i = \mathbf{c}_{C,1} + \frac{1}{6}\frac{d^2-1}{d} = \sum_{p|d} \frac{\varphi(p)}{p^2} + \frac{1}{6}\frac{d^2-1}{d}. \tag{6}$$

During the time this manuscript went through the review process, the author learned that David Aulicino has independently calculated these and other *area Siegel-Veech constants*. For a general evaluation of generic area Siegel-Veech constants for branched cyclic torus covers see the forthcoming article [**AS**]. For a background on Lyapunov exponents, see [**EKZ11**]. In section 10 and section 11 we evaluate quadratic constants for (some) d-symmetric differentials with finite orbit.

Acknowledgments. An early version of this paper, already containing all non-weighted Siegel-Veech constants, would not have been possible without the support of Serge Troubetzkoy (IML-Marseille), Anatole Katok (Penn State) and Marc Burger (ETH). The author thanks the referee for many suggestions and observations that helped to improve the paper.

2. Background on translation surfaces

Below we recall briefly some standard terminologies in the theory of translation surfaces. For a more comprehensive background this topic see [**MT**] and [**Zo**]. In this section we assume X is a Riemann surface, either compact, or compact after adding finitely many points. As in the introduction we consider a metric on X defined by a holomorphic 1-form $\omega \in \Omega(X)$. More generally the 1-form induces a *translation structure* on $\mathring{X} := X \backslash Z(\omega)$, where $Z(\omega)$ is the zero set of ω, as follows. Around every point $x_0 \in \mathring{X}$ $\zeta_0(x) := \int_{x_0}^{x} \omega$ defines *natural* charts. The maximal atlas associated to those charts is called *translation structure*. Indeed, in natural charts the coordinate changes are translations $\zeta_0 = \zeta_1 + c$, $c \in \mathbb{C}$, because $d\zeta_0 = \omega = d\zeta_1$. A translation structure allows us to pull back all translation invariant geometric objects of $\mathbb{R}^2$ to $\mathring{X}$. First, the euclidean metric on $\mathbb{C} \cong \mathbb{R}^2$ is translation invariant and defines a flat metric $|\omega|$ on $\mathring{X}$. Next, the translation invariant line field on $\mathbb{C}$ in a given direction $\theta \in S^1 \subset \mathbb{C}$, viewed as points on the unit circle, defines a directed line field $\mathcal{F}_\theta(\omega)$ on $\mathring{X}$. Note that θ is well-defined on the surface since chart changes are translations. In particular we have a canonical trivialization of the (unit) tangent bundle on $\mathring{X}$. Finally, dz is invariant under translations and defines a 1-form on $\mathring{X}$. That form is ω, since $dz = d\zeta = \omega$. All pullback structures extend to the whole surface X with some care. The metric's $|\omega|$ continuation has cone point singularities in $Z(\omega)$: An order n zero of ω is a metric cone point of total angle $2(n+1)\pi$. Near a zero of order n, ω looks like $z^n dz$. All objects, including the line fields can be defined directly using ω, for example $\mathcal{F}_\theta(\omega) = \ker \operatorname{Im} e^{-i\theta}\omega$ with leaves l directed so that $\int_l \operatorname{Re} e^{-i\theta}\omega > 0$. A pair (X, ω) is called a translation surface. The leaves of any direction foliation of (X, ω) are traces of geodesics and it is convenient to identify them with the respective geodesic. A leaf of a direction foliation is called *regular*, if it does not contain a singular point, or removed point. The leaves that start and terminate at singular points, are called *saddle connections*. For our purpose it is useful to extend that definition to leaves that start and terminate at marked points or removed points of (X, ω). Closed regular leaves on (X, ω)

appear in families of isotopic parallel lines. A maximal family occupies a region on (X, ω) that is isometric to an open cylinder $\mathbb{R}/w\mathbb{Z} \times (0, h)$, where the leaves are represented by the loops given by the level sets of the projection from $\mathbb{R}/w\mathbb{Z} \times (0, h)$ on $(0, h)$. In particular the *width* w of the cylinder is the $|\omega|$-length of any leaf in the family. The boundary of a maximal cylinder consists of saddle connections. The pull back ω along any (branched) covering map $\pi : Y \to X$ of Riemann surfaces, gives a translation structure $(Y, \pi^*\omega)$ on Y. We call the map of pairs a translation covering. Note, that if a direction on (X, ω) contains a cylinder of closed leaves, then this is true for the same direction on all translation covers of (X, ω). The cylinder width on a cover is an integer multiple of the width of its image cylinder on (X, ω). By eventually considering the maximal cylinders of $(X, \{\text{branch points}\})$ with branch points marked, the height of a cylinder and of any of its preimages on a cover are the same.

2.1. Affine group and Veech group. Let us consider the group $\mathrm{SL}_2\mathbb{R}$ acting real linearly on $\mathbb{C}$. Then it acts by post-composition of charts on translation structures to give a new translation structure on X. The translation structure obtained by postcomposition with $A \in \mathrm{SL}_2\mathbb{R}$ is characterized by the 1-form $A\omega$. An *affine map* $\phi : (X, \omega) \to (X, \omega)$ is orientation preserving homeomorphism of X that is affine linear in natural charts (defined by the translation structure). If X is connected, than its derivative $D\phi$ is necessary constant and it is not hard to see $D\phi \in \mathrm{SL}_2\mathbb{R}$. The affine maps of (X, ω) are a group denoted by $\mathrm{Aff}^+(X, \omega)$, its image $\mathrm{SL}(X, \omega) := D\,\mathrm{Aff}^+(X, \omega) \subset \mathrm{SL}_2\mathbb{R}$ is commonly called the *Veech group* of (X, ω).

Two translation structures defined by (X, ω_1) and (X, ω_2) are *equivalent*, if there is an affine map $\phi : X \to X$, with respect to natural charts in each 1-form, so that $\phi^*\omega_2 = D\phi \cdot \omega_1$. Here $D\phi \cdot \omega_1 := (1, i)D\phi(\mathrm{Re}\,\omega, \mathrm{Im}\,\omega)^T$ with $D\phi$ in matrix representation. If in particular $\omega_1 = \omega_2 = \omega$ then we have that (X, ω) and $(X, A\omega)$ induce the same translation structure whenever $A \in \mathrm{SL}(X, \omega)$. The affine group of a fixed translation surface (X, ω) acts on covers of (X, ω) by postcomposition.

We call two translation covers of equivalent, if the diagram

$$\begin{array}{ccc} (Y_1, \tau_1) & \xrightarrow{\ \widetilde{\psi}\ } & (Y_2, \tau_2) \\ {\scriptstyle \pi_1}\downarrow & & \downarrow{\scriptstyle \pi_2} \\ (X, \omega_1) & \xrightarrow{\ \psi\ } & (X, \omega_2) \end{array} \tag{7}$$

commutes. Here $\psi : (X, \omega_1) \to (X, \omega_2)$ is an affine map and consequently $\widetilde{\psi}$ is an affine lift of ψ. All covering maps are translation maps. In other words, the 1-forms in the upper row of the diagram are pull backs. If $\omega = \omega_1 = \omega_2$, we must have $D\psi = D\widetilde{\psi} \in \mathrm{SL}(X, \omega)$ and $D\widetilde{\psi} \cdot \tau_1 = \widetilde{\psi}^*\tau_2$.

The group $\mathrm{SL}_2\mathbb{R}$ acts on the translation structure of a cover of (X, ω) by post-composing cover and base with the same element of $\mathrm{SL}_2\mathbb{R}$. If $A \in \mathrm{SL}_2\mathbb{R}$ is so that $A = D\psi$ for some $\psi \in \mathrm{Aff}^+(X, \omega)$, then post-composition of the deformed translation structure gives a new translation covering. It is equivalent to the old one in the sense above, if ψ lifts.

Lattice surfaces and optimal dynamics. A translation surface (X, ω) is called a *lattice surface*, or *Veech surface*, if $\mathrm{SL}(X, \omega)$ is a lattice in $\mathrm{SL}_2\mathbb{R}$. This property is equivalent for the volume of $\mathrm{SL}_2\mathbb{R}/\mathrm{SL}(X,\omega)$ being finite, or the hyperbolic area of $\mathcal{H}/\mathrm{SL}(X,\omega)$ being finite. A branched cover $f : Y \to X$ of a lattice surface (X,ω) is itself a lattice surface $(Y, f^*\omega)$, if f is ramified over points with finite $\mathrm{Aff}^+(X,\omega)$ orbit, such as the zeros of ω. It is rather generally difficult to find lattice surfaces, that are not already covers of lattice surfaces, for the description of several families see **[V1, C, McM1]** and **[BM]**.

Example. If $\Lambda \subset \mathbb{C}$ is a lattice then the torus $\mathbb{C}/\Lambda$ is a lattice surface with Veech group conjugate to $\mathrm{SL}_2\mathbb{Z}$. The group $\mathrm{SL}_2\mathbb{Z}$ is the Veech group of the standard torus $\mathbb{C}/\mathbb{Z}[i]$. The action of the group $\mathrm{SL}_2\mathbb{Z}$ on the torus is induced by its linear action on $\mathbb{R}^2$, in particular all rational points are periodic.

Lattice surfaces have a property known as *optimal dynamics*. That is, in a given direction on a lattice surface (X,ω) all leafs of the foliation $\mathcal{F}_\theta(X,\omega)$ are either compact, or dense. In fact, for a dense direction the directional (speed 1) flow is ergodic with respect to the flow invariant measure induced by Lebesgue measure on $\mathbb{C}$. In a direction with compact leaves those form maximal open cylinders of parallel periodic leaves bounded by saddle connections. It is common to call those directions completely periodic.

For the calculation of quadratic growth rates it is an important property of a lattice surface (X,ω), that the set of completely periodic directions decomposes into *finitely* many $\mathrm{SL}(X,\omega)$ orbits. The *holonomy vector* $\mathrm{hol}(C) \in \mathbb{R}^2$ is the vector in direction of a cylinder of periodic leaves, that has modulus the width of the cylinder C. Because $\phi \in \mathrm{Aff}^+(X,\omega)$ maps cylinders to cylinder and acts locally linear on $\mathbb{R}^2$ via their constant derivative we have $\mathrm{hol}(\phi C) = D\phi\, \mathrm{hol}(C)$.

For covers it is convenient to use a relative version of the coverings Veech group. For a given cover $(Y,\tau) \to (X,\omega)$, the *relative Veech group* $\mathrm{SL}(Y/X,\tau) \subset \mathrm{SL}(X,\omega)$ is the group of derivatives of affine maps that lift to affine maps of (Y,τ), i.e. lifts so that diagram 7 with $(Y,\tau) = (Y_1,\tau_1) = (Y_2,\tau_2)$ commutes in the sense described.

Most important for our considerations is, if (X,ω) is a *lattice surface*, a quadratic asymptotics $\lim_{T\to\infty} \frac{N_V((X,\omega),T)}{T^2}$ exists for virtually all relevant leaves of finite length and can be calculated by Veech's formula, see **[V2]** and **[GJ]**. More generally, quadratic constants exist for all branched covers of lattice surfaces and can be calculated as in **[EMS]**.

To evaluate quadratic growth rates of d-cyclic covers of a lattice surface (X,ω) equipped with the pullback 1-forms and metrics, which we will call *d-symmetric covers* (of (X,ω)), we involve the parameter space of d-cyclic covers and note:

Finite $\mathrm{SL}_2\mathbb{Z}$ orbits on tori

To calculate quadratic growth rates for d-symmetric torus covers we need an $\mathrm{SL}_2\mathbb{Z}$ orbit classification of their parameter space. Granting the statements in the introduction this $\mathrm{SL}_2\mathbb{Z}$ orbit classification can be extracted from the standard $\mathrm{SL}_2\mathbb{Z}$ action on the translation torus $\mathbb{T}_{[0]} = (\mathbb{C}/\mathbb{Z}[i], [0], dz)$, marked in $[0]$. The discussion generalizes to tori $\mathbb{C}/\Lambda$ defined by other unimodular lattices Λ without adding essential new ideas.

The group of orientation preserving affine diffeomorphisms $\mathrm{Aff}^+(\mathbb{T}, dz)$ of $(\mathbb{T}, dz)$ is isomorphic to the group $\mathrm{SL}_2\mathbb{Z} \ltimes \mathbb{C}/\mathbb{Z}[i]$ with composition rule

$$(A, [a]) \circ (B, [b]) = (A \cdot B, [b + Aa]) \quad \text{where } a, b \in \mathbb{C} \text{ and } A, B \in \mathrm{SL}_2\mathbb{Z}.$$

The two parts of $\mathrm{Aff}^+(\mathbb{T}, dz)$ are seen in

$$0 \longrightarrow \mathbb{C}/\mathbb{Z}[i] \xrightarrow{\ i\ } \mathrm{SL}_2\mathbb{Z} \ltimes \mathbb{C}/\mathbb{Z}[i] \xrightarrow{\ \mathrm{D}\ } \mathrm{SL}_2\mathbb{Z} \longrightarrow 1 \tag{8}$$

We eliminate the continuous subgroup of translations $\mathrm{Aut}(\mathbb{T}, dz) \cong \mathbb{C}/\mathbb{Z}[i]$ by making the origin $[0] \in \mathbb{T}$ a fixed point.

Let us denote the subgroup $\mathrm{SL}(\mathbb{T}_{[x]}, dz) \subset \mathrm{SL}_2\mathbb{Z}$ to be the set of linear maps that stabilize the point $[x] \in \mathbb{T}$. While in many cases the projection of $\mathrm{SL}(\mathbb{T}_{[x]}, dz)$ to $\mathrm{PSL}_2(\mathbb{R})$ is the called the Veech group of the two marked torus $\mathbb{T}_{[x]} = (\mathbb{C}/\mathbb{Z}[i], [0], [x])$, we will regard $\mathrm{SL}(\mathbb{T}_{[x]}, dz)$ as the Veech group of $(\mathbb{T}_{[x]}, dz)$. So we *distinguish* the marked points. The group $\mathrm{SL}(\mathbb{T}_{[x]}, dz)$ is a lattice in $\mathrm{SL}_2\mathbb{R}$, if and only if $[x] = x + \mathbb{Z}[i]$ is rational. More precisely:

PROPOSITION 4. *Given $a, b, n \in \mathbb{N}_0$ with $\gcd(a, b, n) = 1$ then the stabilizer $\mathrm{SL}(\mathbb{T}_{[x]}, dz)$ of $[x] = [\frac{a}{n}, \frac{b}{n}]$ is conjugate to*

$$\Gamma_1(n) := \left\{ \left(\begin{smallmatrix} a & b \\ c & d \end{smallmatrix}\right) \in \mathrm{SL}_2\mathbb{Z} : \left(\begin{smallmatrix} a & b \\ c & d \end{smallmatrix}\right) \equiv \left(\begin{smallmatrix} 1 & b \\ 0 & 1 \end{smallmatrix}\right) \mod n \right\} \subset \mathrm{SL}_2\mathbb{Z}.$$

In particular for $[x] = [\frac{1}{n}, 0]$ we have $\mathrm{SL}(\mathbb{T}_{[x]}, dz) = \Gamma_1(n)$.

This is an easy exercise, see **[S1]**.

For $[\frac{1}{n}, 0] \in \mathbb{T}$ we have

$$\mathrm{SL}_2\mathbb{Z} \cdot \left[\frac{1}{n}, 0\right] = \left\{ \left[\frac{a}{n}, \frac{b}{n}\right] \in \mathbb{T} : \ a, b, n \in \mathbb{Z} \text{ with } \gcd(a, b, n) = 1 \right\} \tag{9}$$

in particular

$$[\Gamma_1(n) : \mathrm{SL}_2\mathbb{Z}] = n^2 \prod_{p|n} \left(1 - \frac{1}{p^2}\right) = \varphi(n)\psi(n).$$

The last product is taken over all prime divisors p of n. The two functions on the right are the well known *Euler φ function* and the *Dedekind ψ function*:

$$\varphi(n) := n \prod_{p|n} \left(1 - \frac{1}{p}\right), \quad \psi(n) := n \prod_{p|n} \left(1 + \frac{1}{p}\right). \tag{10}$$

2.2. Torsion points on tori.

For the torus $\mathbb{T}_d = \mathbb{C}/d\mathbb{Z}[i]$ any $m \in \mathbb{N}$ defines a group homomorphism $\mathbb{T}_d \to \mathbb{T}_d$ by $[z]_d \mapsto m \cdot [z]_d$ ($[z]_d := z + d\mathbb{Z}[i]$). The kernel of this m-homomorphism is

$$\mathbb{T}_d[m] := \ker(\mathbb{T}_d \xrightarrow{m} \mathbb{T}_d) = \frac{d}{m}\mathbb{Z}[i]/d\mathbb{Z}[i].$$

The point in the kernel are called *torsion points.* The *order* of a torsion point $[z]_d \in \mathbb{T}_d$ is the smallest $m \in \mathbb{N}$ such that $m \cdot [z]_d = 0 \in \mathbb{T}_d$. Denote the set of torsion points of order m on $\mathbb{T}_d$ by $\mathbb{T}_d(m)$. On the standard torus $\mathbb{T} = \mathbb{T}_1$ we have by equation 9

$$\mathbb{T}(m) = \mathrm{SL}_2\mathbb{Z} \cdot [1/m, 0]. \tag{11}$$

The rescaling map $\mathbb{T}_d \xrightarrow{d^{-1}} \mathbb{T}$, given by

$$[z]_d = z + d\mathbb{Z}[i] \mapsto d^{-1}(z + d\mathbb{Z}[i]) = d^{-1}z + \mathbb{Z}[i] = [d^{-1}z]$$

is $\mathrm{SL}_2\mathbb{Z}$ equivariant and identifies torsion points of order m.

Let $n = p_1^{l_1} \cdot p_2^{l_2} \dots p_r^{l_r}$ be the *prime-factor decomposition* for $n \in \mathbb{N}$. Then the number of positive *divisors* of n is

$$D(n) := (l_1 + 1) \cdots (l_r + 1).$$

PROPOSITION 5. *The real linear action of* $\mathrm{SL}_2\mathbb{Z}$ *on* $\mathbb{T}_d$ *restricted to* $\mathbb{T}_d[m] = \frac{d}{m}\Lambda/d\Lambda \subset \mathbb{T}_d$ *contains precisely* $D(m)$ *orbits.*

PROOF. After taking the $\mathrm{SL}_2\mathbb{Z}$ equivariant map from $\mathbb{T}_d$ to $\mathbb{T}$, we need to count the $\mathrm{SL}_2\mathbb{Z}$ orbits on the m-torsion points $\mathbb{T}[m]$. Since by the $\mathrm{SL}_2\mathbb{Z}$-orbit classification for $\mathbb{T}$, $\mathrm{SL}_2\mathbb{Z} \cdot [1/n, 0] = \mathbb{T}(n)$ for any $n \in \mathbb{N}$ and $\mathbb{T}[m] = \bigsqcup_{n|m} \mathbb{T}(n)$ the statement follows. □

For every $d > 1$ we consider the map $\pi_d : \mathbb{T}_d \to \mathbb{T}$ defined by taking $[z]_d \in \mathbb{T}_d$ modulo $\mathbb{Z}[i]$. We want to classify the $\mathrm{SL}_2\mathbb{Z}$ orbits containing the preimage $\pi_d^{-1}[1/m, 0] \subset \mathbb{T}_d$ of the point $[1/m, 0] \in \mathbb{T}$.

PROPOSITION 6. *For* $m \in \mathbb{N}$ *consider* $[1/m, 0] \in \mathbb{T}$. *Then for given* $d \in \mathbb{N}$ $\mathrm{SL}_2\mathbb{Z}(\pi_d^{-1}[1/m, 0]) \subset \mathbb{T}_d$ *is the union of* $D(d_m)$ *orbits.*

PROOF. Represent the set $\pi_d^{-1}[1/m, 0]$ by $\{(k, l + 1/m) \in \mathbb{R}^2 : 0 \leq k, l < d\}$. Rescaling with d^{-1} transforms this becomes $\{(km/dm, (lm+1)/dm) \subset \mathbb{R}^2 : 0 \leq k, l < d\}$ this represents points on $\mathbb{T}$ and we need the $\mathrm{SL}_2\mathbb{Z}$ orbits through these points on $\mathbb{T}$. First let us look which values $\gcd(km, lm+1, dm)$ will take when the integers k and l range between 0 and $d-1$. Since the middle term of $\gcd(km, lm+1, dm)$ leaves remainder 1 modulo any divisor of m, $\gcd(km, lm+1, dm)$ is relatively prime to m. We claim that any divisor of d_m is attained in the set $\{\gcd(km, lm+1, dm) : 0 \leq k, l \leq d-1\}$. Indeed, given $p|d_m$, take $k = p$. Since $\gcd(p, m) = 1$, we get a full set of remainders $lm + 1 \mod p$ for $0 \leq l \leq d-1$. Thus, for some $l = l_p$ we have $p|l_p m + 1$ By the $\mathrm{SL}_2\mathbb{Z}$ orbit classification for $\mathbb{T}$ $[p/d, (l_p m + 1)/dm] \in \mathrm{SL}_2\mathbb{Z}[p/dm, 0]$ and there are $D(d_m)$ such orbits. □

It follows from this proposition that the maximal number of $\mathrm{SL}_2\mathbb{Z}$ orbits generated by $\pi_d^{-1}[1/m, 0] \subset \mathbb{T}_d$ is achieved when $\gcd(d, m) = 1$ and it is $D(d)$.

COROLLARY 5. *If* $\gcd(d, m) = 1$, *then the set* $\pi_d^{-1}[1/m, 0] \subset \mathbb{T}_d$ *lies on* $D(d)$ $\mathrm{SL}_2\mathbb{Z}$*-orbits. If* $d|m^n$ *for some* $n \in \mathbb{N}$ *the set* $\pi_d^{-1}[1/m, 0]$ *lies on one* $\mathrm{SL}_2\mathbb{Z}$ *orbit. In particular if* $d = 2$, *then the points in* $\pi_2^{-1}[1/m, 0] \subset \mathbb{T}_2$ *lie on one* $\mathrm{SL}_2\mathbb{Z}$ *orbit if and only if* m *is even, otherwise there are two* $\mathrm{SL}_2\mathbb{Z}$ *orbits through this set.*

Note, that $d|m^n$ for some $n \in \mathbb{N}$ is equivalent to: For any prime p, so that $p|d$ also $p|m$.

3. Parameter spaces of cyclic covers

Cyclic covers defined by homology classes. Given a group G, we call a cover $Y \to X$ *G-cover* if G acts properly discontinuously on Y with quotient space $X = Y/G$. With $\mathring{X} = X \backslash \{x_0, ..., x_n\}$ we consider $\mathbb{Z}$-covers of $\mathring{X}$, X a (compact) Riemann surface. Those covers are parameterized by $\mathrm{Hom}(\pi_1(\mathring{X}, x), \mathbb{Z})$. Using commutativity of $\mathbb{Z}$ and the Hurewicz isomorphism

$$\pi_1(\mathring{X}, x)/[\pi_1(\mathring{X}, x), \pi_1(\mathring{X}, x)] \cong H_1(\mathring{X}; \mathbb{Z}),$$

one finds $\mathrm{Hom}(\pi_1(\mathring{X}, x), \mathbb{Z})$ is $\mathrm{Hom}(H_1(\mathring{X}; \mathbb{Z}), \mathbb{Z})$ as a set. Algebraic intersection defines a non-degenerate pairing

$$\langle \cdot\, , \cdot \rangle \;: H_1(X, \{x_0, ..., x_n\}; \mathbb{Z}) \times H_1(\mathring{X}; \mathbb{Z}) \to \mathbb{Z}$$

using that we can represent any $\mathbb{Z}$-cover by a relative homology class. The same remains true if we replace $\mathbb{Z}$ by $\mathbb{Z}_d$ in the above homology groups and take the intersection modulo d. One assigns an (eventually disconnected) $\mathbb{Z}_d$ (respectively $\mathbb{Z}$) cover to an element of $H_1(X, \{x_0, ..., x_n\}; \mathbb{Z}_d)$ as follows.

A class $\gamma \in H_1(X, \{x_0, ..., x_n\}; \mathbb{Z})$ characterizes a cover $p : X_\gamma \to X$ with deck group $\mathbb{Z}$ by prescribing how loops lift from X to X_γ through intersection with γ. In fact, a lift $\widetilde{\sigma}$ of a loop $\sigma : [t_0, t_1] \to X$ to X_γ is determined by its deck shift $\widetilde{\sigma}(t_1) = \langle \gamma, [\sigma] \rangle \cdot \widetilde{\sigma}(t_0)$. Here $[\sigma] \in H_1(X; \mathbb{Z})$ denotes the homology class defined by σ and $\cdot$ denotes the action of $\mathbb{Z}$ as group of deck-transformations. For such a cover on the other hand, if $\widetilde{\rho} : [t_0, t_1] \to X_\gamma$ is a curve with $\widetilde{\rho}(t_1) = n \cdot \widetilde{\rho}(t_0)$ for some $n \in \mathbb{Z}$, then $n = \langle \gamma, [\pi \circ \widetilde{\rho}] \rangle$ where $[\pi \circ \widetilde{\rho}] \in H_1(X; \mathbb{Z})$ is the homology class of the loop $\pi \circ \widetilde{\rho}$ on X. Covers with deck group $\mathbb{Z}_d$ are obtained as quotients from covers with deck group $\mathbb{Z}$. Equivalently $\mathbb{Z}_d$ covers are defined by considering deck-shifts of lifted curves modulo d. Not all covers characterized by (relative) homology classes in this fashion are connected. The cover associated to $2[\gamma] \in H_1(X; \mathbb{Z})$, where $[\gamma] \in H_1(X; \mathbb{Z})$ is a non-trivial class of a simple loop, for example, is not connected.

PROPOSITION 7. *Let*

$$\{[\gamma_i] : i = 1, ..., 2g\} \cup \{[\gamma_j^r] : j = 1, ..., n\} \subset H_1(X, \{x_0, ..., x_n\}; \mathbb{Z}_d)$$

be a basis. Then the cover defined by the class

$$[\gamma] = \sum_{j=1}^{n} r_j [\gamma_j^r] + \sum_{i=1}^{2g} a_i [\gamma_i] \in H_1(X, \{x_0, x_1\}; \mathbb{Z}_d)$$

is connected, if and only if $\gcd(r_1, ..., r_n, a_1, ..., a_{2g}, d) = 1$. *A relative homology class with coefficients in* $\mathbb{Z}$ *determines a connected cover, if and only if*

$$\gcd(r_1, ..., r_n, a_1, ..., a_{2g}) = 1.$$

For $\mathbb{Z}_d$ classes that means, there is no proper divisor $k|d$ with $k \cdot [\gamma] \equiv 0 \mod d$. If, on the other hand $\gcd(r_1, ..., r_n, a_1, ..., a_{2g}, d) = k > 1$ then $(d/k) \cdot [\gamma] \equiv 0 \mod d$.

Recall that $\mathbb{Z}_d$-covers up to $\mathbb{Z}_d$ equivariant isomorphisms are in one-to-one correspondence to classes in $H_1(X, \{x_0, ..., x_n\}; \mathbb{Z}_d)$. To parameterize d-cyclic covers we will not distinguish covers that have distinct $\mathbb{Z}_d$ actions on the fibers of the cover. In order to preserve the cyclic structure of the cover the maps between fibers must be isomorphisms of $\mathbb{Z}_d$. Those are given by the elements of $\mathbb{Z}_d^*$, the group of units in the ring $\mathbb{Z}_d$, applied multiplicatively. Let $\mathbb{P}H_1(X, \{x_0, ..., x_n\}; \mathbb{Z}_d)$ be the set of $\mathbb{Z}_d^*$ orbits in $H_1(X, \{x_0, ..., x_n\}; \mathbb{Z}_d) \backslash \{0\}$, i.e. the quotient $(H_1(X, \{x_0, ..., x_n\}; \mathbb{Z}_d) \backslash \{0\}) / \mathbb{Z}_d^*$.

PROPOSITION 8. *Isomorphy classes of d-cyclic covers are in one-to-one correspondence to* $\mathbb{Z}_d^*$ *orbits in* $H_1(X, \{x_0, ..., x_n\}; \mathbb{Z}_d) \backslash \{0\}$.

PROOF. We show that the action of $\mathbb{Z}_d^*$ on homology is induced by the multiplicative action of $\mathbb{Z}_d^*$ on the fibers of the cover associated to the class. Given the cover $\pi : X_\gamma \to X$ defined by the class $\gamma \in H_1(X, \{x_0, ..., x_n\}; \mathbb{Z}_d) \backslash \{0\}$, we refer to the elements of a fiber $\pi^{-1}(x) \subset X_\gamma$ as decks. Note that a fiber $\pi^{-1}(x)$ is identified with $\mathbb{Z}_d$. Suppose a lift $\widetilde{c}$ of the curve $c : I \to X \backslash \{x_0, ..., x_n\}$ starts on deck $i \mod d$

of $\pi^{-1}(x) \subset X_\gamma$ and ends on deck $(\langle\gamma, c\rangle + i) \mod d$. Shuffling the d elements of the fiber by a permutation σ, the lift of c that starts on deck $j = \sigma(i)$ will end on deck $\sigma((\langle\gamma, c\rangle + \sigma^{-1}(j)) \mod d)$. In our case σ is multiplication with $k \in \mathbb{Z}_d^*$ and the previous expression becomes:

$$k(\langle\gamma, c\rangle + k^{-1}j) \equiv \langle k\gamma, c\rangle + j \mod d.$$

So the cover $X_{k\gamma}$ defined by the class $k\gamma$ is the cover obtained from X_γ by rearranging the decks via k. □

Consequently the classes in

$$\mathbb{P}H_1(X, \{x_0, ..., x_n\}; \mathbb{Z}_d) := H_1(X, \{x_0, ..., x_n\}; \mathbb{Z}_d)\backslash\{0\}/\mathbb{Z}_d^*$$

parameterize d-cyclic covers. Since the ramification points of a d-symmetric cover are assumed to have maximal order d, the weights $r_k \in \mathbb{Z}_d$ of the relative classes $[\gamma_k^r]$ need to be in $\mathbb{Z}_d^*$. This condition alone ensures connectedness independent from the remaining absolute homology part defining the cover. So we can always assume, that one relative cycle has weight 1. In particular, since d-symmetric covers have two branch points, we pick a class with relative cycle that has weight 1. Let us denote the set of those classes by $\mathbb{P}_1H_1(X, \{x_0, x_1\}; \mathbb{Z}_d)$. As before $[n]_d \in \mathbb{Z}_d$ denotes the residue class of an integer n modulo d, then using $a_0 \equiv 1$ for d-symmetric covers. The following decomposition is obvious:

PROPOSITION 9. *For any* $x_1 \in X\backslash\{x_0\}$

$$\mathbb{P}H_1(X, \{x_0, x_1\}; \mathbb{Z}_d) = \sqcup_{n|d}(H_1(X; \mathbb{Z}_d)) \times \{[n]_d[\gamma_0]\}$$

and

$$\mathbb{P}_1H_1(X, \{x_0, x_1\}; \mathbb{Z}_d) = H_1(X; \mathbb{Z}_d) \times \{[\gamma_0]\}.$$

So after choosing a basis class $[\gamma_0]$ for the relative homology to be a simple curve $\gamma_r(t)$, $t \in [0, 1]$ with $\gamma_0(0) = x_0$ and $\gamma_0(1) = x_1$, we can write any class $[\gamma] \in \mathbb{P}_1H_1(X, \{x_0, x_1\}; \mathbb{Z}_d)$ as

$$[\gamma] = [\gamma_0] + \sum_{i=1}^{2g} a_i[\gamma_i], \quad a_i \in \mathbb{Z}_d$$

where the $[\gamma_i], i = 1, ..., 2g$ are a basis of $H_1(X, \mathbb{Z}_d)$. In particular, as sets

$$\mathbb{P}_1H_1(X, \{x_0, x_1\}; \mathbb{Z}_d) \cong (\mathbb{Z}_d)^{2g} \cong H_1(X; \mathbb{Z}_d).$$

The cover associated to a projective homology class. Let us now construct a branched cover $X_{[\gamma]} \to X$ associated to the class $[\gamma] \in \mathbb{P}H_1(X, \{x_0, x_1\}; \mathbb{Z}_d)$. First pick a homology class in $H_1(X, \{x_0, x_1\}; \mathbb{Z}_d)$ that represents the chosen projective class, here also denoted by $[\gamma]$. Recall that topologically a cover is given by the way (oriented) loops $\beta : [0, 1] \to X\backslash\{x_0, x_1\}$ lift. Suppose $X_{[\gamma]} \to X$ is given by the class $[\gamma] = a_0[\gamma_0] + \sum_{i=1}^{2g} a_i[\gamma_i]$ with $a_i \in \mathbb{Z}_d$ and simple closed and oriented curves γ_i, then every time β crosses γ_i with *positive orientation*, its lift, say $\widetilde{\beta}$, will move a_i decks *up*.

In order to construct $X_{[\gamma]}$ from the marked surface (X, x_0), realize each class $[\gamma_i]$ as simple oriented curve starting in x_0. Then cut X along all γ_i. Denote the resulting surface X^{cut}, along each cut there are two oriented strands, we label those by $+a_i$ and $-a_i$. To get $X_{[\gamma]}$ take d copies $\sqcup_{k=1}^d X_k^{cut}$ of X^{cut} and identify the $+a_i$ strand on X_k^{cut} with the $-a_i$ strand on X_l^{cut}, when $l = k + a_i \mod d$. We denote the surface obtained from $[\gamma]$ by $\#^d_{[\gamma]}X$. Identifying all copied points defines the covering map to X. The group $\mathbb{Z}_d$ acts on $\#^d_{[\gamma]}X$, because identifications are done cyclically.

3.1. Cyclic covers over the universal abelian cover. The *universal abelian cover* $\pi_{ab} : X_{ab} \to X$ of a Riemann surface, also called *homology cover*, is the regular cover associated to the commutator subgroup $K := [\pi_1(X, x_0), \pi_1(X, x_0)]$ of the fundamental group $\pi_1(X, x_0)$ of X. The homology cover has deck group $H_1(X;\mathbb{Z}) \cong \mathbb{Z}^{2g}$ and $H_1(X_{ab};\mathbb{Z}) \cong K/[K,K]$. In particular the π_{ab} image of any loop γ in X_{ab} is trivial in homology:

$$[\pi_{ab} \circ \gamma] = 0 \in H_1(X;\mathbb{Z}).$$

If, as before, $x_0 \in X$ denotes the locus of the first branch point, we pick

$$x_{ab} \in \pi_{ab}^{-1}(x_0) \subset X_{ab},$$

obtaining a marked cover

$$\pi_d : (X_{ab}, x_{ab}) \to (X, x_0).$$

For $\widetilde{x}_1 \in X_{ab}$ consider a path γ from x_{ab} to $\widetilde{x}_1$, and the following map

$$\widetilde{x}_1 \mapsto [\pi_{ab} \circ \gamma] \in \begin{cases} \mathbb{P}_1 H_1(X, \{x_0, x_1\};\mathbb{Z}), & \text{if } x_0 \neq x_1 := \pi_{ab}(\widetilde{x}_1) \\ H_1(X;\mathbb{Z}), & \text{if } x_0 = \pi_{ab}(\widetilde{x}_1) \end{cases} \tag{12}$$

PROPOSITION 10. *The homology class associated to $\widetilde{x}_1 \in X_{ab}$ is well-defined.*

PROOF. If $\widetilde{x}_1 \notin \pi^{-1}(x_0)$, then $x_0 \neq x_1 := \pi_{ab}(\widetilde{x}_1)$. We have to show that two different curves γ_1 and γ_2 connecting x_{ab} with $\widetilde{x}_1 \in X_{ab}$ project to the same homology class in $H_1(X, \{x_0, x_1\};\mathbb{Z})$. This is true, because $\pi_{ab} \circ (\gamma_2^{-1} * \gamma_1)$ is a loop whose homotopy class lies in the commutator subgroup of $\pi_1(X, x_0)$ and therefore defines the trivial class in $H_1(X;\mathbb{Z})$. The argument is the same, if we assume $\widetilde{x}_1 \in \pi^{-1}(x_0)$. □

Let us now denote the above map (into the union of relative homology groups) by π_{ab,x_0}. So, π_{ab,x_0} assigns a $\mathbb{Z}$-cover to each point $\widetilde{x}_1 \in X_{ab}$ in the marked surface (X_{ab}, x_{ab}). This cover is connected, if $\widetilde{x}_1 \notin \pi_{ab}^{-1}(x_0)$ and eventually degenerate, if $\widetilde{x}_1 \in \pi_{ab}^{-1}(x_0)$. In both these cases π_{ab,x_0} is bijective since the points of $\pi_{ab}^{-1}(x_1)$ are in one-to-one correspondence to $\mathbb{P}_1 H_1(X, \{x_0, x_1\};\mathbb{Z})$ if $x_0 \neq x_1$, or to $H_1(X;\mathbb{Z})$ in the other case.

A family of $\mathbb{Z}$ covers over X_{ab}. The construction of covers associated to a relative homology class defines a family $\mathfrak{X}_d$ of $\mathbb{Z}_d$ covers on X_{ab} for each $d \in \mathbb{Z}$. This construction works for $\mathbb{Z}$ covers and the associated family over X_{ab} is denoted by $\mathfrak{X}_\infty$. The alternative construction of the family $\mathfrak{X}_\infty$ below may give a better idea of its regularity.

For $x_1 \in X$ consider a directed simple path, say ϕ, from $x_{ab} \in X_{ab}$ to any point $\widetilde{x}_1 \in \pi_{ab}^{-1}(x_1) \in X_{ab}$. The path can be chosen so that it is disjoint to any of its deck-translates under the deck group $H_1(X,\mathbb{Z})$. We remove all the deck translates of ϕ and fit in two strands, labeled right and left, for each side in each translate. Call the resulting surface $X_{ab}^{cut}(\widetilde{x}_1)$ take $\mathbb{Z}$ labeled copies and identify the copy of a particular right strand on the i-th deck in $\mathbb{Z} \times X_{ab}^{cut}(\widetilde{x}_1)$ with the left strand of the same origin in the $(i+1)$st deck. We obtain a connected surface $\#^\infty_{\widetilde{x}_1} X_{ab}$ with a $\mathbb{Z} \times H_1(X;\mathbb{Z})$ action. Here the $\mathbb{Z}$ action induced by $m \cdot (i, x) = (i + m, x)$ and the

$H_1(X;\mathbb{Z})$ action induced by $\gamma \cdot (i,x) = (i,\gamma x)$ for $(i,x) \in \mathbb{Z} \times X_{ab}^{cut}(\widetilde{x}_1)$ descend to $\#_{\widetilde{x}_1}^{\infty} X_{ab}$. Note that both actions commute. So $\#_{\widetilde{x}_1}^{\infty} X_{ab}$ is a $\mathbb{Z}$-cover of X_{ab} and its quotient by the deck group $H_1(X;\mathbb{Z})$ is the $\mathbb{Z}$-cover $\#_{\widetilde{x}_1}^{\infty} X$ of X.

Varying $\widetilde{x}_1$ over X_{ab} we obtain the family $\mathfrak{X}_\infty \to X_{ab}$ of infinte cyclic covers $\#_{\widetilde{x}_1}^{\infty} X \to X$. The group of integers acts on that family by deck-transformations on each individual surface. For $d > 1$ consider the quotient family $\mathfrak{X}_\infty / d\mathbb{Z} \to X_{ab}$ with respect to the subgroup $d\mathbb{Z}$.

PROPOSITION 11. *If $\#_{\widetilde{x}_1}^{\infty} X \in \mathfrak{X}_\infty$, then $\#_{\widetilde{x}_1}^{\infty} X/d\mathbb{Z} \cong \#_{[\widetilde{x}_1]_d}^{\infty} X \in \mathfrak{X}_d$.*

PROOF. If $[\gamma] \in \mathbb{P}_1 H_1(X, \{x_0, x_1\}; \mathbb{Z})$ is the projective class defining the cover $\pi_{\widetilde{x}_1} : \#_{\widetilde{x}_1}^{\infty} X \to X$, then $\gamma = \pi_{\widetilde{x}_1} \widetilde{\gamma}$ where $\widetilde{\gamma}$ is a lift of γ to X_{ab} that starts in x_{ab} and ends in $\widetilde{x}_1$. Now consider $\#_{\widetilde{x}_1}^{\infty} X/\mathbb{Z}_d$. The lift $\widetilde{\sigma}$ of any loop $\sigma : [0,1] \to X \backslash \{x_0, x_1\}$ to the original $\mathbb{Z}$ cover $\#_{\widetilde{x}_1}^{\infty} X$ projects to the loop $\widetilde{\sigma}_d$ on $\#_{\widetilde{x}_1}^{\infty} X/\mathbb{Z}_d$ given by identifying points whenever their decks differ by an element in $d\mathbb{Z}$. Since deck changes of $\widetilde{\sigma}$ are given by the intersection $\langle \gamma, \sigma \rangle \in \mathbb{Z}$ the deck changes of its projection $\widetilde{\sigma}_d$ to the quotient surface are given by $\langle \gamma, \sigma \rangle \mod d$. Since this is true for any loops γ on X we see that the d cover $\#_{\widetilde{x}_1}^{\infty} X/d\mathbb{Z}$ is given by the class $[\sigma]_d \in \mathbb{P}_1 H_1(X, \{x_0, x_1\}; \mathbb{Z}_d)$. This class is in $\mathbb{P}_1 H_1(X, \{x_0, x_1\}; \mathbb{Z}_d)$ since it is the image of a class in $\mathbb{P}_1 H_1(X, \{x_0, x_1\}; \mathbb{Z})$, so particularly the cover $\#_{\widetilde{x}_1}^{\infty} X/d\mathbb{Z}$ is fully branched. The claim follows. □

As a consequence we have the parameter space of d symmetric covers.

COROLLARY 6. *The points of $(X_{ab} - H_1 x_{ab})/dH_1$ parameterize d-symmetric covers.*

The covers defined by the points $H_1 x_{ab}/dH_1$ are used obtain a family over the compact parameter space. We use polygonal representations for those.

3.2. Polygonal representation of d-symmetric covers. Any (compact and connected) translation surface (X, ω) can be represented by a planar polygon, with pairs of parallel edges that are identified by translations. This allows us to do the previous copy, cut and glue construction of $\mathbb{Z}_d$ covers using representations of homology classes by straight line segments, or concatenations of straight line segments. This is a special case of the *slit construction* that utilizes a cut along a single regular line segment.

3.3. Degenerate covers via absolute homology classes. The space and family of d-symmetric surfaces over (X, x_0, ω) has a natural compactification by adding covers for the lattice points $\pi_d^{-1}(x_0) \subset X_{ab}/(d\mathbb{Z})^{2g}$. Since $\pi_d^{-1}(x_0)$ and $H_1(X;\mathbb{Z}_d)$ can be identified as sets, we may represent the respective points by "d-cyclic covers" defined by absolute homology classes in $H_1(X;\mathbb{Z}_d)$. This can be done as before just using the polygon representation in the plane: By gluing d copies of X cut along line segments representing an absolute homology basis. Along the cuts the copies are cyclically identified according to the values of the homology cycle $[\gamma] \in H_1(X;\mathbb{Z}_d) \cong \mathbb{Z}_d^{2g}$ for the respective classes. These surfaces are not always connected, but nevertheless carry a $\mathbb{Z}_d$ action with quotient (X, ω). The number of connected components is a divisor of d. We obtain a connected surface by identifying the d preimages of x_0 to one point. Using our usual convention, we denote the connected surface by $\#_z^d(X, x_0, \omega)$ for the cover given by $[\gamma] \in H_1(X;\mathbb{Z}_d)$, so that $z = [\gamma] \cdot x_d \in X_{ab}/dH_1$. We call these covers *degenerate covers*. Figure

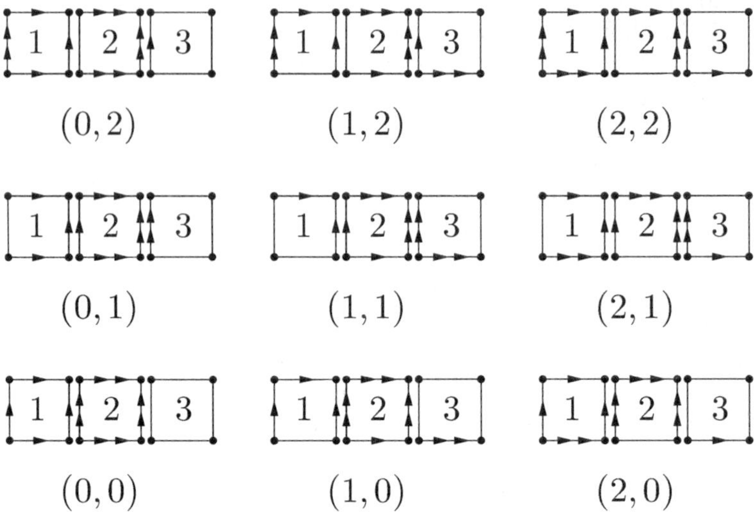

FIGURE 2. Degenerate degree 3 differentials at lattice coordinates of parameter space.

2 shows the nine $\mathbb{Z}_3$ covers of $\mathbb{T}$ defined by absolute homology classes. The coordinates below each cover correspond to their (real) lattice point coordinate in $\mathbb{Z}[i]/3\mathbb{Z}[i] \subset \mathbb{C}/3\mathbb{Z}[i]$. Parallel sides of the squares with the same symbol pattern in each figure are considered identified by a translation. The degenerate covers have a singularity that is not induced by a zero of the pullback of ω, nevertheless they have a natural interpretation as limits obtained from collapsing the two cone points of a d-symmetric differential to one point. Together with the results of the previous section we have:

THEOREM 5. *Let (X, x_0, ω) be a marked abelian differential of genus g. Then the d-symmetric forms over (X, x_0, ω), degenerate forms included, are parameterized by the points in $(X_{ab}/dH_1, x_d, \pi^{-1}\omega)$. The point $x_d \in \pi_d^{-1}(x_0)$ corresponds to the trivial class in $H_1(X;\mathbb{Z}_d)$, i.e. to the degenerate cover $\#_{[0]}^d(X, x_0, \omega)$ defined by the trivial homology class.*

Since the degenerate d symmetric forms represented by points in $\pi_d^{-1}(x_0)$ have only one (artificial) branch point they may have symmetries and so the compactification $(X_{ab}/dH_1, x_d)$ does not necessary classify covers up to isomorphy. This can be seen particularly in the case of torus covers. Tori are hyperelliptic and we can consider a hyperelliptic involution that exchanges the branch points of the cover. That in turn induces an involution of the parameter torus $(\mathbb{C}/dH_1, x_d)$, that is the hyperelliptic involution fixing x_d. Since the hyperelliptic involution fixes the preimage of x_0 on degenerate surfaces, those are symmetric. On the other hand the point in $(\mathbb{C}/dH_1, x_d)$ that represents this cover may not be fixed by the hyperelliptic involution of $(\mathbb{C}/dH_1, x_d)$. So we have two isomorphic copies for those among the d^2 degenerate surfaces in $(\mathbb{C}/dH_1, x_d)$ that are not already fixed by the hyperelliptic involution.

4. Asymptotic formulas for branched covers

Here we recall a formalism to calculate asymptotic quadratic growth rates for (branched covers of) lattice surfaces. In case the covering is itself a lattice surface we derive our formula from a variant of the asymptotic formula presented in Gutkin and Judge **[GJ]** and independently Vorobetz **[Vrb]**. Throughout this section we assume (X,ω) is a lattice surface with Veech-group $\mathrm{SL}(X,\omega)$.

Growth rate of vector distributions. Take a discrete and countable distribution of vectors, say $V\subset\mathbb{R}^2$, and let $B_T\subset\mathbb{R}^2$ be the ball of radius T. If existent, we want to calculate the quadratic asymptotic of $\#(V\cap B_T)$, that is $\pi c(V):=\lim_{T\to\infty}\frac{\#(V\cap B_T)}{T^2}$. For the vector distribution $V=\mathbb{Z}^2$ it is easy to see $c(\mathbb{Z}^2)=1$. Indeed for $T=N\in\mathbb{N}$ considering the rescaled lattice $\frac{1}{N}\mathbb{Z}^2$ we count the number of rational points in the unit disk with denominator N. This converges to the area of the unit disk. So $c(V)$ measures the quadratic growth rate relative to the integer lattice.

Asymptotic quadratic constants have some obvious properties: They are invariant under translations of V and if $A\in\mathrm{GL}_2(\mathbb{R})$, then $c(AV)=\frac{1}{\det(A)}c(V)$. In particular $c(AV)=c(V)$, if $A\in\mathrm{SL}_2\mathbb{R}$. For a given distribution V the *primitive distribution* $\mathbb{P}V\subseteq V$ contains the points of V *visible from the origin* and its *completion* $\mathbb{Z}V:=\{mv\in\mathbb{R}^2: m\in\mathbb{Z}, v\in V\}$. We call a vector-distribution V *complete*, if $V=\mathbb{Z}V$. Given a vector distribution V that has defined quadratic asymptotics for $\mathbb{Z}V$ and $\mathbb{P}V$ then there is the following universal relation:

$$c(\mathbb{Z}V)=\zeta(2)c(\mathbb{P}V)=\frac{\pi^2}{6}c(\mathbb{P}V).$$

Distributions generated by Veech groups. We now take a look at the asymptotic growth rate of the vector distribution $V=\Gamma\cdot v$ that is the orbit of a *lattice* subgroup $\Gamma\subset\mathrm{SL}_2\mathbb{R}$ applied to a vector $v\in\mathbb{R}^2\backslash\{0\}$. If for example $\Gamma=\mathrm{SL}_2\mathbb{Z}$ and $v=(1,0)^T$, then

$$\mathrm{SL}_2\mathbb{Z}(1,0)^T=\{(a,b)\in\mathbb{Z}^2:\gcd(a,b)=1\}\subset\mathbb{Z}^2=\mathrm{GL}_2(\mathbb{Z})(1,0)^T$$

is the set of *visible points* and so is the primitive distribution associated to $\mathbb{Z}^2$. If $v=\lambda(1,0)^T$, where $\lambda\in\mathbb{R}\backslash\{0\}$, the identity

$$c(\mathrm{SL}_2\mathbb{Z}v)=\frac{1}{\|v\|^2}c(\mathrm{SL}_2\mathbb{Z}(1,0)^T)=\frac{1}{\|v\|^2}\frac{6}{\pi^2}c(\mathbb{Z}^2)=\frac{1}{\|v\|^2}\frac{6}{\pi^2}$$

follows because the distribution under consideration is $D_\lambda\mathbb{Z}^2$ where D_λ is the diagonal matrix with diagonal entries λ.

The next result gives the asymptotic constant of a distribution generated by a general *lattice group* $\Gamma\subset\mathrm{SL}_2\mathbb{R}$. Recall that Γ is a lattice, if it is discrete and $\mathbb{H}/\Gamma$ has finite hyperbolic area. Since distributions generated by subgroups of $\mathrm{SL}_2\mathbb{R}$ contain only visible points, we normalize by multiplying with $\pi^2/6$, so that $c(\mathrm{SL}_2\mathbb{Z}(1,0)^T)=1$. A lattice $\Gamma\subset\mathrm{SL}_2\mathbb{Z}$ is called symmetric, if $-\mathrm{id}\in\Gamma$, or equivalently $-\Gamma=\Gamma$. The following fundamental formula ties the asymptotic growth rate for Γ orbits of vectors to geometric properties of its action on the hyperbolic plane. Statement equivalent to the following can be found in Gutkin and Judge **[GJ]**, Vorobetz **[Vrb]**, but also in **[EMM]**.

PROPOSITION 12. *Assume the subgroup of linear maps $N_v \subset \mathrm{SL}_2\mathbb{R}$ stabilizing the vector $v \in \mathbb{R}^2$ has nontrivial intersection with a symmetric lattice $\Gamma \subset \mathrm{SL}_2\mathbb{R}$ and let A be a generator of $N_v \cap \Gamma$, then*

$$c(\Gamma v) = \frac{\pi}{3\,\mathrm{vol}(\mathbb{H}/\Gamma)} \frac{|\langle A u_v^\perp, u_v\rangle|}{|v|^2}, \tag{13}$$

where u_v is the unit vector in direction v and $u_v^\perp \in \mathbb{R}^2\backslash\{0\}$ is a unit vector perpendicular to v.

We apply formula 13 to a holonomy vector v defined by a saddle connection, or by a maximal cylinder on a lattice surface (X,ω). In this case the group Γ is the Veech group $\mathrm{SL}(X,\omega)$ and formula 13 gives the asymptotic quadratic constant of the set of holonomy vectors in the $\mathrm{SL}(X,\omega)$ orbit of v. One easily applies the formula to tori, particularly to $\mathbb{T} := \mathbb{C}/\mathbb{Z}[i]$ marked at $[0]$. This torus has $\mathrm{SL}(X,\omega) = \mathrm{SL}_2\mathbb{Z}$ as Veech group. The horizontal saddle connection and any horizontal loop have holonomy $v = (1,0)^T$. The set $\mathrm{SL}_2\mathbb{Z}\cdot(1,0)^T \subset \mathbb{Z}^2$ is the set of integer points *visible from the origin* for which we calculated $c(\mathrm{SL}_2\mathbb{Z}\cdot(1,0)^T) = 1$. The right hand side of formula 13 gives the same value. Since $(1,0)^\perp = (0,1)$ and the stabilizer of $(1,0)^T$ in $\mathrm{SL}_2\mathbb{Z}$ is generated by (the powers of) $\left[\begin{smallmatrix}1&1\\0&1\end{smallmatrix}\right]$, one obtains the above quadratic constant by taking $\mathrm{vol}(\mathbb{H}/\mathrm{SL}_2\mathbb{Z}) = \pi/3$ into account.

If we assume the horizontal foliation of the Veech surface (X,ω) is periodic, taking $v^\perp = \left[\begin{smallmatrix}0\\h\end{smallmatrix}\right]$ and $v = \left[\begin{smallmatrix}w\\0\end{smallmatrix}\right] \in \mathbb{R}^2$ to be the height h and circumference w of a maximal cylinder, then we have Veech's asymptotic formula **[V1, V2]**:

$$c(\mathrm{SL}(X,\omega)\cdot v) = \frac{\pi}{3\,\mathrm{vol}(\mathbb{H}/\Gamma)} \frac{l}{w^2}, \tag{14}$$

where $l \in \mathbb{Q}\frac{w}{h}$ is the positive minimal, so that $\left[\begin{smallmatrix}1&l\\0&1\end{smallmatrix}\right] \in \mathrm{SL}(X,\omega)$.

Veech's asymptotic formula. We recall Veech's asymptotic formula given formula 14. Assume (X,ω) is a Veech surface with Veech group $\mathrm{SL}(X,\omega)$. Generally picking any periodic direction on a Veech surface we can associate a holonomy vector $\mathrm{hol}(C) = \int_{l_C}\omega \subset \mathbb{C}$ to a maximal periodic cylinder C by integration of any of its core leaves. The length of a cylinder equals the modulus of its holonomy vector. Since A acts real linearly by postcomposition of natural charts taking holonomy commutes with the action of the affine group in the following sense:

$$\mathrm{hol}(\phi C) = D\phi\,\mathrm{hol}(C).$$

Here ϕC denotes the image of C under the affine map ϕ. With regard of the above counting formula we need to know the orbit of a cylinder with respect to $\mathrm{Aff}^+(X,\omega)/\mathrm{Aut}(X,\omega) \cong \mathrm{SL}(X,\omega)$. It is known that the group of parabolic matrices in $\mathrm{SL}_2\mathbb{R}$ with $\mathrm{hol}(C)$ as eigendirection has nontrivial intersection with $\mathrm{SL}(X,\omega)$. Let us denote this subgroup by $N_1 \subset \mathrm{SL}(X,\omega)$. Then $\mathrm{SL}(X,\omega)/N_1$ parameterizes the slopes obtained from $\mathrm{hol}(C)$. Those correspond to cusps of $\mathbb{H}/\mathrm{SL}(X,\omega)$, in group theoretical terms the conjugacy classes of parabolic subgroups in $\mathrm{SL}(X,\omega)$. Since there are only finitely many cusps, we can write down an asymptotic constant as follows: For each cusp, labeled with $j = 1, ..., n$ pick a representative totally periodic direction on (X,ω). The j-th direction has n_j cylinders C_k^j, $k = 1, ..., n_j$.

Let us put $w_k^j := |\operatorname{hol}(C_k^j)|$. Then we have

$$c(X,\omega) = \frac{\pi}{3\operatorname{vol}(\mathbb{H}/\Gamma)} \sum_{j=1}^{n} i_j \sum_{k=1}^{n_j} \frac{1}{(w_k^j)^2}. \tag{15}$$

Here $i_j = \operatorname{lcm}(m_{j1}^{-1}, ..., m_{jn_j}^{-1})$ where $m_{jk} = h_k^j (w_k^j)^{-1}$ and h_k^j is the height of cylinder C_k^j.

Asymptotic formula for covers of lattice surfaces. Consider a branched cover $\pi : (Y,\tau) \to (X,\omega)$ of a lattice surface (X,ω). The Veech-group $\operatorname{SL}(X,\omega)$ of (X,ω) acts on covers and the stabilizer is the relative group $\operatorname{SL}(Y/X,\tau) \subseteq \operatorname{SL}(X,\omega) \cap \operatorname{SL}(Y,\tau)$. If $\operatorname{SL}(Y/X,\tau)$ is a lattice, then (Y,τ) is a Veech surface and $\operatorname{SL}(Y/X,\tau)$ has finite index in both groups $\operatorname{SL}(X,\omega)$ and $\operatorname{SL}(Y,\tau)$. In particular the orbit

$$\mathscr{O}_Y := \operatorname{SL}(X,\omega)[Y \to X] = \operatorname{SL}(X,\omega)/\operatorname{SL}(Y/X,\tau)$$

on covers is finite and has order $|\mathscr{O}_Y| = [\operatorname{SL}(X,\omega) : \operatorname{SL}(Y/X,\tau)]$. To calculate quadratic asymptotics, it is sufficient to use the relative Veech group.

Pick a set of periodic directions labeled by $j = 1, ..., n$ on (X,ω) representing the cusps of $\mathbb{H}/\operatorname{SL}(X,\omega)$. Pick one periodic direction. Without restrictions of generality we can assume this direction is *horizontal*, by rotating the translation structure if necessary.

PROPOSITION 13. *The cusps of* $\mathbb{H}/\operatorname{SL}(Y/X,\tau)$ *in the preimage of the horizontal cusp with respect to the map* $\mathbb{H}/\operatorname{SL}(Y/X,\tau) \to \mathbb{H}/\operatorname{SL}(X,\omega)$ *are in one-to-one correspondence to the* $N_h(X,\omega)/N_h(Y/X,\tau)$ *orbits on* $\operatorname{SL}(X,\omega)/\operatorname{SL}(Y/X,\tau)$.

PROOF. Cusps of $\mathbb{H}/\operatorname{SL}(Y/X,\tau)$, or $\operatorname{SL}(Y/X,\tau)$ are in one-to-one correspondence to $\operatorname{SL}(Y/X,\tau)$ orbits of slopes of periodic directions on (Y,τ). Directions on (Y,τ) are periodic if and only if they are periodic on (X,ω), we fix a periodic direction, say the horizontal direction, and need to see how many $\operatorname{SL}(Y/X,\tau)$ orbits the set of directions $\operatorname{SL}(X,\omega)/N_h$ has. As for the action of $\operatorname{SL}(Y/X,\tau)$ on $\operatorname{SL}(X,\omega)$ the cosets $\operatorname{SL}(X,\omega)/\operatorname{SL}(Y/X,\tau)$ parameterize the orbits. We note that the orbits of $N_h(X,\omega)$ on the cosets correspond to the image of the (horizontal) cusps of $\operatorname{SL}(Y/X,\tau)$. Those orbits are the claimed orbits of $N_h(X,\omega)/N_h(Y/X,\tau)$. □

In a geometric way, one can think of the abstract orbit $\operatorname{SL}(X,\omega)/\operatorname{SL}(Y/X,\tau)$ of cover classes as an actual coset $\operatorname{SL}(Y/X,\tau)\backslash\operatorname{SL}(X,\omega)(Y,\tau)$. Then any periodic direction in the horizontal $\operatorname{SL}(X,\omega)$ cusp on a surface in $\operatorname{SL}(Y/X,\tau)\backslash\operatorname{SL}(X,\omega)(Y,\tau)$ is horizontal on another surface, say S, in this orbit. The horizontal foliations of all surfaces in the $N_h(X,\omega)/N_h(Y/X,\tau)$ orbit of S stay horizontal. So a direction in the $\operatorname{SL}(Y/X,\tau)$ orbit of the horizontal one on any of $N_h(X,\omega)/N_h(Y/X,\tau)S$ belongs to the same cusp. And if there is a surface in that cusp on which the horizontal direction is the image of the horizontal direction on S, then the map from S to that surface has to lie in $N_h(X,\omega)/N_h$ because horizontal leafs are preserved. A similar characterization has been applied in [**HL**].

Let us now consider a general periodic direction $\theta \in S^1$ on (X,ω). By $N_\theta(Y/X,\tau) := \operatorname{SL}(Y/X,\tau) \cap N_\theta(X,\omega)$ we denote the parabolic stabilizer of the direction θ on (Y,τ). Its index

$$i_\theta(Y,X) := [N_\theta(X,\omega) : N_\theta(Y/X,\tau)] = |N_\theta(X,\omega) \cdot [Y \to X]| \tag{16}$$

depends only on the cusp defined by θ and we call it the *relative width* of the cusp.

PROPOSITION 14. *Let $v \in \mathbb{C}\backslash\{0\}$ be a vector parallel to a periodic direction on (X, ω) and $(S, \alpha) \in \mathrm{SL}(X, \omega)(Y, \tau)$, then*

$$c(\mathrm{SL}(S, \alpha)v) = \mathrm{vol}(\mathbb{H}/\mathrm{SL}(X, \omega))^{-1} \frac{i_v(X) \cdot i_v(S, X)}{[\mathrm{SL}(X, \omega) : \mathrm{SL}(Y/X, \tau)]} \frac{1}{|v|^2}. \tag{17}$$

PROOF. We apply formula 14 to the (relative) lattice $\mathrm{SL}(S/X, \alpha)$. This group is a conjugate of $\mathrm{SL}(Y/X, \tau)$, in particular it has the same index in $\mathrm{SL}(X, \omega)$. Now use the following facts: The index of the parabolic stabilizer of v in $\mathrm{SL}(S/X, \alpha)$ is $i_v(X) \cdot i_v(S, X)$ and

$$\mathrm{vol}(\mathbb{H}/\mathrm{SL}(Y/X, \tau)) = \mathrm{vol}(\mathbb{H}/\mathrm{SL}(X, \omega))[\mathrm{SL}(X, \omega) : \mathrm{SL}(Y/X, \tau)].$$

□

To apply this formula to the counting of closed cylinders, let us consider a positive real vector $\vec{\lambda} := (\lambda_1, ..., \lambda_n) \in \mathbb{R}^n_+$ and let $\theta \in \{z \in \mathbb{C} : |z| = 1\} \subset \mathbb{C}$ be a unit vector. Then $\vec{\lambda}\theta \in \mathbb{C}^n$ and we think of this vector as a set of n holonomy vectors of cylinders or saddle connections in direction θ on any translation cover $(S, \alpha) \in \mathrm{SL}(X, \omega)(Y, \tau)$. Let us now evaluate the quadratic asymptotics of the set

$$\mathrm{SL}(S, \alpha) \cdot \vec{\lambda}\theta := \{\vec{\lambda}A\theta \in \mathbb{C}^n : A \in \mathrm{SL}(S, \alpha)\}$$

and describe it in terms of the orbit space of covers $\mathrm{SL}(X, \omega)/\mathrm{SL}(Y/X, \tau)$. Using the orbit description of a relative cusp $\mathscr{U}_S := N_\theta(X, \omega) \cdot [S \to X]$ we obtain from the previous formula:

$$c(\mathrm{SL}(S, \alpha)\vec{\lambda}\theta) = \frac{\pi}{3\,\mathrm{vol}(\mathbb{H}/\mathrm{SL}(X, \omega))} \frac{i_\theta(X)}{|\mathscr{O}_Y|} \sum_{(S,\alpha)\in\mathscr{U}_S} \sum_{i=1}^{n_\theta} \frac{1}{\lambda_i^2} \tag{18}$$

Now consider all directions in the $\mathrm{SL}(X, \omega)$ orbit of the periodic direction θ on (Y, τ). Here we need to sum over the relative cusps on (Y, τ) that appear in $\mathrm{SL}(X, \omega)\theta$. If $C_Y(\theta)$ denotes this set of relative cusps we have

$$c_C(Y, \tau, \theta) = \sum_{\vartheta \in C_Y(\theta)} c(\mathrm{SL}(S, \alpha)\vec{\lambda}_\vartheta \vartheta)$$

where we specialize and take the positive vector to be the holonomy vector of cylinders for each cusp. Putting $\mathscr{O}_\tau := \mathrm{SL}(X, \omega)/\mathrm{SL}(Y/X, \tau)$, the asymptotic constant can be written using the orbit decomposition of cusps:

$$c_C(Y, \tau, \theta) = \frac{\pi}{3\,\mathrm{vol}(\mathbb{H}/\mathrm{SL}(X, \omega))} \frac{i_\theta(X)}{|\mathscr{O}_\tau|} \sum_{(S,\alpha)\in\mathscr{O}_\tau} \sum_{i=1}^{n_\alpha} \frac{1}{|\lambda_{\alpha,i}|^2}. \tag{19}$$

Note that for lattice surfaces saddle connection are always in the boundaries of cylinders. So the asymptotic constant for saddle connections is obtained by changing the entries of the last sum of 19 using saddle connection holonomy vectors.

The Siegel-Veech constants including all periodic directions of (X, ω) is now obtained by summing the previous expression over the cusps C_X of (X, ω), that is the cusps of $\mathrm{SL}(X, \omega)$ here represented by a periodic direction.

$$c_C(Y, \tau) = \sum_{\theta \in C_X} c_C(Y, \tau, \theta) \tag{20}$$

To summarize: Besides the *index and length of cylinders in a particular direction* the only quantities required to evaluate the constants are a *finite set of directions on the base surface corresponding to the cusps of its Veech group*, the number $i_X(\theta)$ associated to those directions and the hyperbolic area (of $\mathbb{H}/\mathrm{SL}(X,\omega)$).

Examples **Torus covers branched over one point.** Recall that arithmetic surfaces are (representable as) torus covers branched over one point. Taking the lattice defined by holonomy vectors of the absolute periods on an arithmetic surface defines a lattice Λ and a translation covering $(Y,\tau) \to (\mathbb{C}/\Lambda, dz)$ (up to translation). Assuming Λ is unimodular, i.e. $\mathbb{C}/\Lambda$ has area 1 with respect to the natural metric, we can use the $\mathrm{SL}_2\mathbb{R}$ action on (Y,τ) to deform the lattice to be $\mathbb{Z}[i]$. This does not change asymptotic quadratic growth rates. The Veech group of $\mathbb{T} = \mathbb{C}/\mathbb{Z}[i]$ is $\mathrm{SL}_2\mathbb{Z}$, having only one cusp, here represented by by the horizontal direction. Because the modulus of the horizontal cylinder is 1, we have $i_h(\mathbb{T}) = 1$. So,

$$c_C(Y,\tau) = \frac{\pi}{3\,\mathrm{vol}(\mathbb{H}/\mathrm{SL}_2\mathbb{Z})} \frac{1}{|\mathscr{O}_\tau|} \sum_{(S,\alpha)\in\mathscr{O}_\tau} \sum_{i=1}^{n_\alpha} \frac{1}{|\lambda_{\alpha,i}|^2}. \tag{21}$$

From this the general orbit formula for covers of Veech surfaces can be developed.

One takes into account, that given a periodic direction θ of (X,ω), there are exactly i_θ different surfaces in the orbit of the stabilizer $N_\theta \subset (X,\omega)$. Since elements in N_θ map cylinders (and saddle connections) in direction θ to cylinders (and saddle connections) of the same length we find

$$c_C(Y,\tau) = \frac{\pi}{3\,\mathrm{vol}(\mathbb{H}/\mathrm{SL}(X,\omega))} \frac{1}{|\mathscr{O}_\omega|} \sum_{(S,\alpha)\in\mathscr{O}_\omega} \sum_{i=1}^{n_\alpha} \frac{1}{|\lambda_{\alpha,i}|^2}. \tag{22}$$

Here $\mathscr{O}_\omega$ denotes the orbit of the cover $(Y,\tau) \to (X,\omega)$ under the full Veech group $\mathrm{SL}(X,\omega)$.

One cusp surfaces. Let us further mention the class of lattice surfaces having properties very similar to arithmetic ones. Those are Veech surfaces with *one cusp*, more precisely:

- *all periodic directions are in the orbit of a direction with k cylinders and*
- *the moduli $w_1/h_1 = \ldots = w_k/h_k$ of the cylinders are identical.*

As before w_i denotes the width and h_i the height of the ith cylinder. So $i_v(X) = m$ and the Siegel-Veech constant for cylinders is

$$c_C(X,\omega) = \frac{\pi}{3\,\mathrm{vol}(\mathbb{H}/\mathrm{SL}(X,\omega))} \sum_{i=1}^{k} \frac{1}{h_i w_i}. \tag{23}$$

A paper by Eskin, Marklof and Morris [**EMM**] contains a discussion of the Veech surfaces X_n obtained by linearization of the billiard in the triangle with angles $(\pi/n, \pi/n, (n-2)\pi/n)$. Veech showed in [**V3**] that X_n has one cusp that stabilizes a direction whose moduli are all the same. Thus formula 23 applies, for mere see [**EMM**] pages 26-28. In this case the w_i and h_i are explicit, see [**V2, EMM**]. Both of the above examples were used to study branched covers. The asymptotic growth rates for 2-fold branched covers of X_n for example have been evaluated in [**EMM**].

Flat geometry of parameter spaces and counting. We apply the asymptotic formula 20 in combination with the parameter space for d-symmetric covers. To do that we generally assume (X,ω) is a Veech surface, even though some of the following statements will hold in more generality. Let us fix a base point $x_0 \in Z(\omega) \subset X$ that will serve as the basic branch point of any d-symmetric to be considered. Then the set of d-symmetric covers branched over x_0 and $x \in X \backslash \{x_0\}$ is given by the points in the translation cover

$$\pi_d : (X_{ab}/dH_1, \pi_d^{-1}(x_0), \pi_d^* \omega) \to (X, \omega).$$

PROPOSITION 15. *Any affine map $\phi : (X, x_0, \omega) \to (X, x_0, \omega)$ lifts to $(X_{ab}/dH_1, x_d, \pi_d^* \omega)$ where $x_d \in \pi_d^{-1}(x_0)$.*

PROOF. We need to check the homotopy lifting criterion. Any homomorphism of groups $\psi : G \to H$ maps the commutator subgroup of G into the commutator subgroup of H. Since $\pi_1(X_{ab}) = [\pi_1(X, x_0), \pi_1(X, x_0)]$ we conclude $(\phi \circ \pi_{ab})_* \pi_1(X_{ab}, x_0) = C = (\pi_{ab})_* \pi_1(X_{ab}, x_0)$ for any affine map $\phi \in \mathrm{Aff}^+(X, x_0, \omega)$. Here $C = [\pi_1(X, x_0), \pi_1(X, x_0)] = \pi_1(X_{ab}, x_0)$. If we consider the lift $\widetilde{\phi}$ of ϕ to X_{ab} that fixes $x_{ab} \in \pi^{-1}(x_0)$, then the points in $\pi^{-1}(x_0)$ can be canonically identified with homology classes x_{ab} being the trivial class. Then the action of $\widetilde{\phi}$ restricted to $\pi^{-1}(x_0)$ is the action of ϕ on homology. Since the homology action is a $\mathbb{Z}$-module homomorphism it preserves the submodule $dH(X; \mathbb{Z})$ for any $d \geq 1$. That means $\widetilde{\phi}$ descends to an affine map, say ϕ_d of X_{ab}/dH_1 fixing $x_d := x_{ab} + dH_1$. By construction the cover $\pi_{ab} : (X_{ab}, x_{ab}) \to (X, x_0)$ factors over $(X_{ab}/dH_1, x_d)$ and ϕ_d is a lift of ϕ. □

Under the assumption that (X, x_0, ω) is a Veech surface all the covers

$$\pi_d : (X_{ab}/dH_1, \pi_d^{-1}(x_0), \pi_d^* \omega) \to (X, x_0, \omega).$$

are Veech surfaces and all their Veech groups contain the lattice $\mathrm{SL}(X, x_0, \omega)$ of (X, x_0, ω). Whenever a point on a translation surface is stabilized by all elements of a group of affine maps, we identify the affine maps with their derivatives. With respect to this identification we can make the following statement.

PROPOSITION 16. *For all $z \in (X_{ab}/dH_1, \pi_d^* \omega)$ and $A \in \mathrm{SL}(X, x_0, \omega)$*

$$A \cdot \#_z^d(X, x_0, \omega) = \#_{Az}^d(X, x_0, \omega).$$

PROOF. The dotted product is the action of $\mathrm{SL}(X, x_0, \omega)$ on d-symmetric covers. Since $A \in \mathrm{SL}(X, x_0, \omega)$ there is an affine map ϕ of (X, ω) that fixes x_0 with derivative A. The cover $A \cdot \#_z^d(X, x_0, \omega)$ is the cover associated to the homology class $\phi(\gamma)$, if γ is a path representing the relative class $[\gamma]$. This γ has starting point x_0 and by definition lifts to a path $\widetilde{\gamma}$ on X_{ab}/dH_1 that terminates in z. Because ϕ lifts to an affine map $\widetilde{\phi}$ that stabilizes x_d. By construction $\widetilde{\phi}$ maps the lift of $\widetilde{\gamma}$ to a lift of $\phi(\gamma)$ that terminates in $\widetilde{\phi}(z)$. Since all the affine maps have derivative A and fix the initial points of the respective curves, we have $Az = \widetilde{\phi}(z)$ and so the claim for d-symmetric covers follows. □

Any translation cover of (X, x_0, ω) is completely periodic in every direction where (X, x_0, ω) is completely periodic. Let us consider a periodic direction $\theta \in S^1$ on (X, x_0, ω). Then θ is a completely periodic direction for any d-symmetric cover $\#_x^d(X, x_0, \omega)$ given by a point $x \in (X_{ab} - \pi_{ab}^{-1}(x_0))/dH_1$. In particular this x lies on a saddle connection or in an open cylinder of the cylinder decomposition of $(X_{ab} - \pi_{ab}^{-1}(x_0))/dH_1$ in direction θ. The holonomy vectors (with multiplicity) of all cylinders in a given completely periodic direction on a translation surface is called cylinder datum.

LEMMA 1. *Given a completely periodic direction $\theta \in S^1$ on (X, x_0, ω). Then the cylinder datum on a d-symmetric cover in direction θ depends only on the cylinder, or saddle connection of $(X_{ab} - \pi_{ab}^{-1}(x_0))/dH_1$ that contains the point representing the cover.*

PROOF. Again we fix a periodic direction, say $\theta \in S^1$, on (X, x_0, ω). Let us consider a maximal cylinder $\mathscr{C} \subset (X_{ab} - \pi_{ab}^{-1}(x_0))/dH_1$ and its closure $\overline{\mathscr{C}} := \mathscr{C} \cup \partial\mathscr{C}$. Then $\overline{\mathscr{C}_X} := \pi_d(\overline{\mathscr{C}}) \subset X$ is a single closed cylinder on (X, x_0, ω). Indeed $\mathscr{C}_X$ is a cylinder, since regular leaves are mapped to regular leaves. Because $\partial\mathscr{C}$ contains preimages of $Z(\omega)$ or x_0 on each connected component, $\mathscr{C}_X$ has one of those points on its boundary and hence is maximal.

Let us take two points $z_i \in \mathscr{C}$ $i = 1, 2$, and path $\widetilde{\gamma}_{x_d}^{z_i}$ on $(X_{ab} - \pi^{-1}(x_0))/dH_1$ from x_d to z_i. Since both points $z_i \in \mathscr{C}$ lie in the same cylinder we can connect z_1 with z_2 with a line segment l and obtain a loop $\widetilde{\gamma}$ on $(X_{ab} - \pi^{-1}(x_0))/dH_1$. Since loops in $(X_{ab} - \pi^{-1}(x_0))/dH_1$ map to homologically trivial loops on X the intersection number $\pi_d\widetilde{\gamma}$ with any loop on X is trivial. This is particularly true for all loops defined by regular closed leaves in cylinders parallel to $\mathscr{C}$ on (X, ω). Consequently as long as those loops do not intersect $\pi_d(l)$ they have the same intersection number with both relative classes $[\gamma_{x_0}^{x_i}] \in H_1(X, \{x_0, x_i\}; \mathbb{Z}_d)$ defined by $\gamma_{x_0}^{x_i} = \pi_d \circ \widetilde{\gamma}_{x_d}^{z_i}$, $(x_i = \pi_d(z_i), i = 1, 2)$. This is trivially true for all leaves that do not intersect $\mathscr{C}_X$. Since the straight line segment $\pi_d(l)$ connects the points x_i in the interior of $\mathscr{C}_X$ there are cylinder leaves of $\mathscr{C}_X$ that do not intersect $\pi_d(l)$. Indeed both points x_i have positive distance from the boundary and any cylinder leaf that is closer to the boundary than the minimum distance of the two points does not intersect $\pi_d(l)$. So the intersections of such leaves with either class $[\gamma_{x_0}^{x_i}]$ are the same.

Now the intersection numbers of the cylinder leaves in a cylinder decomposition of (X, ω) in direction θ with $\gamma_{x_0}^{x_i}$ determine the circumferences of the direction θ cylinders in $\#_{z_i}^d(X, x_0, \omega)$. Since the covers are cyclic that also determines the multiplicity of the cylinders of $\#_{z_i}^d(X, x_0, \omega)$. That shows the claim for all cylinders on either surface $\#_{z_i}^d(X, x_0, \omega)$ in the preimage of any direction θ cylinder on (X, ω) besides $\mathscr{C}_X$. For $\mathscr{C}_X$, that contains the branching points x_i, let us look at x_1 and pretend the cylinder $\mathscr{C}_X$ is horizontal, in order to use geometric phrases. Then the lifts of cylinder loops from $\mathscr{C}_X$ to $\#_{z_1}^d(X, x_0, \omega)$ are determined by a loop below x_1 and by a second loop above x_1. Both can be represented by loops that avoid $\pi_d(l)$ since all cylinder loops of $\mathscr{C}_X$ above, respectively below x_1 have the same lift.

If two points $z_i \in \partial\mathscr{C}$ lie in the same boundary components of $\mathscr{C}$ we can connect them using part of a direction θ leaf in $\mathscr{C}$ as close as we wish to $\partial\mathscr{C}$ together with two appropriate line segments perpendicular to θ that will have z_i as endpoints. Then we apply the same intersection argument as before. □

Because holonomy data are constant on cylinders and saddle connections of the parameter space the orbit type asymptotic formulas can be simplified.

5. Asymptotic constants and parameter space geometry.

The following discussion together with the asymptotic formulas 19 and 20 will provide Theorem 6 and Theorem 7.
Consider the parameter space $(X_{ab}/dH_1, \pi_d^{-1}(x_0), \pi_d^* \omega)$ as the parameter space of d-symmetric covers and assume for simplicity the horizontal direction is completely periodic. Let $\mathscr{C}_i$, $i = 1, ..., n_h$ denote the maximal horizontal cylinders on the marked surface $(X_{ab}/dH_1, \pi_d^{-1}(x_0), \pi_d^* \omega)$ and their top boundary components by $\partial^{top}\mathscr{C}_i$. The previous Lemma shows, that the length spectrum of horizontal cylinders with multiplicity is constant on cylinders, such as $\mathscr{C}_i$ and their (top) boundaries. Each top boundary $\partial^{top}\mathscr{C}_i$ is a union of saddle connections and the Lemma shows that the d-symmetric covers parameterized by $\partial^{top}\mathscr{C}_i$ have the same length spectrum.

Associated to each horizontal cylinder $\mathscr{C}_i$ there is a datum $w_{i,1}, ..., w_{i,n_i}$ for the length of the horizontal cylinders on any d-symmetric cover defined by a point on $\mathscr{C}_i$. The same holds for the length of horizontal cylinders on d-symmetric surfaces parameterized by $\partial^{top}\mathscr{C}_i$, let us denote their total number by m_i and their length datum by $c_{i,1}, ..., c_{i,m_i}$. If $\mathscr{O}_z$ denotes the $\mathrm{SL}(X, x_0, \omega)$-orbit of a $z \in X_{ab}/dH_1$ with finite orbit, then the asymptotic formula for the horizontal cusp θ_h is:

$$c_C(\theta_h) = C \cdot i_h \sum_{i=1}^{n_h} \left[\frac{|\mathscr{O}_z \cap \mathscr{C}_i|}{|\mathscr{O}_z|} \sum_{k=1}^{n_i} \frac{1}{w_{i,k}^2} + \frac{|\mathscr{O}_z \cap \partial^{top}\mathscr{C}_i|}{|\mathscr{O}_z|} \sum_{k=1}^{m_i} \frac{1}{c_{i,k}^2} \right]. \tag{24}$$

Here $C := \frac{\pi}{3 \operatorname{vol}(\mathbb{H}/\mathrm{SL}(X, x_0, \omega))}$ denotes the constant in formula 18. Recall that i_h is the upper right entry of the minimal parabolic stabilizer of the (periodic) horizontal direction. For tori we have $i_h = 1$. To obtain the general version of the formula stated in the introduction assume $\{\theta_1, ..., \theta_{n_c}\}$ is a set of directions corresponding to the cusps of $\mathrm{SL}(X, x_0, \omega)$. Pick a cylinder decomposition $\mathscr{C}_j^l$ $j = 1, ..., m_l$ of $(X_{ab} - dH_1 x_{ab})/dH_1$ in direction θ_l. Then any $\#_z^d(X, x_0, \omega)$ with $z \in \mathscr{C}_j^l$ has precisely n_j^l maximal cylinders of circumference $w_{j1}^l, ..., w_{jn_j^l}^l$ and height $h_{j1}^l, ..., h_{jn_j^l}^l$ in direction θ_l. Any $\#_z^d(X, x_0, \omega)$ with $z \in \partial^{top}\mathscr{C}_j^l$ has s_j^l maximal cylinders of circumference $c_{j1}^l, ..., c_{js_j^l}^l$ in direction θ_l.

THEOREM 6. **[Cylinders]** *Under the previous assumptions and conventions, if $\#_z^d(X, x_0, \omega)$ has infinite $\mathrm{SL}(X, x_0, \omega)$ orbit, the asymptotic constant for periodic cylinders in the $\mathrm{SL}(X, x_0, \omega)$ orbit of the direction θ_l is:*

$$\mathbf{c}_C^l(z) = \frac{1}{d^2 \operatorname{area}(X, \omega)} \sum_{j=1}^{m_l} \sum_{k=1}^{n_j^l} \frac{\operatorname{area}(\mathscr{C}_j^l)}{(w_{jk}^l)^2}. \tag{25}$$

If $\#_z^d(X, x_0, \omega)$ has a finite $\mathrm{SL}(X, x_0, \omega)$ orbit $\mathscr{O}_z$, then:

$$\mathbf{c}_C^l(z) = \frac{C \cdot i_l}{|\mathscr{O}_z|} \sum_{j=1}^{m_l} \left(\sum_{k=1}^{n_j^l} \frac{|\mathscr{O}_z \cap \mathscr{C}_j^l|}{(w_{jk}^l)^2} + \sum_{k=1}^{s_j^l} \frac{|\mathscr{O}_z \cap \partial^{top}\mathscr{C}_j^l|}{(c_{jk}^l)^2} \right). \tag{26}$$

In either case the asymptotic formula for cylinders on (S, α) is given by

$$\mathbf{c}_C(z) = \sum_{l=1}^{n_c} \mathbf{c}_C^l(z).$$

Saddle connections. The asymptotic formula for saddle connections on $\#_z^d(X, x_0, \omega)$ is analogous to the one for cylinders as long as $z \in X_{ab}/dH_1$ has a finite $\mathrm{SL}(X, x_0, \omega)$-orbit, i.e. if the d-symmetric surface $\#_z^d(X, x_0, \omega)$ is a Veech surface. In fact, one only needs to replace the holonomy vector datum corresponding to widths of horizontal cylinders with the one for the length of the saddle connections s_j bounding the horizontal cylinders of $\#_z^d(X, x_0, \omega)$ in formula 24.

Saddle connections between the two branch points. Let $V(\widehat{x}_0, \widehat{z}_0)$ denote the set of saddle connections on the d-symmetric cover $\pi_z : \#_z^d(X, x_0, \omega) \to (X, x_0, \omega)$ connecting the two cone-points $\widehat{z}_0 = \pi^{-1}\pi_d(z) \in \#_z^d(X, x_0, \omega)$ and $\widehat{x}_0 = \pi^{-1}(x_0) \in \#_z^d(X, x_0, \omega)$. The covering map π_z induces a surjective and isometric map

$$\pi_{d*} : V(\widehat{x}_0, \widehat{z}_0) \to V(x_0, z_0)$$

to the set of regular line segments connecting the marked points x_0 and z_0 on (X, x_0, z_0, ω). Since every regular segment in $V(x_0, z_0)$ has d preimages on $\#_z^d(X, x_0, \omega)$, the map π_* has degree d. In particular

$$c(V(\widehat{x}_0, \widehat{z}_0)) = d \cdot c(V(x_0, z_0))$$

and we only need to describe the respective counting formula for (X, x_0, z_0, ω).

Given this, it seems more interesting to look at certain subsets of saddle connections related to the saddle connections of $(X_{ab}/dH_1, \pi_d^{-1}(x_0), \pi_d^*\omega)$, particularly when $\#_z^d(X, x_0, \omega)$ is a lattice surface, or equivalently, $z \in X_{ab}/dH_1$ has finite orbit with respect to the Veech group $\mathrm{SL}(X, x_0, \omega)$ acting on X_{ab}/dH_1. We are mainly interested in the torus case, so let us further assume $x_d \in \pi_d^{-1}(x_0)$ is fixed under the action of $\mathrm{SL}(X, \omega)$ and that every periodic direction is in the $\mathrm{SL}(X, \omega)$ orbit of the periodic horizontal direction on (X, x_0, ω). Then, since z_0 is periodic, the marked surface (X, x_0, z_0, ω) is a Veech surface and so any $s \in V(x_0, z_0)$ is parallel to a completely periodic direction. Given our assumptions, there is $A \in \mathrm{SL}(X, x_0, \omega)$ so that $s_h = As$ is a horizontal saddle connection on (X, x_0, Az_0, ω) and hence all preimages of s_h on $\#_{Az}(X, x_0, \omega)$ are isometric to s_h. Since $s_h \in V_h(x_0, z_0)$ is horizontal, it is a subset of the π_d image of the horizontal saddle connection $(X_{ab}/dH_1, \pi_d^{-1}(x_0))$ that contains Az. This saddle connection, $\widetilde{s}_h$, emanates from, say $x_s \in \pi_d^{-1}(Ax_0)$ and contains the horizontal segment s_{Az}^+ from x_s to Az that maps isometrically onto s_h under π_d. If the base surface X is a torus then $\widetilde{s}_h$ connects two preimages of x_0. Then the set theoretical complement of $s_{Az}^+ \cup \{Az\}$ in $\widetilde{s}_h$ is also a horizontal saddle connection, say s_{Az}^- that corresponds isometrically to d saddle connections on $\#_{Az}(X, x_0, \omega)$.

We now state an orbit version of the counting formula for saddle connections of the type just described for *torus covers*. If $\#_z^d(X, x_0, \omega)$ is a lattice torus cover, any saddle connection in $V(\widehat{x}_0, \widehat{z}_0)$ is up to isometry (and multiplicity by a factor d) in the $\mathrm{SL}(X, x_0, \omega)$ orbit of a horizontal saddle connection on the marked parameter space $(X_{ab}/dH_1, \pi_d^{-1}(x_0), \mathscr{O}_z \cap SC_h(d))$. Here $\mathscr{O}_z := \mathrm{SL}(X, x_0, \omega) \cdot z$ and $SC_h(d)$ is the set of horizontal saddle connections on $(X_{ab}/dH_1, \pi_d^{-1}(x_0))$. For the formula we only need to know the multiplicity of each horizontal saddle connection on the actual covers. For d-symmetric covers it is always d.

THEOREM 7 (Saddle connections on d-symmetric torus covers). *Let $SC_h(d)$ denote the horizontal saddle connections on the parameter space* $(\mathbb{C}/d\mathbb{Z}[i], \mathbb{Z}[i], dz)$

of d-symmetric torus covers $(\mathbb{C}/\mathbb{Z}[i], [0], dz)$. *If* $\#_z^d(\mathbb{C}/\mathbb{Z}[i], [0], dz)$ *is a lattice surface, then the asymptotic quadratic growth rate of saddle connections connecting two cone points of* $\#_z^d(\mathbb{C}/\mathbb{Z}[i], [0], dz)$ *is:*

$$c_{\pm}(z) = \frac{d}{|\mathscr{O}_z|} \sum_{s \in SC_h(d)} \sum_{y \in \mathscr{O}_z \cap s} \left[\frac{1}{|s_y^-|^2} + \frac{1}{|s_y^+|^2} \right]. \tag{27}$$

Since on the torus $(\mathbb{C}/d\mathbb{Z}[i], \mathbb{Z}[i], dz)$ the affine involution induced by $-\operatorname{id}$ on $\mathbb{C}$ preserves $\mathbb{Z}[i]$, the formula simplifies:

$$c_{\pm}(z) = \frac{2d}{|\mathscr{O}_z|} \sum_{s \in SC_h(d)} \sum_{y \in \mathscr{O}_z \cap s} \frac{1}{|s_y^+|^2}. \tag{28}$$

The formula is easily derived using the same method as for the counting formulas for cylinders. Instead of square reciprocals of cylinder circumferences assigned to each saddle connection in $SC_h(d)$, one considers the pair of (horizontal reciprocal square) distances of a point in $\mathscr{O}_z \cap SC_h(d)$ to the relevant points in $\pi_d^{-1}(x_0)$.

6. Teichmüller curves and moduli space

If (Y, τ) is a lattice surface with Veech group $\mathrm{SL}(Y, \tau)$, then the image of $\mathrm{SL}_2\mathbb{R}(Y, \tau)$ in $\Omega\mathcal{M}_g$ projects to a Teichmüller curve $\mathcal{C} \subset \mathcal{M}_g$ in moduli space of Riemann surfaces of genus $g = g(Y)$. This curve is the image of the algebraic immersion

$$\mathbb{H}/\mathrm{SL}(Y, \tau) \to \mathcal{M}_g,$$

that is an isometry with respect to the Teichmüller metric on $\mathcal{M}_g$. For more on Teichmüller curves particularly in genus 2, see McMullen [**McM1**]–[**McM5**], as well as Bouw and Möller [**BM**]. The first examples of Teichmüller curves that do not arise as torus covers were discovered by Veech [**V1**].

All d-symmetric (X, x_0, ω) covers of a lattice surface (X, x_0, ω) that have finite $\mathrm{SL}(X, x_0, \omega)$ orbit in their respective parameter space define Teichmüller curves. Since we may not use the full group of affine maps the curves we obtain in the global description below may only be covers of the actual Teichmüller curve.

The family of d-symmetric differentials. To obtain Teichmüller curves we consider the family of all d-symmetric torus covers over any (normalized) base torus. That is, we look at all (unimodular) lattices and include degenerate covers for simplicity. All those covers are parameterized by points in

$$\mathrm{SL}_2\mathbb{R} \ltimes \mathbb{C}/\mathrm{SL}_2\mathbb{Z} \ltimes d\mathbb{Z}[i].$$

If $(Y, \widetilde{x}, \tau) \to (\mathbb{C}/\Lambda, [0], dz)$ is a d-symmetric cover with lattice stabilizer $\mathrm{SL}(Y, \widetilde{x}, \tau)$ then its orbit

$$\mathrm{SL}_2\mathbb{R} \cdot (Y, \widetilde{x}, \tau) \hookrightarrow \mathrm{SL}_2\mathbb{R} \ltimes \mathbb{C}/\mathrm{SL}_2\mathbb{Z} \ltimes d\mathbb{Z}[i].$$

is given by $\mathrm{SL}_2\mathbb{R}/\mathrm{SL}(Y, \widetilde{x}, \tau)$. The quotient with respect to $\mathrm{SO}_2\mathbb{R}$,

$$\mathrm{SL}_2\mathbb{R}/\mathrm{SL}(Y, \widetilde{x}, \tau) \to \mathbb{H}/\mathrm{SL}(Y, \widetilde{x}, \tau)$$

is a finite cover of the Teichmüller curve (Y, τ) defines in moduli space.

7. Cylinder decompositions of d-symmetric forms

We calculate the cylinder decompositions for d-symmetric forms over the fixed lattice $\Lambda := \mathbb{Z}[i]$. Note, that the knowledge of the parameter space and its geometry simplifies some arguments. In this section we will write $\mathbb{T}_d := \mathbb{C}/d\Lambda$, $\dot{\mathbb{T}}_d := (\mathbb{C} - \Lambda)/d\Lambda$ and denote marked tori as $\mathbb{T}_{d,m} := (\mathbb{C}/d\Lambda, \Lambda/d\Lambda)$. Further, let us denote the integer part of $y \in \mathbb{R}_+$ by $\lfloor y \rfloor$ and if $[0, z] \subset \mathbb{C}$ is a line segment, then $[\![0, z]\!] \in H_1(\mathbb{T}, \{[0], [z]\}; \mathbb{Z})$ denotes the indicated relative homology class of its image on $\mathbb{T}$. Whenever it does not lead to confusion we will use the abbreviation $(a_1, ..., a_n) = \gcd(a_1, ..., a_n)$.

Proposition 17. *Let $\#_z^d(\mathbb{T}, dz)$ be a d-symmetric form. Then its horizontal foliation is completely periodic and it contains*

- $(\lfloor \operatorname{Im} z \rfloor, d)$ *cylinders of width* $\frac{d}{(\lfloor \operatorname{Im} z \rfloor, d)}$ *and*
- $(\lfloor \operatorname{Im} z \rfloor + 1, d)$ *cylinders of width* $\frac{d}{(\lfloor Im(z) \rfloor + 1, d)}$, *if* $\operatorname{Im} z \notin \mathbb{Z}_d$
- $(\operatorname{Im} z, d)$ *cylinders of width* $\frac{d}{(\operatorname{Im} z, d)}$, *if* $\operatorname{Im} z \in \mathbb{Z}_d$.

Proof. Given $\#_z^d(\mathbb{T}, dz)$, denote the projection of $z \in \mathbb{T}_d$ to $\mathbb{T}$ by z. First we assume $\operatorname{Im} z \notin \mathbb{Z}_d$. If l is a horizontal leaf on the two marked torus $(\mathbb{T}, [0], [z])$ that lies on the cylinder having $[z]$ in its upper boundary, then we have an intersection number $\langle l, [\![0, z]\!] \rangle \equiv i + 1 \mod d$ where $i \equiv \lfloor \operatorname{Im} z \rfloor \mod d$. By cyclicity every preimage of l on $\#_z^d(\mathbb{T}, dz)$ has length $d/(\lfloor \operatorname{Im} z \rfloor + 1, d)$ and so there must be $(\lfloor \operatorname{Im} z \rfloor + 1, d)$ of those. The same way, if l is a horizontal leaf on $(\mathbb{T}, [0], [z])$, that lies on the cylinder with $[z]$ on its lower boundary, then $\langle l, [\![0, z]\!] \rangle = i$ where i is as above. By cyclicity every preimage of l on $\#_z^d(\mathbb{T}, dz)$ has length $d/(\lfloor \operatorname{Im} z \rfloor, d)$ and then there must be $(\lfloor \operatorname{Im} z \rfloor, d)$ of those.

If $\operatorname{Im} z \in \mathbb{Z}_d$, then $(\mathbb{T}, [0], [z])$ has only one horizontal cylinder and a horizontal loop l intersects with $[\![0, z]\!]$ exactly $\operatorname{Im} z$ times. As before we obtain $(\operatorname{Im} z, d)$ lifted loops of length $d/(\operatorname{Im} z, d)$.

Since every horizontal cylinder of $\#_z^d(\mathbb{T}, dz)$ maps to a horizontal cylinder of $(\mathbb{T}, [0], [z])$ the claim follows. □

The proof of the previous proposition allows us to specify the area of each horizontal cylinder on $\#_z^d(\mathbb{T}, dz)$. We record this for later use.

Corollary 7. *If $\operatorname{Im} z \notin \mathbb{Z}_d$ then the horizontal cylinders of $\#_z^d(\mathbb{T}, dz)$ mapped to the cylinder on $(\mathbb{T}, [0], [z])$ that has $[z]$ on its top boundary have area $\operatorname{Im} z \cdot d/\gcd(\lfloor \operatorname{Im} z \rfloor + 1, d)$. The cylinders mapped to the cylinder on $(\mathbb{T}, [0], [z])$ that has $[z]$ on its lower boundary have area $(1 - \operatorname{Im} z) \cdot d/\gcd(\lfloor \operatorname{Im} z \rfloor, d)$. If $\operatorname{Im} z \in \mathbb{Z}_d$ then the horizontal cylinders on $\#_z^d(\mathbb{T}, dz)$ have (integer) area $d/\gcd(\operatorname{Im} z, d)$.*

Divisibility properties for the numbers of cylinders calculated in Proposition 17 give:

Corollary 8. *Assume $d > 2$ is prime, then for any j the only possible numbers of maximal* horizontal cylinders *of d-symmetric covers parameterized by $\mathscr{C}_j$ are* 2 *and* $d + 1$.

Forms parameterized by lattice coordinates. Now we describe cylinder decompositions of surfaces with lattice twist coordinates, i.e. those $\#_z^d(\mathbb{T}, dz)$ with

$$z + d\Lambda \in \Lambda/d\Lambda.$$

Because the hyperelliptic involution of $\mathbb{T}_d$ sends the surface $\#_z^d(\mathbb{T}, dz)$ to the translation equivalent surface $\#_{-z}^d(\mathbb{T}, dz)$, $\Lambda/d\Lambda \subset \mathbb{T}_d$ does not provide a classification space for those. Here we study their horizontal cylinder decompositions and their $\mathrm{SL}_2\mathbb{Z}$-orbits in $\mathbb{T}_d$.

With the convention $(a, b) = \gcd(a, b)$, given a Gaussian integer $z \in \Lambda$ we write $(z) := (\operatorname{Re} z, \operatorname{Im} z)$. If furthermore $d \in \mathbb{N}$ we set $(z, d) := ((\operatorname{Re} z, \operatorname{Im} z), d)$.

COROLLARY 9. *For $z \in \Lambda/d\Lambda$ the d-symmetric differential $\#_z^d(\mathbb{T}, dz)\backslash\{\pi^{-1}[0]\}$ is a disjoint union of (z, d) tori tiled by unit squares. Each such torus has area $d/(z, d)$ and decomposes in $(\operatorname{Im} z, d)/(z, d)$ horizontal cylinders of width $d/(\operatorname{Im} z, d)$ and $(\operatorname{Re} z, d)/(z, d)$ vertical cylinders of width $d/(\operatorname{Re} z, d)$. For any given divisor $k|d$ the surfaces that are unions of k tori are all on one $\mathrm{SL}_2\mathbb{Z}$ orbit.*

PROOF. If $j \in \mathbb{Z}$, the horizontal foliation of $\#_j^d(\mathbb{T}, dz)$ has d cylinders of width 1 while its vertical foliation has (j, d) cylinders of width $d/(j, d)$. Thus $\#_j^d(\mathbb{T}, dz)$ consists of (j, d) tori each of width 1 and height $d/(j, d)$. By the $\mathrm{SL}_2\mathbb{Z}$ orbit classification every form represented by a point in $\Lambda/d\Lambda$ is on the $\mathrm{SL}_2\mathbb{Z}$ orbit of a form $\#_a^d(\mathbb{T}, dz)$ for some $a \in \mathbb{Z}_d$. The shape of those $\mathrm{SL}_2\mathbb{Z}$ orbits imply that $\#_z^d(\mathbb{T}, dz) \in \mathrm{SL}_2\mathbb{Z} \cdot \#_a^d(\mathbb{T}, dz)$ if and only if $(z, d) = (a, d)$. In particular for $z + d\Lambda \in \Lambda/d\Lambda$ the forms $\#_z^d(\mathbb{T}, dz)$ with fixed (z, d) are unions of (z, d) tori, each of area $d/(z, d)$.

It follows from Proposition 17 that $\#_z^d(\mathbb{T}, dz)$ has $(\operatorname{Im} z, d)$ horizontal cylinders of width $d/(\operatorname{Im} z, d)$ and $(\operatorname{Re} z, d)$ vertical cylinders of width $d/(\operatorname{Re} z, d)$. □

8. Illumination and non illumination

On most d-symmetric torus covers there is a family of parallel saddle connections that, when removed together with its endpoints, leaves exactly d connected components. Since the only degenerate d-symmetric surface that, after removing one point, falls into d connected components is $\#_0^d(\mathbb{T}, dz)$, i.e. the surface determined by the origin of $\mathbb{C}/d\Lambda$, we can formulate that property in terms of the parameter space geometry as follows. Can we see a given point in $\mathbb{C}\backslash\Lambda/d\Lambda$ from the origin? We will see that there are finitely many points for which this is not possible, at least if $d > 2$. Another way to characterize such a d-symmetric cover is, that one cannot represent its defining relative homology class by a straight line segment that does not intersect itself. In case of a d-symmetric torus cover with second branch point $[z]_\Lambda \in \mathbb{C}/\Lambda$ the existence of a homology representing line segment is the same as the existence of an open line segment in $\{(0, w) : w \in z + d\Lambda\} \subset \mathbb{C}$ that lies in $\mathbb{C} - \Lambda$.

Let $[a, b] \subset \mathbb{C}$ denote the (closed) line segment in $\mathbb{C}$ and $[a, b]_\Lambda$ its image (modulo Λ) on $\mathbb{C}/\Lambda$. Recall the isogeny

$$(\mathbb{C}\backslash\Lambda)/d\Lambda \to (\mathbb{C}\backslash d^{-1}\Lambda)/\Lambda = (\mathbb{C}/\Lambda)\backslash T(d)$$

given by multiplication with d^{-1}. Since the set of d-torsion points $T(d) \subset \mathbb{C}/\Lambda$ is invariant under the action of $\mathrm{SL}(\Lambda)$, this action is well-defined on both tori with the respective points removed. The isogeny d^{-1} is $\mathrm{SL}(\Lambda)$-equivariant.

LEMMA 2. *For $d > 0$ fixed let $n \in \{2, ..., d\}$. Then the d-symmetric torus cover $\#_z^d(\mathbb{C}/\Lambda, dz)$ is representable by a regular slit construction, if and only if z is not a torsion point of order $n \nmid d$. In particular those covers appear only for $d > 2$.*

PROOF. By $\mathrm{SL}_2\mathbb{R}$ invariance of geodesic properties it is enough to show the claim for $\Lambda = \mathbb{Z}[i]$, nevertheless (for readability) we continue to denote the integer lattice by Λ. Representability of a regular d-symmetric form $\#_z^d(\mathbb{T}, dz)$ by a line segment $[0, w]$, for some $w \in z + d\Lambda$ follows, if w is visible from the origin, that is if the line segment $[0, w]$ lies in $\mathbb{C}\backslash\Lambda$. If we can find $A \in \mathrm{SL}(\Lambda)$ mapping the point $[z]$ into the open disk

$$D := \{[w] \in (\mathbb{C}\backslash\Lambda)/d\Lambda : |w| < 1\} \subset (\mathbb{C}\backslash\Lambda)/d\Lambda,$$

we are done. First D is a genuine disk since no point of $\Lambda/d\Lambda$ has distance less than one from the origin. But then the interval $[0, Az]_\Lambda$ lies in $(\mathbb{C}\backslash\Lambda)/d\Lambda$, and by $\mathrm{SL}(\Lambda)$ invariance of $(\mathbb{C}\backslash\Lambda)/d\Lambda$ $A^{-1}[0, A[z]]$ is an interval in $(\mathbb{C}\backslash\Lambda)/d\Lambda$ with endpoints $[0]$ and $[z]$. It is always possible to find a transformation $A \in \mathrm{SL}_2\mathbb{Z} = \mathrm{SL}(\Lambda)$ mapping an irrational $z \in \mathbb{C}$ into D, since $\mathrm{SL}_2\mathbb{Z}$ orbits of irrational numbers are dense on $\mathbb{T}$. For rational $[z]$ we use the isogeny to transform the torus $(\mathbb{C}\backslash\Lambda)/d\Lambda$ to the marked torus $\mathbb{T}\backslash(\frac{1}{d}\Lambda)/\Lambda$, which maps rational points to rational points. This is an $\mathrm{SL}_2\mathbb{Z}$-equivariant map, in particular the action of $\mathrm{SL}_2\mathbb{Z}$ is well-defined on $\mathbb{T}\backslash(\frac{1}{d}\Lambda)/\Lambda$. The image of D under this map is

$$D_d := \{[w] \in \mathbb{C}/d\Lambda : |w| < d^{-1}\} \subset (\mathbb{C}\backslash\frac{1}{d}\Lambda)/\Lambda.$$

By the $\mathrm{SL}_2\mathbb{Z}$ orbit classification we there is an $A \in \mathrm{SL}_2\mathbb{Z}$ that maps a given rational coordinate $z \in (\mathbb{Q}\oplus\mathbb{Q}i)/\Lambda \subset \mathbb{C}/\Lambda$ of denominator n to the point $[1/n] \in \mathbb{C}/\Lambda$. Since $[1/n] \in U_d$ if and only if $n > d$, the existence of a regular slit representation follows by transforming back to $(\mathbb{C}\backslash\Lambda)/d\Lambda$.

The remaining points to consider are in $n^{-1}\Lambda/\Lambda$ with $0 < n < d$. If n is a d-torsion point, then n must divide d and $n^{-1}\Lambda/\Lambda$ is a subset of the set of removed points $d^{-1}\Lambda/\Lambda$. If on the other hand $n \nmid d$, consider a line segment $I \subset \mathbb{C}/\Lambda$ between $[0] \in \mathbb{C}/\Lambda$ and some point $[z] \in n^{-1}\Lambda/\Lambda$. Then there is an $A \in \mathrm{SL}_2\mathbb{Z}$, such that $A \cdot I = [0, 1/n]$. Since $n < d$, the point $[1/d]$ lies on the segment $[0, 1/n] = A \cdot I$. Since $\mathrm{SL}_2\mathbb{Z}$ preserves the order of points the d-torsion point $A[1/d]$ lies on I and so $I \not\subseteq (\mathbb{C}\backslash d^{-1}\Lambda)/\Lambda$ showing the claim. □

We close this section with some remarks: For any representation of the cover $\pi : \#_w^d(\mathbb{C}/\Lambda, dz) \to (\mathbb{C}/\Lambda, dz)$ by a slit construction along $[0, w] \subset \mathbb{C} - \Lambda$ the preimage $\pi^{-1}([0, w])$ consists of d *saddle connections* of length $|w|$. If $\#_z^d(\mathbb{C}/\Lambda, dz)$ is on the $\mathrm{SL}_2\mathbb{Z}$ orbit of $\#_w^d(\mathbb{C}/\Lambda, dz)$ then is representable by a slit construction. Because then there is $A \in \mathrm{SL}_2\mathbb{Z}$, so that $A[w] = [z] \in (\mathbb{C} - \Lambda)/d\Lambda$. Viewing $[0, w] \subset (\mathbb{C} - \Lambda)/d\Lambda$ as a line segment in parameter space and noticing $A[0] = [0]$ we obtain a line segment $A[0, w] \subset (\mathbb{C} - \Lambda)/d\Lambda$ from $[0]$ to $[z]$. Now $\#_z^d(\mathbb{T}, dz)$ can be deformed into the trivial d-symmetric cover $\#_{[0]}^d(\mathbb{T}, dz)$ inside $(\mathbb{C} - \Lambda)/d\Lambda$ by moving $[z]$ into $[0]$ along the line segment $A[0, w] \subset (\mathbb{C} - \Lambda)/d\Lambda$.

The lattice $\Lambda \in \mathbb{C}/d\Lambda$ is a *blocking set* for the finite set $\bigsqcup_{n\nmid d}(n^{-1}d\Lambda)/d\Lambda$ relative to the point $[0] \in \mathbb{C}/d\Lambda$. Informally, a rotating laser installed at the point $[0]$ will never illuminate any point in $\bigsqcup_{n\nmid d}(n^{-1}d\Lambda)/d\Lambda$, because the rays that would illuminate will be swallowed by the holes $\Lambda \in \mathbb{C}/d\Lambda$.

9. Asymptotic constants — cylinders, generic case

Siegel-Veech constants for generic forms. We have all information to evaluate the Siegel-Veech constants as function of the point in the space of d-symmetric covers. As before we restrict our considerations to the covers with absolute period lattice Λ.

Proof of Theorem 4. Since we assume $\#_z^d(\mathbb{T}, dz)$ is not arithmetic, i.e. $[z]_d \in \mathbb{T}_d$ is not a torsion point, it has infinite $\mathrm{SL}_2\mathbb{Z}$ orbit. In that case formula 25 of Theorem 6 is applied. This formula is the integral formula in [**EMS**]. It is itself derived from the Siegel-Veech formula in [**V3**] and states that the the $\mathrm{SL}_2\mathbb{Z}$ orbit of a generic surface can be treated as uniform distribution with respect to Lebesgue measure. The only relevant part of the Siegel-Veech formula is the integral over the horizontal cylinders of the surface parameterizing covers over a fixed base surface, here $\mathbb{T} = \mathbb{C}/\mathbb{Z}[i]$. The measure on the parameter space is the standard Lebesgue measure induced by the euclidian structure. Given this, each point z, representing the cover $\#_z^d(\mathbb{T}, dz)$, is weighted by the areas of the covers horizontal cylinders $C = C(z)$. Recall that the numbers of horizontal cylinders and their width is constant for all surfaces $\#_z^d(\mathbb{T}, dz)$ in a horizontal cylinder $\mathscr{C}$ of parameter space. Written in differential form for any horizontal cylinder in parameter space one obtains:

$$dc = \frac{1}{\mathrm{area}(\mathbb{C}/d\mathbb{Z}[i])} \frac{(\mathrm{area}\, C(z))^\alpha}{w_C^2} \frac{i}{2} dz \wedge d\bar{z}$$

Using the data from Proposition 17 and Corollary 7 this differential becomes:

$$dc = d^{\alpha-4}[(\lfloor \mathrm{Im}\, z \rfloor + 1, d)^{3-\alpha} (\mathrm{Im}\, z)^\alpha + (\lfloor \mathrm{Im}\, z \rfloor, d)^{3-\alpha}(1 - \mathrm{Im}\, z)^\alpha] \frac{i}{2} dz \wedge d\bar{z}$$

Putting $h = \mathrm{Im}\, z$, and $l = \mathrm{Re}\, z$, for the i-th cylinder, with respect of our chosen cylinder parameterization of parameter space one has:

$$dc = d^{\alpha-4}[(i+1, d)^{3-\alpha} h^\alpha + (i, d)^{3-\alpha}(1-h)^\alpha] dh\, dl$$

Integrating over a horizontal cylinder in the bounds $0 \leq l \leq d$ and $0 \leq h \leq 1$ gives:

$$\mathbf{c} = 2 \frac{d^{\alpha-3}}{1+\alpha} \sum_{i=1}^{d} (i, d)^{3-\alpha}$$

Using the Euler totient $\varphi(p)$, as in the introduction, brings the formula into its final form:

$$\mathbf{c}_{C,\alpha}^d = \frac{2}{1+\alpha} \sum_{i=1}^{d} \left(\frac{(i,d)}{d} \right)^{3-\alpha} = \frac{2}{1+\alpha} \sum_{p|d} \frac{\varphi(p)}{p^{3-\alpha}}. \tag{29}$$

□

10. Asymptotic constants — cylinders, finite orbit case

We calculate the Siegel-Veech constants for d-symmetric forms represented by torsion points of order n in $\mathbb{T}_d$. For fixed $n \in \mathbb{N}$ take a natural number $1 \leq a \leq n$ and define

$$c_{d,n}(a) := \frac{\pi}{6} \lim_{T \to \infty} \frac{N(V_n(a), T)}{T^2}$$

to be the quadratic growth rate of the vector distribution

$$V_n(a) := \{\mathrm{SL}_2\mathbb{Z} \cdot \mathrm{hol}(l) : l \text{ regular horizontal leaf on } \#_z^d(\mathbb{T}, dz) \text{ with } z \in \mathcal{L}_{a/n} \cap \mathbb{T}_d(n)\}. \tag{30}$$

Here $\mathcal{L}_{a/n}$ denotes the horizontal leaf on $(\mathbb{T}_d, dz)$, that contains the point $[id \cdot a/n]_d$. As before let $\lfloor ad/n \rfloor \in \mathbb{Z}$ be the integer part of ad/n

LEMMA 3. *For* $1 \leq a \leq n$

$$c_{d,n}(a) = \frac{n}{\varphi(n)\psi(n)} \frac{\varphi((a,n))}{(a,n)} \left(\frac{(\lfloor ad/n \rfloor, d)^3}{d^2} + \frac{(\lfloor ad/n \rfloor + 1, d)^3}{d^2} \right). \tag{31}$$

PROOF. Recall that $\mathbb{T}_d(n)$ is the number of primitive torsion points of order n. Consider the horizontal leaf $\mathcal{L}_{a/n} \subset \mathbb{T}_d$ as defined already. Then for fixed a and n:

$$|\{b \in \mathbb{Z}_n : (b, a, n) = 1\}| = |\{b \in \mathbb{Z}_n : (b, (a, n)) = 1\}| = \\ = \begin{cases} |\Psi^{-1}((\mathbb{Z}_{(a,n)})^*)|, & \text{if } (a, n) \geq 2 \\ n, & \text{if } (a, n) = 1 \end{cases}$$

where $\Psi : \mathbb{Z}_n \to \mathbb{Z}_{(a,n)}$ is given by taking classes modulo (a, n). Thus

$$|\mathcal{L}_{a/n} \cap \mathbb{T}_d(n)| = n \frac{\varphi((a, n))}{(a, n)}. \tag{32}$$

By Proposition 17 the horizontal foliation of $\#_z^d(\mathbb{T}, dz)$ with $z \in \mathcal{L}_{a/n} \cap \mathbb{T}_d(n)$

- *always has* $(\lfloor ad/n \rfloor, d)$ *cylinders of width* $d/(\lfloor ad/n \rfloor, d)$
- *and it has* $(\lfloor ad/n \rfloor + 1, d)$ *cylinders of width* $d/(\lfloor ad/n \rfloor + 1, d)$, *if* $ad/n \notin \mathbb{Z}$.

With $|\mathbb{T}_d(n)| = \varphi(n)\psi(n)$ we find the quadratic growth constants above. □

The first part of the proof of the Lemma implies that the Teichmüller disk through the torsion points $\mathbb{T}(n)$ of order n has

$$cu(n) = \frac{1}{2} \sum_{a=1}^{n} \varphi((a, n)) = \frac{1}{2} \sum_{l|n} \varphi\left(\frac{n}{l}\right) \varphi(l) \tag{33}$$

cusps if $n \geq 3$, and 2 cusps if $n = 2$.

Now the Siegel-Veech constant $\mathbf{c}_{d,n} = \sum_{a=1}^{n} c_{d,n}(a)$ for periodic cylinders on d-symmetric differentials $\#_z^d(\mathbb{T}, dz)$ with $[z]_d \in \mathbb{T}_d(n)$ is

$$\mathbf{c}_{d,n} = \frac{n}{\varphi(n)\psi(n)} \left(\sum_{a=1}^{n} \frac{\varphi((a, n))}{(a, n)} \frac{(\lfloor ad/n \rfloor, d)^3}{d^2} + \right. \\ \left. + \sum_{ad/n \notin \mathbb{Z}} \frac{\varphi((a, n))}{(a, n)} \frac{(\lfloor ad/n \rfloor + 1, d)^3}{d^2} \right). \tag{34}$$

To simplify this expression further, we *consider torsion points of order* n *with* $(n, d) = 1$. Then all a with $ad/n \in \mathbb{Z}$ are multiples of n and consequently

$$|\mathbb{T}_d(n) \cap \partial^{btm}\mathscr{C}_i| = \begin{cases} \varphi(n) & \text{if } i = 0, \\ 0 & \text{if } i \neq 0. \end{cases} \tag{35}$$

and:

$$\mathbf{c}_d(n) = \frac{n}{\varphi(n)\psi(n)}\left(d\frac{\varphi(n)}{n} + 2\sum_{a\not\equiv 0 \bmod n}\frac{\varphi((a,n))}{(a,n)}\frac{(\lfloor ad/n\rfloor, d)^3}{d^2}\right). \tag{36}$$

We always assume $a \in \{-n+1, -n+2, ... -1, 0, 1, ..., n-1\}$ represents a class in $\mathbb{Z}_n$.

Prime d**.** If in addition to the previous assumptions d *is prime* the numbers of maximal cylinders in a given direction is either $1, 2, d$, or $d+1$ and the quadratic growth rates are:

# of cylinders	$\mathbf{c}_d(n) =$
1	$\frac{n}{\varphi(n)\psi(n)}\frac{1}{d^2}\sum_{da/n\in\{1,\dots,d-1\}}\frac{\varphi((a,n))}{(a,n)}$
2	$\frac{n}{\varphi(n)\psi(n)}\frac{2}{d^2}\sum_{d-1>\frac{a}{n}d>1}\frac{\varphi((a,n))}{(a,n)}$
d	$\frac{n}{\varphi(n)\psi(n)}\frac{d\varphi(n)}{n} = \frac{d}{\psi(n)}$
d+1	$\frac{n}{\varphi(n)\psi(n)}2\left(d+\frac{1}{d^2}\right)\sum_{0<\lvert a\rvert<n/d}\frac{\varphi((a,n))}{(a,n)}$

In particular there are two possibilities for prime d:

— *either* $(n,d) = 1$, then there are *no* directions with only *one maximal cylinder* on the forms located on the $\mathrm{SL}_2\mathbb{Z}$ orbit of $[1/n]$ and we find

$$\begin{aligned}\mathbf{c}_d(n) = \frac{n}{\varphi(n)\psi(n)}\Bigg(&\frac{2}{d^2}\sum_{1<\frac{a}{n}d<d-1}\frac{\varphi((a,n))}{(a,n)} + \frac{d\varphi(n)}{n} + \\ &+ 2\left(d+\frac{1}{d^2}\right)\sum_{0<\frac{\lvert a\rvert}{n}d<1}\frac{\varphi((a,n))}{(a,n)}\Bigg),\end{aligned} \tag{37}$$

— *or* $(d,n) = d$, then the asymptotic constant is the sum of all four terms in the table above. In particular, there are *always* directions containing precisely *one maximal cylinder* on forms contained in the orbit of $[1/n]$.

Explicit constants for 2-symmetric covers. For small d the above formulæ simplify. In particular for $d = 2$ and n *odd*, formula (37) reads

$$\begin{aligned}c_2(n) &= \frac{1}{\varphi(n)\psi(n)}\left(2\varphi(n) + \frac{9n}{4}\sum_{a\not\equiv 0 \mod n}\frac{\varphi((a,n))}{(a,n)}\right) = \\ &= \frac{2}{\psi(n)} + \frac{9}{4}\left(1 - \frac{1}{\psi(n)}\right) = \frac{9}{4} - \frac{1}{4\psi(n)}.\end{aligned} \tag{38}$$

The only thing we have used at this point, is the following identity derived from equation (32):

$$\varphi(n)\psi(n) = n\sum_{a=1}^{n}\frac{\varphi((a,n))}{(a,n)} = \varphi(n) + n\sum_{a=1}^{n-1}\frac{\varphi((a,n))}{(a,n)}.$$

If $d = 2$ the d-cylinder directions are simply 2-cylinder directions The parameter space $\mathbb{T}_{2,m}$ has two horizontal cylinders, and the horizontal foliation of surfaces parameterized by either one of these cylinders contains $3 = 2+1$ maximal cylinders.

For *even* n the parameter choice $a = n/d = n/2$ gives points on horizontal leaf going through $[i]_2 \in \mathbb{T}_2$. This horizontal leaf (see table 10 above) reflects all directions with exactly one closed cylinder, thus:

(39)
$$\begin{aligned}\mathbf{c}_2(n) &= \frac{1}{\varphi(n)\psi(n)}\left(2\varphi(n) + \frac{1}{4}\varphi(\frac{n}{2}) + \frac{9n}{2}\sum_{2a/n\neq 1,2}\frac{\varphi((a,n))}{(a,n)}\right) = \\ &= \begin{cases}\frac{9}{4\psi(n)} + \frac{9}{4}\left(1 - \frac{2}{\psi(n)}\right) = \frac{9}{4} - \frac{9}{4\psi(n)} & \text{if } 4 \nmid n \\ \frac{17}{8\psi(n)} + \frac{9}{4}\left(1 - \frac{3}{2\psi(n)}\right) = \frac{9}{4} - \frac{5}{4\psi(n)} & \text{if } 4 \mid n.\end{cases}\end{aligned}$$

The two cases appear because:

(40)
$$\varphi\left(\frac{n}{2}\right) = \begin{cases}\varphi(n), & \text{if } 4 \nmid n \\ \frac{1}{2}\varphi(n), & \text{if } 4 \mid n.\end{cases}$$

11. Asymptotic constants — saddle connections

A cover $\pi : (Y, \tau) \to (X, \omega)$ is called *balanced*, if

$$\pi^{-1}(\pi(Z(\omega))) = Z(\omega),$$

where $Z(\omega)$ is the set of zeros of ω. Since d-symmetric covers are fully branched they are *balanced*.

Since $\#_z^d(\mathbb{T}, dz)$ is a $\mathbb{Z}_d$ cover every saddle connection $s \in SC(\mathbb{T}, [0], [z])$ on the marked torus $(\mathbb{T}, [0], [z])$ has d saddle connections as preimages

$$\pi^{-1}(s) = \{s_1, ..., s_d\} \subset SC(\#_z^d(\mathbb{T}, dz).$$

Using that it is easy to calculate the quadratic growth rate of saddle connections on $\#_z^d(\mathbb{T}, dz)$ once the growth rates on the two marked torus $(\mathbb{T}, [0], [z])$ are known. Those are calculated in **[S1]**. As the two marked torus can be viewed as a 1-symmetric cover, the results in **[S1]** on the other hand easily follow from the formulas given in the introduction. All we need to do is to determine the quadratic constants for saddle connections between $[0]$ and $[z]$ on $(\mathbb{T}, [0], [z])$. We may use that $(\mathbb{T}, [0], dz)$ is the parameter space for 1-symmetric forms and apply the formula for saddle connections developed earlier.

To consider parameters $[z]$, that are primitive n-torsion points, we only need the pairs of distances to primitive the points in $\mathrm{SL}_2\mathbb{Z} \cdot [1/n] \cap \mathcal{L}_0 = \{[i/n] \in \mathbb{T} :$

$\gcd(i,n)=1\}$ to the origin and add the summed squares of their reciprocals. So

$$\mathbf{c}_{\pm}(n) = \frac{2n^2}{\varphi(n)\psi(n)} \sum_{(i,n)=1} \frac{1}{i^2} \tag{41}$$

and the limit $n \to \infty$ is, up to multiplicity, the Siegel-Veech constant for saddle connections starting at irrational points, i.e. when $[z] \notin \mathbb{Q}[i]/\mathbb{Z}[i]$:

$$\mathbf{c}_{\pm} = \frac{\pi^2}{3}. \tag{42}$$

As a consequence we obtain:

COROLLARY 10. *The quadratic constant* $c^d_{\pm}(n)$ *for saddle connections* $SC_{\pm}(\#^d_z\mathbb{T}, dz)$ *connecting distinct cone points on a d-symmetric cover* $\#^d_z(\mathbb{T}, dz)$ *so that* $z \in \mathbb{T}_d(n)$ *is*

$$\mathbf{c}^d_{\pm}(n) = d\frac{2n^2}{\varphi(n)\psi(n)} \sum_{(i,n)=1} \frac{1}{i^2}. \tag{43}$$

For d-symmetric covers with infinite orbit:

$$\mathbf{c}^d_{\pm} = d\frac{\pi^2}{3}. \tag{44}$$

While this is elementary, there are quadratic growth rates of certain saddle connections between the two cone points that tell us more about the geometry of the surface.

Let us fix a d-symmetric cover $\pi : \#^d_z(\mathbb{T}, dz) \to (\mathbb{T}, dz)$ and remove the d saddle connections including their endpoints, say $\{s_1, ...s_d\}$, in the preimage of a saddle connection on $\mathbb{T}$ connecting the ramification points of the cover. Then the number of connected components of $\#^d_z(\mathbb{T}, dz)\backslash\{\overline{s}_1, ..., \overline{s}_d\}$ is a divisor m of d. Indeed, moving the parameter z of $\#^d_z(\mathbb{T}, dz)$ in (shrinking) direction of any of the d-saddle connection in parameter space gives a path in parameter space. Along that path the d saddle connections get shrink and after going the distance $|s_i|$ in parameter space they disappear. We hit a degenerate differential $\#^d_{z_0}(\mathbb{T}, dz)$, $z_0 \in \Lambda/d\Lambda$, that is homeomorphic to $\#^d_z(\mathbb{T}, dz)\backslash\{s_1, ...s_d\}$ and the number of connected components of $\#^d_{z_0}(\mathbb{T}, dz)\backslash\{\pi^{-1}[0]\}$ divides d. This motivates the following definition.

DEFINITION 1. *Let us call a family* S_d *of d parallel saddle connections between the cone points of the d-symmetric cover* $\#^d_z(\mathbb{T}, dz)$ m-homologous, *if collapsing the cone points of* $\#^d_z(\mathbb{T}, dz)$ *along the family* S_d *gives a degenerate cover* $\#^d_{z_0}(\mathbb{T}, dz)$ *that has* d/m *connected components after its singular point* $[0]$ *is removed.*

The name m-homologous is inspired by the fact, that a chain of m loops, each formed by a pair of saddle connections, bounds a torus tiled by d/m squares. Let us denote the set of m-homologous saddle connections on $\#^d_z(\mathbb{T}, dz)$ by $SC_m(\#^d_z(\mathbb{T}, dz))$. It is clear that in general

$$\bigsqcup_{m|d} SC_m(\#^d_z(\mathbb{T}, dz)) = SC_{cp}(\#^d_z(\mathbb{T}, dz))$$

where SC_{cp} denotes the set of all saddle connections between distinct cone points. Recall, that the divisors m of d are in one-to-one correspondence to orbits

$$\mathscr{O}_m = \mathrm{SL}_2\mathbb{Z} \cdot [\#_{d/m}(\mathbb{T}, dz)] \subset \Lambda/d\Lambda$$

and that

$$\bigsqcup_{m|d} \mathscr{O}_m = \Lambda/d\Lambda.$$

It is possible to reformulate the counting problem as a counting problem on parameter space $(\mathbb{T}_d, \Lambda/d\Lambda, dz)$. Namely on a d-symmetric cover $\#_z^d(\mathbb{T}, dz)$ consider those m-homologous saddle connections that terminate in a particular degenerate cover, say $\#_{z_m}^d(\mathbb{T}, dz)$. A parallel family S_d of d such saddle connections corresponds to a line segment l from z to the point z_m on $(\mathbb{T}_d, \Lambda/d\Lambda, dz)$, of the same length and direction as any saddle connection in S_d. If $SC(z_m, z)$ denotes the set of all line segments connecting the point z with z_m on $(\mathbb{T}_d, \Lambda/d\Lambda, dz)$. Since $\pi_d : \mathbb{T}_d \to \mathbb{T}$ is a translation map of degree d^2 that maps all lattice points on $\mathbb{T}_d$ to the origin of $\mathbb{T}$ the saddle connections in $SC(z_m, z)$ can be viewed as a set of saddle connections on $\mathbb{T}$. In fact, any saddle connection $s \in SC(z_m, z)$ is mapped isometrically to a saddle connection $(\pi_d)_*(s) \in SC([0], [z])$. So we can identify $SC(z_m, z)$ with a subset of $SC([0], [z])$. However, we need to identify $SC(z_m, z)$ in $SC([0], [z])$ and this is not completely trivial. Instead we apply our counting formula and consider the standard problem of finding all saddle connections in

$$SC(\mathscr{O}_m, z) = \bigsqcup_{z_m \in \mathbb{T}_d(m)} SC(z_m, z).$$

Recall, that all primitive m-torsion points are on one $\mathrm{SL}_2\mathbb{Z}$ orbit denoted by $\mathscr{O}_m$. First consider the case z is a torsion point. Then any saddle connection $s \in SC(\mathscr{O}_m, z)$ has horizontal ones in its $\mathrm{SL}_2\mathbb{Z}$-orbit. If As is horizontal than $[Az]_d$ lies on one of the horizontal leaves through the points $[ki]_d$, where $k = 1, ..., d$. In fact, the horizontal saddle connection of the marked torus $(\mathbb{T}_d, \mathbb{Z}[i]/d\mathbb{Z}[i])$ that contains $[Az]_d$ is bounded by at least one element from $\mathscr{O}_m$, since As is a horizontal line segment from $[Az]_d$ to a point in $\mathscr{O}_m$ that cannot cross any integer lattice point on $\mathbb{T}_d$. By the counting formula we need to record the length of As and this is either $|\{\mathrm{Re}\, Az\}| = |\mathrm{Re}\, Az| \mod 1$, or $|1 - \{\mathrm{Re}\, Az\}| = |\mathrm{Re}\, Az| \mod 1$. Here we representing any class modulo d by $(0, d]^2$. To write down the counting formula let $\mathscr{S}_{h,m} \subset \mathbb{T}_d$ denote the set of horizontal saddle connections that emanate or terminate in a primitive m-torsion point.

THEOREM 8. *Assume $\#_z^d(\mathbb{T}, dz) \to (\mathbb{T}, dz)$ is a d-symmetric differential with $z \in (\mathbb{C} - \mathbb{Z}[i])/d\mathbb{Z}[i]$. Then, if $[z]_d \in \mathbb{T}(n)$ the m-homologous saddle connections on $\#_z^d(\mathbb{T}, dz)$ have Siegel-Veech constant:*

$$\mathbf{c}_m^d(n) = \frac{2 \cdot d}{|\mathscr{O}_n|} \sum_{[y]_d \in \mathscr{S}_{h,m} \cap \mathscr{O}_n} \frac{1}{|\{Re(y)\}|^2}. \tag{45}$$

If $[z]_d \in \mathbb{T}$ is not a torsion point we have:

$$\mathbf{c}_m^d(gen) = \frac{\pi^2}{3} \frac{\varphi(m)\psi(m)}{d}.$$

PROOF. For torsion points this is a straight forward application of the parameter space growth rate formula. The fact, that only distances of the form $|\{\mathrm{Re}(y)\}|$ appear is because of the hyperelliptic involution of $\mathbb{T}_d$. This causes the factor 2 in the formula. For generic points we have the standard growth rate between two points when one is generic multiplied by the number of terminal points, here $|\mathscr{O}_m|$. □

12. Questions and Comments

The flat geometry of parameter spaces of covers has not been used in depth to study and particularly classify lattice covers, with the exception of Duryev's recent work [**D18**]. The growth rate formulas generalize directly to other parameter spaces of covers. The description of d-symmetric covers given in this paper can be generalized to cyclic and abelian covers, see [**AS**].

References

[AS] Aulicino, David; Schmoll, Martin. Weighted Siegel-Veech constants for branched cyclic covers. In final preparation.

[Ba] Matt Bainbridge, *Euler characteristics of Teichmüller curves in genus two*, Geom. Topol. **11** (2007), 1887–2073, DOI 10.2140/gt.2007.11.1887. MR2350471

[BM] Irene I. Bouw and Martin Möller, *Teichmüller curves, triangle groups, and Lyapunov exponents*, Ann. of Math. (2) **172** (2010), no. 1, 139–185, DOI 10.4007/annals.2010.172.139. MR2680418

[C] Kariane Calta, *Veech surfaces and complete periodicity in genus two*, J. Amer. Math. Soc. **17** (2004), no. 4, 871–908, DOI 10.1090/S0894-0347-04-00461-8. MR2083470

[D18] Duryev, Eduard. Teichmüller curves in genus two: Square-tiled surfaces and modular curves. Thesis, Harvard University 2018.

[E98] Alex Eskin, *Counting problems and semisimple groups*, Proceedings of the International Congress of Mathematicians, Vol. II (Berlin, 1998), Doc. Math. **Extra Vol. II** (1998), 539–552. MR1648103

[E05] Alex Eskin, *Counting problems in moduli space*, Handbook of dynamical systems. Vol. 1B, Elsevier B. V., Amsterdam, 2006, pp. 581–595, DOI 10.1016/S1874-575X(06)80034-2. MR2186249

[EKZ11] Alex Eskin, Maxim Kontsevich, and Anton Zorich, *Lyapunov spectrum of square-tiled cyclic covers*, J. Mod. Dyn. **5** (2011), no. 2, 319–353, DOI 10.3934/jmd.2011.5.319. MR2820564

[EKZ14] Alex Eskin, Maxim Kontsevich, and Anton Zorich, *Sum of Lyapunov exponents of the Hodge bundle with respect to the Teichmüller geodesic flow*, Publ. Math. Inst. Hautes Études Sci. **120** (2014), 207–333, DOI 10.1007/s10240-013-0060-3. MR3270590

[EM98] Alex Eskin and Howard Masur, *Asymptotic formulas on flat surfaces*, Ergodic Theory Dynam. Systems **21** (2001), no. 2, 443–478, DOI 10.1017/S0143385701001225. MR1827113

[EMM] Alex Eskin, Jens Marklof, and Dave Witte Morris, *Unipotent flows on the space of branched covers of Veech surfaces*, Ergodic Theory Dynam. Systems **26** (2006), no. 1, 129–162, DOI 10.1017/S0143385705000234. MR2201941

[EMS] Alex Eskin, Howard Masur, and Martin Schmoll, *Billiards in rectangles with barriers*, Duke Math. J. **118** (2003), no. 3, 427–463, DOI 10.1215/S0012-7094-03-11832-3. MR1983037

[EMZ] Alex Eskin, Howard Masur, and Anton Zorich, *Moduli spaces of abelian differentials: the principal boundary, counting problems, and the Siegel-Veech constants*, Publ. Math. Inst. Hautes Études Sci. **97** (2003), 61–179, DOI 10.1007/s10240-003-0015-1. MR2010740

[EO] Alex Eskin and Andrei Okounkov, *Asymptotics of numbers of branched coverings of a torus and volumes of moduli spaces of holomorphic differentials*, Invent. Math. **145** (2001), no. 1, 59–103, DOI 10.1007/s002220100142. MR1839286

[FS] Krzysztof Frączek and Martin Schmoll, *On ergodicity of foliations on $\mathbb{Z}^d$-covers of half-translation surfaces and some applications to periodic systems of Eaton lenses*, Comm. Math. Phys. **362** (2018), no. 2, 609–657, DOI 10.1007/s00220-018-3186-9. MR3843424

[FMZ1] Giovanni Forni, Carlos Matheus, and Anton Zorich, *Square-tiled cyclic covers*, J. Mod. Dyn. **5** (2011), no. 2, 285–318, DOI 10.3934/jmd.2011.5.285. MR2820563

[FMZ2] Giovanni Forni, Carlos Matheus, and Anton Zorich, *Lyapunov spectrum of invariant subbundles of the Hodge bundle*, Ergodic Theory Dynam. Systems **34** (2014), no. 2, 353–408, DOI 10.1017/etds.2012.148. MR3233697

[GJ] Eugene Gutkin and Chris Judge, *Affine mappings of translation surfaces: geometry and arithmetic*, Duke Math. J. **103** (2000), no. 2, 191–213, DOI 10.1215/S0012-7094-00-10321-3. MR1760625

[GHS] Eugene Gutkin, Pascal Hubert, and Thomas A. Schmidt, *Affine diffeomorphisms of translation surfaces: periodic points, Fuchsian groups, and arithmeticity* (English, with English and French summaries), Ann. Sci. École Norm. Sup. (4) **36** (2003), no. 6, 847–866 (2004), DOI 10.1016/j.ansens.2003.05.001. MR2032528

[HBJ] Friedrich Hirzebruch, Thomas Berger, and Rainer Jung, *Manifolds and modular forms*, Aspects of Mathematics, E20, Friedr. Vieweg & Sohn, Braunschweig, 1992. With appendices by Nils-Peter Skoruppa and by Paul Baum. MR1189136

[H] W. Patrick Hooper, *Grid graphs and lattice surfaces*, Int. Math. Res. Not. IMRN **12** (2013), 2657–2698, DOI 10.1093/imrn/rns124. MR3071661

[HW] W. Patrick Hooper and Barak Weiss, *Generalized staircases: recurrence and symmetry* (English, with English and French summaries), Ann. Inst. Fourier (Grenoble) **62** (2012), no. 4, 1581–1600. MR3025751

[HL] Pascal Hubert and Samuel Lelièvre, *Prime arithmetic Teichmüller discs in $\mathcal{H}(2)$*, Israel J. Math. **151** (2006), 281–321, DOI 10.1007/BF02777365. MR2214127

[HST] P. Hubert, M. Schmoll, and S. Troubetzkoy, *Modular fibers and illumination problems*, Int. Math. Res. Not. IMRN **8** (2008), Art. ID rnn011, 42, DOI 10.1093/imrn/rnn011. MR2428145

[Ka1] Ernst Kani, *The number of curves of genus two with elliptic differentials*, J. Reine Angew. Math. **485** (1997), 93–121, DOI 10.1515/crll.1997.485.93. MR1442190

[Ka2] Ernst Kani, *Hurwitz spaces of genus 2 covers of an elliptic curve*, Collect. Math. **54** (2003), no. 1, 1–51. MR1962943

[KZo] Maxim Kontsevich and Anton Zorich, *Connected components of the moduli spaces of Abelian differentials with prescribed singularities*, Invent. Math. **153** (2003), no. 3, 631–678, DOI 10.1007/s00222-003-0303-x. MR2000471

[Ma1] Howard Masur, *The growth rate of trajectories of a quadratic differential*, Ergodic Theory Dynam. Systems **10** (1990), no. 1, 151–176, DOI 10.1017/S0143385700005459. MR1053805

[Ma2] Howard Masur, *Lower bounds for the number of saddle connections and closed trajectories of a quadratic differential*, Holomorphic functions and moduli, Vol. I (Berkeley, CA, 1986), Math. Sci. Res. Inst. Publ., vol. 10, Springer, New York, 1988, pp. 215–228, DOI 10.1007/978-1-4613-9602-4_20. MR955824

[Ma3] Howard Masur, *Closed trajectories for quadratic differentials with an application to billiards*, Duke Math. J. **53** (1986), no. 2, 307–314, DOI 10.1215/S0012-7094-86-05319-6. MR850537

[MT] Howard Masur and Serge Tabachnikov, *Rational billiards and flat structures*, Handbook of dynamical systems, Vol. 1A, North-Holland, Amsterdam, 2002, pp. 1015–1089, DOI 10.1016/S1874-575X(02)80015-7. MR1928530

[McM1] Curtis T. McMullen, *Billiards and Teichmüller curves on Hilbert modular surfaces*, J. Amer. Math. Soc. **16** (2003), no. 4, 857–885, DOI 10.1090/S0894-0347-03-00432-6. MR1992827

[McM2] Curtis T. McMullen, *Teichmüller curves in genus two: discriminant and spin*, Math. Ann. **333** (2005), no. 1, 87–130, DOI 10.1007/s00208-005-0666-y. MR2169830

[McM3] Curtis T. McMullen, *Teichmüller curves in genus two: the decagon and beyond*, J. Reine Angew. Math. **582** (2005), 173–199, DOI 10.1515/crll.2005.2005.582.173. MR2139715

[McM4] Curtis T. McMullen, *Teichmüller curves in genus two: torsion divisors and ratios of sines*, Invent. Math. **165** (2006), no. 3, 651–672, DOI 10.1007/s00222-006-0511-2. MR2242630

[McM5] Curtis T. McMullen, *Dynamics of $\mathrm{SL}_2(\mathbb{R})$ over moduli space in genus two*, Ann. of Math. (2) **165** (2007), no. 2, 397–456, DOI 10.4007/annals.2007.165.397. MR2299738

[McM6] Curtis T. McMullen, *Foliations of Hilbert modular surfaces*, Amer. J. Math. **129** (2007), no. 1, 183–215, DOI 10.1353/ajm.2007.0002. MR2288740

[GS] Gabriela Schmithüsen, *An algorithm for finding the Veech group of an origami*, Experiment. Math. **13** (2004), no. 4, 459–472. MR2118271

[Sa] Savic, Predrag; Thesis UI-Chicago, 2007.

[S1] M. Schmoll, *On the asymptotic quadratic growth rate of saddle connections and periodic orbits on marked flat tori*, Geom. Funct. Anal. **12** (2002), no. 3, 622–649, DOI 10.1007/s00039-002-8260-x. MR1924375

[S2] M. Schmoll. Asymptotic growth rates for d-symmetric forms. Preprint 2005.

[S3] Martin Schmoll, *Spaces of elliptic differentials*, Algebraic and topological dynamics, Contemp. Math., vol. 385, Amer. Math. Soc., Providence, RI, 2005, pp. 303–320, DOI 10.1090/conm/385/07203. MR2180242

[S4] Martin Schmoll, *Veech groups for holonomy-free torus covers*, J. Topol. Anal. **3** (2011), no. 4, 521–554, DOI 10.1142/S1793525311000647. MR2887675

[V1] W. A. Veech, *Teichmüller curves in moduli space, Eisenstein series and an application to triangular billiards*, Invent. Math. **97** (1989), no. 3, 553–583, DOI 10.1007/BF01388890. MR1005006

[V2] William A. Veech, *The billiard in a regular polygon*, Geom. Funct. Anal. **2** (1992), no. 3, 341–379, DOI 10.1007/BF01896876. MR1177316

[V3] William A. Veech, *Siegel measures*, Ann. of Math. (2) **148** (1998), no. 3, 895–944, DOI 10.2307/121033. MR1670061

[Vrb] Ya. B. Vorobets, *Plane structures and billiards in rational polygons: the Veech alternative* (Russian), Uspekhi Mat. Nauk **51** (1996), no. 5(311), 3–42, DOI 10.1070/RM1996v051n05ABEH002993; English transl., Russian Math. Surveys **51** (1996), no. 5, 779–817. MR1436653

[Zo] Anton Zorich, *Flat surfaces*, Frontiers in number theory, physics, and geometry. I, Springer, Berlin, 2006, pp. 437–583, DOI 10.1007/978-3-540-31347-2_13. MR2261104

Clemson University, O-229 Martin Hall, Clemson, SC 29634

Email address: schmoll@clemson.edu

Contemporary Mathematics
Volume **736**, 2019
https://doi.org/10.1090/conm/736/14835

On the collision manifold of coorbital moons

Kimberly Stubbs and Samuel R. Kaplan

ABSTRACT. This paper develops coordinates for near-collision phenomena of the coorbital problem. The observation of pairs of moons around Saturn which display differing, but stable, orbital properties led to the development of the coorbital model. The model assumes a central mass with two small moons on the same order of mass and with nearly equal radius from the central body. Cors and Hall identified a parameter that determines if the moons simply pass each other or if they switch orbits (also called a horseshoe orbit). There is another range of values for this parameter where the two moons can collide. This paper employs the parameter in finding coordinates near collision.

1. Coorbital Model

Spacecraft Voyager I and Voyager II were launched in August and September of 1977. In 1980, Voyager I discovered two new moons around Saturn, Janus and Epimethus. What is remarkable about these moons is that they are on nearly identical orbits around Saturn. One moon has an average orbital radius of 150,472 km and the other has an average orbital radius of 150,422km. The difference of their average radii is only 50 km; they are virtually on the same orbit (only a 0.03% difference). The mathematical model for such configurations is called coorbital.

Over time, astronomers noticed that as these two moons approach one another, their interaction prevents collision and they switch orbits. This phenomenon is also called a horseshoe orbit for the shape the orbit takes on in rotating coordinates. What was particularly intriguing was that other known coorbital moons, such as Pandora and Prometheus, do not switch orbits; they simply pass one another.

A number of papers began to appear from astronomers and mathematicians trying to understand the mechanism behind this newly discovered behavior. In particular, what distinguishes pairs of coorbital moons that switch orbits from pairs which never switch?

To determine what physical properties are connected with switching versus passing orbits, Cors and Hall [**CH**] carried out a power series analysis. They identified a single parameter that indicated if a given pair of moons always switch orbits or if they always pass.

To understand the assumptions in their model, let us review the set up:

For a planar coorbital system, let M denote the mass of the large central body and denote the mass of the two moons by m_1 and m_2.

2010 *Mathematics Subject Classification.* Primary 70F15, 70F07.

The authors thank the reviewers for their valuable comments and insights.

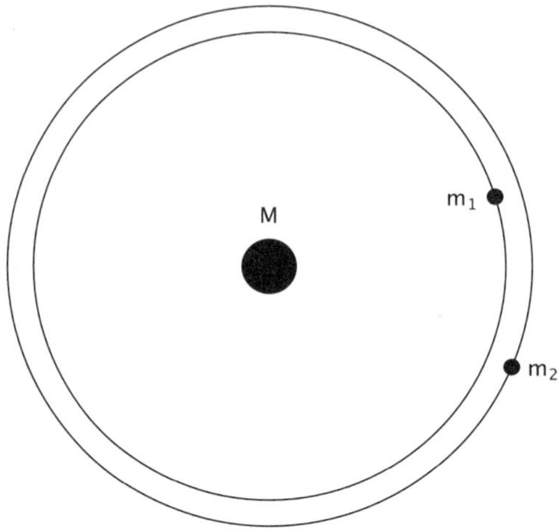

FIGURE 1. A coorbital configuration in the plane.

The next step Cors and Hall took was a clever way to compare several small quantities.

Denote by ϵ the ratio of the gap between the two lunar orbits and the average radius of the two moons.

We are assuming ϵ is small. For Janus and Epimetheus, $\epsilon \approx 3.32 \times 10^{-4}$. For Pandora and Prometheus, $\epsilon \approx 1.60 \times 10^{-2}$.

We also want to measure the ratio of the masses of the moons to the central mass. This will also be small. For Janus and Epimetheus, the ratio is $\approx 3 \times 10^{-9}$. For Pandora and Prometheus, the ratio is $\approx 3 \times 10^{-10}$.

Cors and Hall introduce a parameter, a, that compares smallness of the ratio of the mass of the moons to the central mass relative to the ratio of the gap between the moons to the average radius to the central mass.

$$\epsilon = \frac{\text{gap}}{\text{average radius}}$$

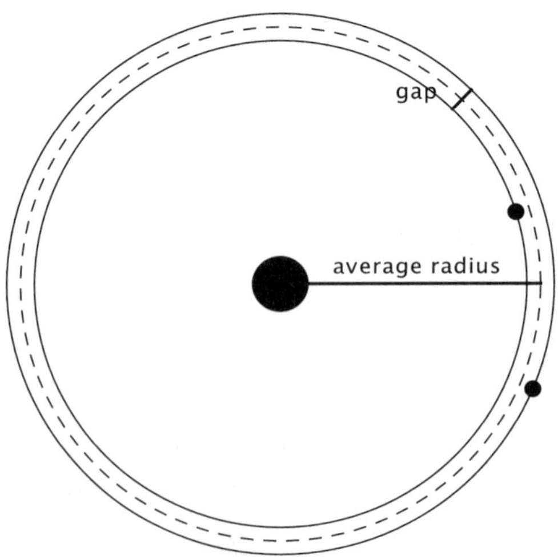

FIGURE 2. Defining the Cors-Hall parameter.

Specifically, we write the masses as:

$$\begin{aligned} m_1 &= \epsilon^a \mu_1 \\ m_2 &= \epsilon^a \mu_2 \end{aligned}$$

where μ_1 and μ_2 are $O(1)$.

By way of example, for Janus and Epimetheus, the ratio of the masses of the moons to Saturn was $\approx 3 \times 10^{-9} = \epsilon^{2.44}$, while $\epsilon \approx 3.32 \times 10^{-4}$, so $a = 2.44$ in this case.

For Pandora and Prometheus, the ratio of the masses of the moons to Saturn was $\approx 3 \times 10^{-10} = \epsilon^{5.3}$, while $\epsilon \approx 1.60 \times 10^{-2}$, so $a = 5.3$ in this case.

Using various techniques, including power series expansions and changes of coordinates, Cors and Hall are able to say that orbit switching occurs for $2 < a < 2.5$ and that only passings occur for $a > 4$. Note that for $a \leq 2$, the moons are massive enough that the full three-body problem is a better model.

However, for a subrange of a between 2.5 and 4, binary collisions between the moons are possible.

2. Solar-system formation

While the motivation for studying coorbital dynamics is based on observed behaviors, it has a possible application to solar-system formation. Observed eccentricity of exoplanets has a wide distribution of values from 0 to 1 (nearly circular to nearly collinear, respectively). The standard model for solar system formation imagines a gaseous cloud slowly rotating, flattening with mass accreting in the center to form a sun. Remaining mass in the rotating disk forms planetoids and eventually planets. The planets inherit a circular orbit from the rotation of the disk. The new sun eventually pushes out remaining light-weight gaseous matter, leaving a solar system in place.

The standard model predicts circular orbits for planets. However, observations of exoplanets include many highly eccentric orbits. The authors posit a modified model based on coorbital behavior. Water is prevalent in the early solar system. Water near the center is vapor, further out is water and eventually, far enough away from the center, water is in the form of ice. This radius where liquid water turns to ice is called the *snow line*. Kretke and Lin showed that planetesimals accumulate near the snow line, drawn to that radius by "magnetorotational instability." [**KL**] This then, is a mechanism for early planetoids to migrate to roughly the same radius, that is to say, coorbital configurations.

As these coorbital planetoids grow in mass, the effective value of a, the Cors-Hall parameter, also grows, eventually passing through a region of a where binary collisions are possible. For a fixed ϵ, the parameter of the largest planetoids grows like $a(t) = \frac{1}{\ln(\epsilon)} \ln(\text{mass}(t))$. Using growth values from Mordasini we can estimate that $a(t)$ is in the region between 2.5 and 4 for about 2,000 rotations. [**Mo**] That is enough time to have several near-collision interactions.

After a near-collision, one possible outcome is for one body to escape the system and the remaining body to move to a Kepler orbit with higher eccentricity. Since the remaining body is also one of the largest masses in the forming solar system, it will guide other growing planets into higher eccentricities as well. This could be one mechanism for solar systems to form with planets of high eccentricity. This model does not force a reexamination of solar system formation in principle, but, if correct, augments our understanding of the standard model by including coorbital dynamics.

However, a detailed analysis or even numerical studies of this hypothesis require the study of near-collision phenomena in the coorbital problem that has no particular relevance to the original motivation for the study of coorbital dynamics. Without specifying a initially, we develop near-collision coordinates. While some papers have included numerical computations near collision [**LO**], our goal is to systematically study orbits which come close to collision.

3. Cors and Hall Coordinates

To reconstruct Cors and Hall's coordinate for the convenience of the reader, we begin with the position, q_0, and momentum, p_0, of a large body of mass M and the positions, q_1, q_2, and momenta, p_1, p_2, of the moons with masses m_1 and m_2 respectively. All three bodies are in the plane, $\mathbb{R}^2$. The Hamiltonian corresponding to the Newtonian three-body problem is given by

$$H(q,p) = \frac{1}{2M}||p_0||^2 + \frac{1}{2m_1}||p_1||^2 + \frac{1}{2m_2}||p_2||^2 - \frac{GMm_1}{||q_1 - q_0||} - \frac{GMm_2}{||q_2 - q_0||} - \frac{Gm_1m_2}{||q_2 - q_1||}.$$

For convenience, we choose units of distance so that $G = 1$ and units of mass so that $M = 1$. [**M**] Our Hamiltonian reduces to

$$H(q,p) = \frac{1}{2}||p_0||^2 + \frac{1}{2m_1}||p_1||^2 + \frac{1}{2m_2}||p_2||^2 - \frac{m_1}{||q_1 - q_0||} - \frac{m_2}{||q_2 - q_0||} - \frac{m_1m_2}{||q_2 - q_1||}.$$

We next move to a different set of coordinates to isolate the linear momentum and center of mass. We accomplish this by using mutual distances between each moon and the large body as two coordinates and the center of mass as the third coordinate. This defines the change of variables:

$$\begin{aligned} u_1 &= q_1 - q_0 \\ u_2 &= q_2 - q_0 \\ u_3 &= \frac{q_1 + m_1q_1 + m_2q_2}{T} \end{aligned}$$

where T is the total mass of the system, $T = 1 + m_1 + m_2$. To keep the new system Hamiltonian, we require a symplectic change of variables. This forces the momenta to be defined as

$$\begin{aligned} v_1 &= -\frac{m_1}{T}p_0 + \frac{1+m_2}{T}p_1 - \frac{m_1}{T}p_2 \\ v_2 &= -\frac{m_2}{T}p_0 - \frac{m_2}{T}p_1 + \frac{1+m_1}{T}p_2 \\ v_3 &= p_0 + p_1 + p_2. \end{aligned}$$

The new Hamiltonian is:

$$H(u,v) = \frac{1+m_1}{2m_1}||v_1||^2 + \frac{1+m_2}{2m_2}||v_2||^2 + v_1^T v_2 + \frac{1}{2T}||v_3||^2 - \frac{m_1}{||u_1||} - \frac{m_2}{||u_2||} - \frac{m_1m_2}{||u_2 - u_1||}.$$

Since the new variables, u_3 and v_3 represent the center of mass and total momentum, we will consider only u_1, u_2, v_1, and v_2 dynamically. To this point,

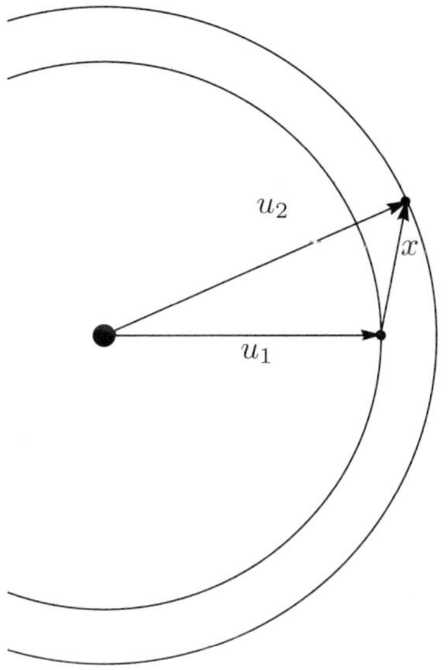

these are standard coordinate changes. We now introduce the assumption that the moons are small by setting

$$\begin{aligned} m_1 &= \epsilon^a \mu_1 \\ m_2 &= \epsilon^a \mu_2 \end{aligned}$$

where the real numbers μ_1 and μ_2 are both $O(1)$. We also rescale the momentum variables by setting

$$\begin{aligned} w_1 &= \epsilon^{-a} v_1 \\ w_2 &= \epsilon^{-a} v_2. \end{aligned}$$

Rescaling by ϵ^{-a}, the new Hamiltonian is given by:

$$H(u,w) = \frac{1}{2}\left(\frac{1}{\mu_1}+\epsilon^a\right)||w_1||^2 + \frac{1}{2}\left(\frac{1}{\mu_2}+\epsilon^a\right)||w_2||^2 + \epsilon^a w_1^T w_2 - \frac{\mu_1}{||u_1||} - \frac{\mu_2}{||u_2||} - \frac{\epsilon^a \mu_1 \mu_2}{||u_2-u_1||}.$$

The corresponding equations of motion are:

$$\begin{aligned} \dot{u}_1 &= \left(\frac{1}{\mu_1}+\epsilon^a\right) w_1 + \epsilon^a w_2 \\ \dot{u}_2 &= \epsilon^a w_1 + \left(\frac{1}{\mu_2}+\epsilon^a\right) w_2 \\ \dot{w}_1 &= -\frac{\mu_1}{||u_1||^3} u_1 + \frac{\epsilon^a \mu_1 \mu_2}{||u_2-u_1||^3}(u_2-u_1) \\ \dot{w}_2 &= -\frac{\mu_2}{||u_2||^3} u_2 - \frac{\epsilon^a \mu_1 \mu_2}{||u_2-u_1||^3}(u_2-u_1). \end{aligned} \tag{3.1}$$

4. Change of Variables into Polar Coordinates

To focus on near-collision of the two small moons, we introduce a mutual distance coordinate, $x = u_2 - u_1$.

We introduce $y = \frac{1}{2}u_1 + \frac{1}{2}u_2$, the midpoint of the two small bodies. In the x, y coordinates, we have $u_1 = -\frac{1}{2}x + y$ and $u_2 = \frac{1}{2}x + y$. This change of coordinates is not symplectic, so the new Hamiltonian only expresses the energy of the system.

$$\begin{aligned} H(x, y, w_1, w_2) &= \frac{1}{2}\left(\frac{1}{\mu_1} + \epsilon^a\right)||w_1||^2 + \frac{1}{2}\left(\frac{1}{\mu_2} + \epsilon^a\right)||w_2||^2 \\ &\quad + \epsilon^a w_1^T w_2 - \frac{\mu_1}{||-\frac{1}{2}x + y||} - \frac{\mu_2}{||\frac{1}{2}x + y||} - \frac{\epsilon^a \mu_1 \mu_2}{||x||}. \end{aligned}$$

The system of equations below is obtained by applying the change of coordinates above directly to the system (3.1).

$$\begin{aligned} \dot{x} &= -\frac{1}{\mu_1}w_1 + \frac{1}{\mu_2}w_2 \\ \dot{y} &= \left(\frac{1}{2\mu_1} + \epsilon^a\right)w_1 + \left(\frac{1}{2\mu_2} + \epsilon^a\right)w_2 \\ \dot{w}_1 &= -\frac{\mu_1}{||-\frac{1}{2}x + y||^3}\left(-\frac{1}{2}x + y\right) + \frac{\epsilon^a \mu_1 \mu_2}{||x||^3}x \\ \dot{w}_2 &= -\frac{\mu_2}{||\frac{1}{2}x + y||^3}\left(\frac{1}{2}x + y\right) - \frac{\epsilon^a \mu_1 \mu_2}{||x||^3}x. \end{aligned}$$

In order to distinguish different orbits that might terminate at binary collision, we move to polar coordinates in the style of McGehee [**Mc**]. In this way, $x = 0$, binary collision, will correspond to the radial coordinate vanishing but allow the angular coordinate to distinguish solutions ejecting from or terminating in collision of the moons.

$$x = r(\cos(\theta), \sin(\theta)).$$

For convenience we will denote $s = (\cos(\theta), \sin(\theta))$, that is, $x = rs$. In these new coordinates we have, in place of x, the scalar differential equations:

$$\begin{aligned} \dot{r} &= \left(\frac{1}{\mu_2}w_2 - \frac{1}{\mu_1}w_1\right)^T s \\ \dot{\theta} &= \frac{1}{r}\left(\frac{1}{\mu_2}w_2 + \frac{1}{\mu_1}w_1\right)^T Ks \end{aligned}$$

where $K = \begin{pmatrix} 0 & -1 \\ 1 & 0 \end{pmatrix}$. Note that $\dot{s} = Ks$ and that K is the negative of the more typical matrix J used in Hamiltonian dynamics.

Near collision, the momentum coordinates grow without bound. In order to expose these rates of growth, we introduce scaled coordinates defined by

$$\begin{aligned} w_1 &= r^{-1/2}W_1 \\ w_2 &= r^{-1/2}W_2. \end{aligned}$$

In order to slow solutions near collision, we scale time as well. We define a fictitious time, τ by

$$\frac{d}{dt} = r^{3/2}\frac{d}{d\tau}.$$

In this new time scale, we will denote the derivative with respect to τ using a prime, $'$. Our new equations of motion are:

$$\begin{aligned}
r' &= r\left(\frac{1}{\mu_2}W_2 - \frac{1}{\mu_1}W_1\right)^T s \\
\theta' &= \left(\frac{1}{\mu_2}W_2 - \frac{1}{\mu_1}W_1\right)^T Ks \\
y' &= r\left(\left(\frac{1}{2\mu_1} + \epsilon^a\right)W_1 + \left(\frac{1}{2\mu_2} + \epsilon^a\right)W_2\right) \\
W_1' &= -\frac{\mu_1 r^2}{||-\frac{1}{2}rs + y||^3}\left(-\frac{1}{2}rs + y\right) + \epsilon^a\mu_1\mu_2 s + \frac{1}{2}W_1^T\left(\frac{1}{\mu_2}W_2 - \frac{1}{\mu_1}W_1\right)s \\
W_2' &= -\frac{\mu_2 r^2}{||\frac{1}{2}rs + y||^3}\left(\frac{1}{2}rs + y\right) - \epsilon^a\mu_1\mu_2 s + \frac{1}{2}W_2^T\left(\frac{1}{\mu_2}W_2 - \frac{1}{\mu_1}W_1\right)s.
\end{aligned}$$

5. Collision Manifold

In the new variables and time-scale, our equations of motion are well-defined at $r = 0$, that is, at binary collision of the two moons. Returning to our Hamiltonian equation, if we fix $H(r, \theta, y, W_1, W_2) = h$ and compute rh, we have

$$\begin{aligned}
rh &= \frac{1}{2}\left(\frac{1}{\mu_1} + \epsilon^a\right)||W_1||^2 + \frac{1}{2}\left(\frac{1}{\mu_2} + \epsilon^a\right)||W_2||^2 \\
&\quad + \epsilon^a W_1^T W_2 - \frac{\mu_1 r}{||-\frac{1}{2}rs + y||} - \frac{\mu_2 r}{||\frac{1}{2}rs + y||} - \epsilon^a\mu_1\mu_2.
\end{aligned}$$

This integral of motion is also well-defined at $r = 0$. Our collision manifold, then, is all θ, y, W_1, and W_2 where

$$0 = \frac{1}{2}\left(\frac{1}{\mu_1} + \epsilon^a\right)||W_1||^2 + \frac{1}{2}\left(\frac{1}{\mu_2} + \epsilon^a\right)||W_2||^2 + \epsilon^a W_1^T W_2 - \epsilon^a\mu_1\mu_2.$$

Note that $\theta \in S^1$, on the circle, while y, W_1, and W_2 are each in $\mathbb{R}^2$. Observe the equations of motion at $r = 0$ reduce to

$$\begin{aligned}
r' &= 0 \\
\theta' &= \left(\frac{1}{\mu_2}W_2 - \frac{1}{\mu_1}W_1\right)^T Ks \\
y' &= 0 \\
W_1' &= \epsilon^a\mu_1\mu_2 s + \frac{1}{2}W_1^T\left(\frac{1}{\mu_2}W_2 - \frac{1}{\mu_1}W_1\right)s \\
W_2' &= -\epsilon^a\mu_1\mu_2 s + \frac{1}{2}W_2^T\left(\frac{1}{\mu_2}W_2 - \frac{1}{\mu_1}W_1\right)s.
\end{aligned}$$

Since $r' = 0$ at $r = 0$, we see that $r = 0$ is an invariant manifold. Since $y' = 0$ at $r = 0$, we see that y is a constant of motion, so can be ignored when integrating the equations. So, the collision manifold reduces to five coordinates with one additional relation from the energy equation:

$$\epsilon^a\mu_1\mu_2 = \frac{1}{2}\left(\frac{1}{\mu_1} + \epsilon^a\right)||W_1||^2 + \frac{1}{2}\left(\frac{1}{\mu_2} + \epsilon^a\right)||W_2||^2 + \epsilon^a W_1^T W_2 \tag{5.1}$$

which is quadratic in W_1 and W_2. The collision manifold is a five-dimensional variety.

THEOREM 5.1. *The collision manifold for binary collision of the moons in the coorbital problem is topologically a four torus,* $\mathbb{T}^4$.

PROOF. Notice equation (5.1) is independent of θ so $\theta \in S^1$.

We next decompose W_1 and W_2 into radial and angular components.

$$\begin{aligned} W_1 &= X \begin{pmatrix} \cos(\beta) \\ \sin(\beta) \end{pmatrix} \\ W_2 &= Y \begin{pmatrix} \cos(\beta+\phi) \\ \sin(\beta+\phi) \end{pmatrix} \end{aligned}$$

where $X = ||W_1||$ and $Y = ||W_2||$. Note that ϕ is the angle between W_1 and W_2. Both β and ϕ are in S^1.

In these coordinates, equation (5.1) becomes:

$$\epsilon^a \mu_1 \mu_2 = \frac{1}{2}\left(\frac{1}{\mu_1} + \epsilon^a\right) X^2 + \frac{1}{2}\left(\frac{1}{\mu_2} + \epsilon^a\right) Y^2 + \epsilon^a XY \cos(\phi). \tag{5.2}$$

We think of equation (5.2) as a family of quadratic relations on X and Y parameterized by ϕ.

The determinant of this conic is

$$-(1-\cos^2(\phi))\epsilon^{2a} - \frac{\epsilon^a}{\mu_1} - \frac{\epsilon^a}{\mu_2} - \frac{1}{\mu_1\mu_2}.$$

The determinant is bounded above by $-\frac{1}{\mu_1\mu_2}$ so is negative for all ϕ.

The discriminant of this conic is

$$-2(1-\cos^2(\phi))\mu_1\mu_2\epsilon^{3a} - 2\mu_1\epsilon^{2a} - 2\mu_2\epsilon^{2a} - 2\epsilon^a$$

which is non-zero for all ϕ. That means, for all $\phi \in S^1$, the conic determined by equation (5.2) is an ellipse.

Since the center of the ellipse is at the origin, a ray from the origin with angle α will intersect the ellipse exactly once. This means we can parameterize the ellipse with an angle $\alpha \in S^1$. In particular, we can set $Y = \tan(\alpha)X$ and solve equation (5.2) to get

$$\begin{aligned} X^2 &= \frac{F}{A + B\tan(\alpha) + C\tan^2(\alpha)} \\ Y^2 &= \frac{F}{C + B\cot(\alpha) + A\cot^2(\alpha)} \end{aligned}$$

where $A = \frac{1}{2}\left(\frac{1}{\mu_1} + \epsilon^a\right)$, $B = \epsilon^a\cos(\phi)$, $C = \frac{1}{2}\left(\frac{1}{\mu_2} + \epsilon^a\right)$ and $F = \epsilon^a\mu_1\mu_2$. We then compute the appropriate sign of the square root for X and Y depending on the quadrant α lies within.

Given α and ϕ, we can determine X and Y. With X, Y, ϕ and β, we can determine W_1 and W_2. So α, ϕ, β and θ uniquely determine a point in the collision manifold. Since each of these variables is in S^1, our collision manifold is $\mathbb{T}^4$. $\square$

6. Flow at Binary Collision

For convenience, we will denote the vectors $b = \begin{pmatrix} \cos(\beta) \\ \sin(\beta) \end{pmatrix}$ and $c = \begin{pmatrix} \cos(\beta+\phi) \\ \sin(\beta+\phi) \end{pmatrix}$ so that we can write $W_1 = Xb$ and $W_2 = Yc$.

In this notation

$$\theta' = \left(\frac{1}{\mu_2} Yc - \frac{1}{\mu_1} Xb\right)^T Ks.$$

This equation simplifies to

$$\theta' = \frac{1}{\mu_2} Y \sin(\beta - \theta + \phi) - \frac{1}{\mu_1} X \sin(\beta - \theta).$$

Taking the τ derivative of $W_1 = Xb$ we can compute β'. Similarly, we can use $W_2 = Yc$ to compute $\beta' + \phi'$ (since c has the argument $\beta + \phi$). From there, we can compute ϕ'. Finally, using $Y = \tan(\alpha)X$ we can compute $Y' = \frac{\alpha'}{1+\alpha^2} X + \tan(\alpha) X'$ and from here α'. Together, we have the system:

$$\begin{aligned}
\theta' &= \frac{1}{\mu_2} Y \sin(\beta - \theta + \phi) - \frac{1}{\mu_1} X \sin(\beta - \theta) \\
\beta' &= -\left(\frac{\epsilon^a \mu_1 \mu_2}{X} + \frac{1}{2\mu_1} Y \cos(\phi) - \frac{1}{2\mu_1} X\right) \sin(\beta - \theta) \\
\phi' &= \left(\frac{\epsilon^a \mu_1 \mu_2}{Y} - \frac{1}{2\mu_2} Y + \frac{1}{2\mu_1} X \cos(\phi)\right) \sin(\beta - \theta + \phi) \\
&\quad + \left(\frac{\epsilon^a \mu_1 \mu_2}{X} + \frac{1}{2\mu_1} Y \cos(\phi) - \frac{1}{2\mu_1} X\right) \sin(\beta - \theta) \\
\alpha' &= \frac{1+\alpha^2}{X} \left\{ \left(-\epsilon^a \mu_1 \mu_2 + \frac{1}{2\mu_2} Y^2 - \frac{1}{2\mu_1} XY \cos(\phi)\right) \cos(\beta - \theta + \phi) \right. \\
&\quad \left. - \left(\epsilon^a \mu_1 \mu_2 \tan(\alpha) + \frac{1}{2\mu_2} Y^2 \cos(\phi) - \frac{1}{2\mu_1} XY\right) \cos(\beta - \theta) \right\}.
\end{aligned}$$

These equations are given with terms X and Y with the understanding that we can compute $X(\alpha)$ and $Y(\alpha)$. We are suppressing the dependence on α.

THEOREM 6.1. *On the collision manifold, the set of equilibria are distributed on eight circles, S^1.*

PROOF. To begin looking for equilibria in this system, we recall that $W_1 = Xb$ which gives $W_1' Kb = W_1 \beta'$. When $\beta' = 0$ then either $W_1' = 0$ or W_1' is perpendicular to Kb. So W_1' is parallel to b. Recall that when $r = 0$, W_1 is parallel to s. For s and b to be parallel, we have $\theta = \beta$ or $\theta = \beta + \pi$.

For the case when W_1' vanishes, we must have $\epsilon^a \mu_1 \mu_2 + \frac{1}{2} W_1^T \left(\frac{1}{\mu_2} W_2 - \frac{1}{\mu_1} W_1\right) = 0$. However, when we substitute this relation in W_2' we see it is not zero. So $\beta' = 0$ forces $\theta = \beta$ or $\theta = \beta + \pi$.

In either case, we can substitute into θ' to get $\theta' = 0$ when $\phi = 0$ or $\phi = \pi$.

The conditions that $\theta = \beta$ and $\phi = 0$ or π forces $\phi' = 0$ for each case.

To explore when α' vanishes, we can ignore the factor $1 + \alpha^2$ since this is always non-zero.

For the case $\phi = 0$, we have

$$0 = -\epsilon^a \mu_1\mu_2 + \frac{1}{2\mu_2}Y^2 - \frac{1}{2\mu_1}XY - \left(\epsilon^a \mu_1\mu_2 + \frac{1}{2\mu_2}XY - \frac{1}{2\mu_1}X^2\right)\tan(\alpha).$$

Recall that $Y = \tan(\alpha)X$. Distributing the $\tan(\alpha)$ and reducing gives

$$0 = -\epsilon^a - \epsilon^a \tan(\alpha)$$

or $\tan(\alpha) = -1$. That is $\alpha = -\frac{1}{4}\pi$ or $\frac{3}{4}\pi$.

For the case $\phi = \pi$, we can reduce as before and find

$$0 = -\epsilon^a + \epsilon^a \tan(\alpha)$$

or $\tan(\alpha) = 1$. That is $\alpha = \frac{1}{4}\pi$ or $-\frac{3}{4}\pi$.

So, we have an infinite number of equilibria. We can reduce the coordinate conditions to those shown in the table.

For each of the eight combinations of θ, α and ϕ there is a circle's worth of equilibria since $\beta \in S^1$. □

THEOREM 6.2. *Of the eight circles of equilibria in the collision manifold, four are the termination sites of binary collision solutions in the full flow and four are the origin of ejection solutions.*

PROOF. Consider the linearization of the system of differential equations at the equilibria. When the two moons come near collision, the direction of their interaction depends on their relative masses. We simplify the computation with the assumption $m_1 > m_2$ and we get eigenvalues that are positive, negative, zero or complex conjugate pairs for all $\epsilon > 0$ and $a > 0$ as indicated in the chart below.

	angle coordinates			signs of eigenvalues		
label	θ	ϕ	α	r-direction	α-direction	plane in $\alpha = r = 0$
C_1	β	0	$-\frac{1}{4}\pi$	-	-	+, -
C_2	β	0	$\frac{3}{4}\pi$	-	-	+, -
C_3	β	π	$\frac{1}{4}\pi$	+	-	complex, Re < 0
C_4	β	π	$-\frac{3}{4}\pi$	+	-	complex, Re < 0
C_5	$\beta+\pi$	0	$-\frac{1}{4}\pi$	+	+	+, -
C_6	$\beta+\pi$	0	$\frac{3}{4}\pi$	+	+	+, -
C_7	$\beta+\pi$	π	$\frac{1}{4}\pi$	-	+	complex, Re > 0
C_8	$\beta+\pi$	π	$-\frac{3}{4}\pi$	-	+	complex, Re > 0

We designate the eight circles of equilibria as C_1 to C_8. On each circle, there is a zero eigenvalue along the circle. This is not included in the chart. One eigenvector is along the r-direction, perpendicular to the collision manifold. The sign of the associated eigenvalue is negative along C_1, C_2, C_7 and C_8 and positive along the other four. This means that solutions of the full flow that end in collision (as $\tau \to \infty$), must approach the collision manifold at an equilibrium on one of C_1, C_2, C_7 or C_8. Similarly, solutions that leave binary collision (that is, approach binary collision as $\tau \to -\infty$) must be have come from an equilibrium on one of C_3, C_4, C_5 or C_6. □

References

[CH] J. M. Cors and G. R. Hall, *Coorbital periodic orbits in the three body problem*, SIAM J. Appl. Dyn. Syst. **2** (2003), no. 2, 219–237, DOI 10.1137/S1111111102411304. MR1998699

[KL] K. A. Kretke and. D. N. C. Lin, *Grain Retention and Formation of Planetesimals near the Snow Line in MRI-driven Turbulent Protoplanetary Disks*, Astrophysical Journal Letters, **664** (2007), pp. L55–L58.

[LO] J. Llibre and M. Ollé, *The motion of Suturn coorbital satellites in the restricted three-body problem,* Astron. Astrophys., **378** (2001), pp. 1087–1099.

[Mc] R. McGehee, *Triple collision in the collinear three-body problem*, Invent. Math. **27** (1974), 191–227, DOI 10.1007/BF01390175. MR0359459

[M] K. R. Meyer and G. R. Hall, *Introduction to Hamiltonian dynamical systems and the N-body problem*, Applied Mathematical Sciences, vol. 90, Springer-Verlag, New York, 1992. MR1140006

[Mo] C. Mordasini, P. Molliere, K. Dittkrist,, S. Jin, and Yann Alibert, Global Models of Planet Formation and Evolution, International Journal of Astrobiology. **14** (2014), pp. 201–232.

University of California at Santa Cruz, Santa Cruz, CA 95064
Email address: `kistubbs@ucsc.edu`

Department of Mathematics, UNC Asheville, Asheville NC 28804
Email address: `skaplan@unca.edu`

Contemporary Mathematics
Volume **736**, 2019
https://doi.org/10.1090/conm/736/14849

Multivariate random fields and their zero sets

Michael Taylor

ABSTRACT. We consider families of frequency-limited random sections of a vector bundle over a compact Riemannian manifold. We focus on the zero sets of such random fields, estimating the mean value of their Hausdorff measures in the appropriate dimension, and obtaining an asymptotic formula as the frequency bound is allowed to increase.

Contents

1. Introduction

This paper deals with random fields, which are basically families of random variables. We start with a brief description of the latter object. Let (Ω,μ) be a probability space. A (real valued) random variable on Ω with finite variance is an element of $L^2(\Omega,\mu)$. A multivariate random variable on Ω (with finite variance) is an $\mathbb{R}^k$-valued L^2 function, i.e., an element f of $L^2(\Omega,\mu;\mathbb{R}^k)$. One says that f is a Gaussian random variable if the push-forward $f_*\mu$ is a Gaussian probability measure on $\mathbb{R}^k$. As is usual, we call the mean value of a random variable on Ω the expectation, and write

$$\mathbb{E}(f) = \int_{\Omega} f(\omega)\, d\mu(\omega). \tag{1.1}$$

We will want to have a probability space (Ω,μ) on which there is a sequence $\{X_j\}$ of independent, identically distributed Gaussian random variables, of mean 0

2010 *Mathematics Subject Classification.* 35J47, 35K45, 35L10, 35P05, 35S05, 60G60.
Key words and phrases. random fields, Gaussian fields, elliptic operators, vector bundles.
Work supported by NSF grant DMS-1500817.

and variance 1, so $\mathbb{E}(X_j) = 0$ and $\mathbb{E}(X_j^2) = 1$. For this, we can take

$$\Omega = \prod_{j=0}^{\infty} I_j, \tag{1.2}$$

where $I_j = [-\infty, \infty]$, with probability measure $\mu_j = (2\pi)^{-1/2} e^{-x^2/2}\, dx$, and the measure μ on Ω is the product measure. Then we can take $X_j(\omega) = \omega_j$ for $\omega = (\omega_0, \omega_1, \omega_2, \dots) \in \Omega$.

The objects of our study will be various random fields on a compact Riemannian manifold M. To introduce these objects, we start with a class of scalar (i.e., real valued) random fields. Let $\{f_j : j \in \mathbb{Z}^+\}$ be an orthonormal basis of $L^2(M)$. Take a continuous function

$$\varphi : [0, \infty) \longrightarrow \mathbb{R}, \tag{1.3}$$

and set

$$F_\omega^\varphi(x) = \sum_{j=0}^{\infty} \varphi(j) X_j(\omega) f_j(x), \tag{1.4}$$

for $\omega \in \Omega$. As above, $\{X_j\}$ is a sequence of independent, identically distributed Gaussian random variables, of mean 0 and variance 1. Note that, for $\omega \in \Omega$,

$$\|F_\omega^\varphi\|_{L^2(M)}^2 = \sum_{j \ge 0} \varphi(j)^2 |X_j(\omega)|^2, \tag{1.5}$$

and hence

$$\mathbb{E}\Big(\|F^\varphi\|_{L^2(M)}^2\Big) = \sum_j \varphi(j)^2, \tag{1.6}$$

which is finite provided

$$\sum_j \varphi(j)^2 < \infty. \tag{1.7}$$

As long as (1.7) holds, $F_\omega^\varphi \in L^2(M)$ for μ-a.e. $\omega \in \Omega$. We say F^φ is a random L^2 field on M. One can loosen the restriction (1.7), and consider more general classes of random fields, but we will not proceed in that direction. (Some discussion and references can be found in [**16**].)

Rather, we will impose further structure. Let Δ denote the Laplace-Beltrami operator on the compact, n-dimensional Riemannian manifold M. Then $L^2(M)$ has an orthonormal basis of eigenfunctions of Δ:

$$-\Delta f_j = \lambda_j^2 f_j, \quad \lambda_j \nearrow +\infty. \tag{1.8}$$

We take φ as in (1.3), and, in place of (1.4), form

$$F_\omega^\varphi(x) = \sum_{j \ge 0} \varphi(\lambda_j) X_j(\omega) f_j(x), \tag{1.9}$$

with $\{X_j\}$ as before, independent, identically distributed Gaussian random variables, of mean 0, variance 1. We strengthen (1.7) to the hypothesis that φ is rapidly decreasing at infinity:

$$\varphi(\lambda) \le C_N (1 + \lambda)^{-N}, \quad \forall\, N \in \mathbb{N},\ \lambda \in \mathbb{R}^+. \tag{1.10}$$

Given $s \in \mathbb{R}$, we can evaluate the H^s-Sobolev norm of F_ω as

$$\|F_\omega^\varphi\|_{H^s(M)}^2 = \sum_{j\geq 0}(1+\lambda_j^2)^s\varphi(\lambda_j)^2|X_j(\omega)|^2, \tag{1.11}$$

parallel to (1.8), and hence

$$\mathbb{E}\Big(\|F^\varphi\|_{H^s(M)}^2\Big) = \sum_{j\geq 0}(1+\lambda_j^2)^s\varphi(\lambda_j)^2 < \infty, \tag{1.12}$$

given (1.10), so ω-a.e. F_ω^φ is in $C^\infty(M)$. In other words, F^φ is a random C^∞ field on M. Such smooth random fields are much studied objects, whose literature includes [**1**], [**17**], [**12**], [**5**], and [**4**], and references given there. We mention particularly investigations of the $(n-1)$-dimensional measure of zero sets of such random fields in [**4**].

The random fields described above are real-valued fields (i.e., scalar fields). It is also of interest to consider multivariate random fields, such as fields on M taking values in $\mathbb{R}^k$. Then the zero set of such a random field is the intersection of the zero sets of each of its components. This paper studies such multivariate random fields.

In fact, we place our study in the following more general, but geometrically natural, setting. As before, let M be a compact, n-dimensional Riemannian manifold. Let $E \to M$ be a smooth, rank k, real vector bundle, such that the fibers E_x are equipped with a smoothly varying inner product. Let

$$L : C^\infty(M,E) \longrightarrow C^\infty(M,E) \tag{1.13}$$

be a strongly elliptic, self-adjoint differential operator. We assume L has order 2 and is positive semi-definite (though other assumptions can be used). The space $L^2(M,E)$ has an orthonormal basis $\{f_j : j \geq 0\}$ consisting of eigenfunctions of L:

$$Lf_j = \lambda_j^2 f_j, \quad \lambda_j \nearrow +\infty. \tag{1.14}$$

We take $\varphi : [0,\infty) \to \mathbb{R}$, continuous and satisfying (1.10), and form the random field $F_\omega^\varphi(x)$ just as in (1.9). Again we have (1.11)–(1.12), so ω-a.e. F_ω^φ is in $C^\infty(M,E)$.

Our goal is to study the set

$$\mathcal{Z}(F_\omega^\varphi) = \{x \in M : F_\omega^\varphi(x) = 0\}. \tag{1.15}$$

We will show that, for a.e. $\omega \in \Omega$, this has Hausdorff dimension $n-k$, and we will produce a formula for its $(n-k)$-dimensional Hausdorff measure. For this, we assume that

$$0 \leq k = \dim E_x \leq n. \tag{1.16}$$

One example where this analysis applies is to $L = -\Delta_1$, where Δ_1 is the Hodge Laplacian, acting on 1-forms. In this setting, random fields are naturally equivalent to random vector fields on M, and we are evaluating the expected value of the number of zeros of such a random vector field.

In §§2–3 we establish the following variant of a Kac-Rice formula:

$$\mathbb{E}\Big[\mathcal{H}^{n-k}\,\mathcal{Z}(F^\varphi)\Big] = \int_M \int_{\mathcal{L}(T_x,E_x)} c_\varphi(x)e^{-\gamma_{\varphi,x}(0,A)}L(A)\,dA\,dV(x), \tag{1.17}$$

where $L(A) = (\det AA^t)^{1/2}$, $\gamma_{\varphi,x}(0,A)$ is a positive definite quadratic form in $A \in \mathcal{L}(T_x, E_x)$, and $c_\varphi(x)$ is a coefficient defined to make (2.8) a probability measure. One key ingredient in this calculation will be the identity

$$\sum_{j\geq 0} \psi(\lambda_j) f_j(x) \otimes f_j(y) = K_\psi(x,y), \tag{1.18}$$

where $K_\psi(x,y)$ is the integral kernel of the operator $\psi(\sqrt{L})$, i.e.,

$$\psi(\sqrt{L})g(x) = \int_M K_\psi(x,y) g(y)\, dV(y). \tag{1.19}$$

Here $K_\psi(x,y) \in E_x \otimes E_y \approx \mathcal{L}(E_y, E_x)$, the latter isomorphism via the inner product on E_y. This figures in the derivation of (1.17), with $\psi(\lambda) = \varphi(\lambda)^2$.

In §4 we take

$$\varphi(\lambda) = \varphi_t(\lambda) = e^{-t\lambda^2/2}, \tag{1.20}$$

and obtain an asymptotic expansion as $t \searrow 0$ of (1.17), with $\varphi = \varphi_t$, under the hypothesis that L has scalar principal symbol. In such a case, the operator (1.19) becomes

$$e^{-tL}, \tag{1.21}$$

the solution operator to the "heat equation"

$$\frac{\partial u}{\partial t} = -Lu, \quad u(0) = g, \tag{1.22}$$

and classical parametrix constructions provide an essential tool to pass from (1.17) to the conclusion, established in Theorem 4.1, that

$$\mathbb{E}\Big[\mathcal{H}^{n-k}\,\mathcal{Z}(F^{\varphi_t})\Big] = (2\pi)^{-m/2} 2^{\nu/2} \gamma(n,k) t^{-k/2} (\mathrm{Vol}\, M)\big(1 + O(t)\big), \tag{1.23}$$

with m and ν given by (4.23) and $\gamma(n,k)$ by (4.28).

In §5 we indicate further directions in which these studies might be pursued. These include more general classes of strongly elliptic operators L, other sorts of spectral cut-offs, beyond (1.20), and the possibility to allow M to have a nonempty boundary. In Appendix A we look at the coefficients $\gamma(n,k)$ that appear in (1.23), and evaluate the endpoint cases $k = 1$ and $k = n$, the latter case making contact with integrals arising in random matrix theory.

2. Formulas for the expected $(n-k)$-dimensional area of $\mathcal{Z}(F^\varphi_\omega)$

Here we tackle the Kac-Rice formula (1.17). We start with the following.

PROPOSITION 2.1. *Assume $F^\varphi_\omega \in C^\infty(M,E)$ and that 0 is a regular value of F^φ_ω. Then the $(n-k)$-dimensional Hausdorff measure of $\mathcal{Z}(F^\varphi_\omega)$ satisfies*

$$\mathcal{H}^{n-k}\mathcal{Z}(F^\varphi_\omega) = \lim_{\varepsilon\to 0} \int_M \eta_\varepsilon(F^\varphi_\omega(x)) L(\nabla F^\varphi_\omega(x))\, dV(x), \tag{2.1}$$

where, for $v \in E_x$,

$$\begin{aligned} \eta_\varepsilon(v) &= V_k^{-1}\varepsilon^{-k} \quad && \text{if } |v| \le \varepsilon, \\ &\quad 0 && \text{if } |v| > \varepsilon, \end{aligned} \tag{2.2}$$

with V_k *the volume of the unit ball in* $\mathbb{R}^k$, *and, for* $A \in \mathcal{L}(T_x, E_x)$,

$$L(A) = (\det AA^t)^{1/2}. \tag{2.3}$$

Here ∇F^φ_ω is defined by a choice of connection on E. Note however that $\nabla F^\varphi_\omega(x_0)$ is independent of the choice of such a connection for $x_0 \in \mathcal{Z}(F^\varphi_\omega)$, so two such connections yield close results for x close to $\mathcal{Z}(F^\varphi_\omega)$. Hence the right side of (2.1) is independent of such a choice.

Proof of Proposition 2.1. Take $x_0 \in \mathcal{Z}(F^\varphi_\omega)$ and pick geodesic coordinates centered at x_0. Identify $T_{x_0}\mathcal{Z}(F^\varphi_\omega)$ with $\mathbb{R}^{n-k}$ and its orthogonal complement $N_{x_0}\mathcal{Z}(F^\varphi_\omega)$ with $\mathbb{R}^k$. The key is to identify, to leading order in ε, the k-dimensional measure of

$$\{x \in N_{x_0}\mathcal{Z}(F^\varphi_\omega) : |F_\omega(x)| \le \varepsilon\}, \tag{2.4}$$

or equivalently (to leading order) the k-dimensional measure of

$$\{x \in N_{x_0}\mathcal{Z}(F^\varphi_\omega) : |Ax| \le \varepsilon\}, \tag{2.5}$$

where

$$A = \nabla F^\varphi_\omega(x_0) : T_{x_0}M \longrightarrow E_{x_0}, \tag{2.6}$$

can be identified with

$$A : \mathbb{R}^n \longrightarrow \mathbb{R}^k, \quad A = (0\ B), \quad B : \mathbb{R}^k \to \mathbb{R}^k, \tag{2.7}$$

and we want to evaluate the k-dimensional volume of

$$\{u \in \mathbb{R}^k : |Bu| \le \varepsilon\}. \tag{2.8}$$

Now applying B multiplies volumes of subsets of $\mathbb{R}^k$ by a factor of

$$|\det B| = (\det AA^t)^{1/2}, \tag{2.9}$$

so the volume of (2.8), hence of (2.5), is $V_k\varepsilon^k|\det B|^{-1}$, and to leading order this is the volume of (2.4). The factor $L(\nabla F^\varphi_\omega(x))$ needs to cancel out the extra factor of $|\det B|^{-1}$, to leading order, and this leads to (2.3). □

Let us denote the integral on the right side of (2.1) by

$$Z_\varepsilon(F^\varphi_\omega) = \int_M \eta_\varepsilon(F^\varphi_\omega(x))L(\nabla F^\varphi_\omega(x))\,dV(x). \tag{2.10}$$

From here, we have

$$\mathbb{E}\,Z_\varepsilon(F^\varphi) = \int_M \mathbb{E}\Big[\eta_\varepsilon(F^\varphi(x))L(\nabla F^\varphi(x))\Big]\,dV(x). \tag{2.11}$$

By (1.9),

$$\begin{aligned} G^\varphi_\omega(x) &= (F^\varphi_\omega(x), \nabla F^\varphi_\omega(x)) \\ &= \sum_k \varphi(\lambda_k)X_k(\omega)(f_k(x), \nabla f_k(x)) \end{aligned} \tag{2.12}$$

is, for each x, a Gaussian random variable, taking values in $E_x \oplus \mathcal{L}(T_x, E_x)$, with mean zero. This Gaussian random variable hence induces a Gaussian probability measure Γ^φ_x on $E_x \oplus \mathcal{L}(T_x, E_x)$, and

$$\mathbb{E}\Big[\eta_\varepsilon(F^\varphi(x))L(\nabla F^\varphi(x))\Big] = \int_{E_x\oplus\mathcal{L}(T_x,E_x)} \eta_\varepsilon(v)L(A)\,d\Gamma^\varphi_x(v, A). \tag{2.13}$$

In §§3–4, we will show that this Gaussian measure has the form

$$d\Gamma_x^\varphi(v,A) = c_\varphi(x) e^{-\gamma_{\varphi,x}(v,A)}\, dv\, dA, \tag{2.14}$$

where $\gamma_{\varphi,x}(v,A)$ is a positive definite quadratic form in (v,A). Consequently,

$$\begin{aligned} &\lim_{\varepsilon\to 0} \mathbb{E}\Big[\eta_\varepsilon(F^\varphi(x)) L(\nabla F^\varphi(x))\Big] \\ &\qquad = c_\varphi(x) \int\limits_{\mathcal{L}(T_x,E_x)} e^{-\gamma_{\varphi,x}(0,A)} L(A)\, dA. \end{aligned} \tag{2.15}$$

Combining this with (2.1) and (2.11) gives the following variant of the *Kac-Rice formula*:

PROPOSITION 2.2. *Given* (2.14), *we have*

$$\mathbb{E}\Big[\mathcal{H}^{n-k} \mathcal{Z}(F^\varphi)\Big] = \int\limits_M \int\limits_{\mathcal{L}(T_x,E_x)} c_\varphi(x) e^{-\gamma_{\varphi,x}(0,A)} L(A)\, dA\, dV(x). \tag{2.16}$$

Our next task, pursued in §§3–4, is to derive information on the integrand on the right side of (2.16), which will follow from information on the Gaussian measure (2.14).

REMARK. These results can be localized. If $U \subset M$ is open and smoothly bounded, then

$$\mathbb{E}\Big[\mathcal{H}^{n-k}(U \cap \mathcal{Z}(F^\varphi))\Big] = \int\limits_U \int\limits_{\mathcal{L}(T_x,E_x)} c_\varphi(x) e^{-\gamma_{\varphi,x}(0,A)} L(A)\, dA\, dV(x). \tag{2.17}$$

3. The Gaussian measure Γ_x^φ on $E_x \oplus \mathcal{L}(T_x, E_x)$

As seen in §2, for each $x \in M$,

$$G_\omega^\varphi(x) = \sum_k \varphi(\lambda_k) X_k(\omega) u_k(x), \quad u_k(x) = (f_k(x), \nabla f_k(x)), \tag{3.1}$$

is a Gaussian random variable, taking values in $E_x \oplus \mathcal{L}(T_x, E_x)$, with mean 0, and this random variable then induces a Gaussian probability measure Γ_x^φ on $E_x \oplus \mathcal{L}(T_x, E_x)$. Our next goal is to see when Γ_x^φ has the form

$$d\Gamma_x^\varphi(v,A) = c_\varphi(x) e^{-\gamma_{\varphi,x}(v,A)}\, dv\, dA, \tag{3.2}$$

as advertised in (2.14), and analyze $c_\varphi(x)$ and $\gamma_{\varphi,x}(v,A)$, which is a quadratic form in (v,A). We use the fact that Γ_x^φ is uniquely determined by the covariance of $G_\omega^\varphi(x)$, which we proceed to analyze. We have

$$\begin{aligned} \mathbb{E}(G^\varphi(x) \otimes G^\varphi(y)) &= \sum_{j,k} \mathbb{E}(X_j, X_k) \varphi(\lambda_j)\varphi(\lambda_k)\, u_j(x) \otimes u_k(y) \\ &= \sum_k \varphi(\lambda_k)^2\, u_k(x) \otimes u_k(y). \end{aligned} \tag{3.3}$$

We can expand out $u_k(x) \otimes u_k(y)$ as

$$u_k(x) \otimes u_k(y) = \begin{pmatrix} f_k(x) \otimes f_k(y) & f_k(x) \otimes \nabla f_k(y) \\ \nabla f_k(x) \otimes f_k(y) & \nabla f_k(x) \otimes \nabla f_k(y) \end{pmatrix}. \tag{3.4}$$

Now, as noted in (1.18),

$$\sum_k \varphi(\lambda_k)^2 f_k(x) \otimes f_k(y) = K_{\varphi^2}(x, y), \tag{3.5}$$

the integral kernel of $\varphi(\sqrt{L})^2$. It follows that

$$\mathbb{E}(G^\varphi(x) \otimes C^\varphi(x)) - \begin{pmatrix} K_{\varphi^2}(x, x) & \nabla_2 K_{\varphi^2}(x, x) \\ \nabla_1 K_{\varphi^2}(x, x) & \nabla_1 \nabla_2 K_{\varphi^2}(x, x) \end{pmatrix}, \tag{3.6}$$

where $\nabla_1 K_\psi(x, y) = \nabla_x K_\psi(x, y)$, $\nabla_2 K_\psi(x, y) = \nabla_y K_\psi(x, y)$, etc. Note that (3.6) is an element of

$$\begin{aligned} \operatorname{End}(E_x \oplus \mathcal{L}(T_x, E_x)) \approx {} & \operatorname{End} E_x \oplus \mathcal{L}(\mathcal{L}(T_x, E_x), E_x) \\ & \oplus \mathcal{L}(E_x, \mathcal{L}(T_x, E_x)) \oplus \operatorname{End} \mathcal{L}(T_x, E_x). \end{aligned} \tag{3.7}$$

We proceed from (3.6) to a formula for the Gaussian measure Γ_x^φ. First, we place the calculation in a more general setting. Let V be an m-dimensional real inner product space, (Ω, μ) a probability space, and $G : \Omega \to V$ a V-valued random variable, yielding the probability measure $G_*\mu = \Gamma$ on V. Let us assume that G is a Gaussian random variable with mean zero. As is well known, Γ is a Gaussian measure, and it is uniquely determined by the covariance

$$\mathbb{E}(G \otimes G) = \int_\Omega G(\omega) \otimes G(\omega)\, d\mu(\omega) = \mathcal{C} \in V \otimes V. \tag{3.8}$$

We can also regard

$$\mathcal{C} \in \mathcal{L}(V), \quad \text{via } V \otimes V \approx \mathcal{L}(V), \tag{3.9}$$

this isomorphism arising from the inner product on V. As such, $\mathcal{C}$ is symmetric and positive semidefinite. If $\mathcal{C}$ is positive definite, then Γ has the form

$$d\Gamma(y) = \alpha(C) e^{-y \cdot Cy}\, dy, \tag{3.10}$$

for some positive definite $C \in \mathcal{L}(V)$, with $\alpha(C)$ chosen so that the right side of (3.10) has mass one. Using orthonormal coordinates on V such that C is diagonal, and computing the Gaussian integrals, via

$$\int_{-\infty}^{\infty} e^{-y^2}\, dy = \sqrt{\pi}, \tag{3.11}$$

we obtain

$$\alpha(C) = \pi^{-m/2} (\det C)^{1/2}. \tag{3.12}$$

Now $\Gamma = G_*\mu$ if and only if

$$\int_V y \otimes y\, d\Gamma(y) = \mathcal{C}. \tag{3.13}$$

To calculate

$$\int_V e^{-y \cdot Cy} y \otimes y\, dy, \tag{3.14}$$

we take an orthonormal basis $\{e_j\}$ of V such that $Ce_j = c_j e_j$, $c_j > 0$. Then $y \otimes y = \sum_{j,k} y_j y_k \, e_j \otimes e_k$, and (3.14) is

$$\sum_{j,k} \int_V e^{-y\cdot Cy} y_j y_k \, dy \, e_j \otimes e_k. \tag{3.15}$$

Symmetry considerations show that each term for which $j \neq k$ vanishes, and we are left to calculate

$$\int_V e^{-y\cdot Cy} y_k^2 \, dy = \prod_{j\neq k} \Bigl(\frac{\pi}{c_j}\Bigr)^{1/2} \int_{-\infty}^{\infty} e^{-c_k y^2} y^2 \, dy, \tag{3.16}$$

making use of the following consequence of (3.11):

$$\int_{-\infty}^{\infty} e^{-cy^2} \, dy = \sqrt{\frac{\pi}{c}}, \tag{3.17}$$

for $c > 0$. Taking the c-derivative of (3.17) yields

$$\int_{-\infty}^{\infty} e^{-cy^2} y^2 \, dy = \frac{\sqrt{\pi}}{2} c^{-3/2}, \tag{3.18}$$

so (3.15)–(3.16) yield

$$\begin{aligned} \int_V e^{-y\cdot Cy} y \otimes y \, dy &= \sum_k \prod_{j\neq k} \Bigl(\frac{\pi}{c_j}\Bigr)^{1/2} \frac{\sqrt{\pi}}{2c_k^{3/2}} \, e_k \otimes e_k \\ &= \frac{1}{2} \frac{\pi^{m/2}}{(\det C)^{1/2}} \sum_k c_k^{-1} \, e_k \otimes e_k \\ &= \frac{1}{2} \frac{\pi^{m/2}}{(\det C)^{1/2}} \sum_k C^{-1} e_k \otimes e_k. \end{aligned} \tag{3.19}$$

Using (3.12) and taking into account the isomorphism $V \otimes V \approx \mathcal{L}(V)$, we have from (3.13) that

$$\mathcal{C} = \frac{1}{2} C^{-1}, \quad \text{hence } C = \frac{1}{2} \mathcal{C}^{-1}. \tag{3.20}$$

We record the (well known) conclusion.

PROPOSITION 3.1. *If $G : \Omega \longrightarrow V$ is a Gaussian random variable with mean 0 and covariance $\mathcal{C}$, given by* (3.8), *and if $\mathcal{C}$ is positive definite, then $\Gamma = G_* \mu$ has the form* (3.10), *with C given by* (3.20) *and $\alpha(C)$ by* (3.12).

Regarding the condition that $\mathcal{C}$ be positive definite, note from (3.8) that, for $v \in V$,

$$v \cdot \mathcal{C} v = \mathbb{E}(|G \cdot v|^2) = \int_\Omega |G(\omega) \cdot v|^2 \, d\mu(\omega). \tag{3.21}$$

Thus $\mathcal{C}$ is positive definite unless there is a proper linear subspace $V_0 \subset V$ such that

$$G(\omega) \in V_0, \quad \text{for } \mu\text{-a.e. } \omega \in \Omega. \tag{3.22}$$

In the case of main interest to us, $\mathcal{C} = \mathcal{C}_x^{\varphi}$ is given by (3.5), as a continuous section of $\mathrm{End}(E \oplus \mathcal{L}(TM, E))$. As long as this is positive definite on $E_x \oplus \mathcal{L}(T_x, E_x)$

for each $x \in M$, we have the results (2.14)–(2.16). We turn to a closer look at such $\mathcal{C}_x^\varphi$ in the next section, for

$$\varphi(\lambda) = \varphi_t(\lambda) = e^{-t\lambda^2/2}, \tag{3.23}$$

and examine asymptotics as $t \searrow 0$.

4. Heat asymptotics and zero set asymptotics

Here we assume that the second order differential operator L has a scalar principal symbol, equal to that of $-\Delta$, where Δ is the Laplace-Beltrami operator on M. Such holds when L is the negative of the Hodge Laplacian on ℓ-forms. Then, for $t \searrow 0$,

$$e^{-tL}u(x) = \int\limits_M K_t(x,y)u(y)\,dV(y), \tag{4.1}$$

where $K_t(x,y) \in \mathcal{L}(E_y, E_x)$ has the form, for x and y close,

$$K_t(x,y) \sim (4\pi t)^{-n/2} e^{-\rho(x,y)/4t}\Big(A_0(x,y) + A_1(x,y)t + \cdots\Big), \tag{4.2}$$

with $A_k \in \mathcal{L}(E_y, E_x)$, depending smoothly on x and y, and

$$A_0(x,x) = I. \tag{4.3}$$

Here,

$$\rho(x,y) = \operatorname{dist}(x,y)^2. \tag{4.4}$$

In particular, if we pick exponential coordinates centered at x,

$$\rho(x,y) = |x-y|^2, \tag{4.5}$$

the square norm being determined by the inner product on T_xM. See [**2**], pp. 204–214, for a derivation of (4.2). The treatment there takes $L = -\Delta$, but the analysis works for the class of operators L described above.

Now, if we take

$$\varphi(\lambda) = \varphi_t(\lambda) = e^{-t\lambda^2/2}, \tag{4.6}$$

then (3.6)–(3.9) give $\mathcal{C} = \mathcal{C}_{t,x}$, with

$$\mathcal{C}_{t,x} = \begin{pmatrix} K_t(x,x) & \nabla_2 K_t(x,x) \\ \nabla_1 K_t(x,x) & \nabla_1\nabla_2 K_t(x,x) \end{pmatrix}. \tag{4.7}$$

We have

$$K_t(x,x) \sim (4\pi t)^{-n/2}\Big(I + A_1(x,x)t + \cdots\Big). \tag{4.8}$$

Since

$$\nabla_1 e^{-|x-y|^2/4t} = -\frac{x-y}{2t} e^{-|x-y|^2/4t}, \tag{4.9}$$

we have

$$\nabla_1 K_t(x,x) = (4\pi t)^{-n/2}\Big(\nabla_1 A_0(x,x) + O(t)\Big). \tag{4.10}$$

Similarly,

$$\nabla_2 K_t(x,x) = (4\pi t)^{-n/2}\Big(\nabla_2 A_0(x,x) + O(t)\Big). \tag{4.11}$$

Furthermore, since

$$\nabla_1\nabla_2 e^{-|x-y|^2/4t} = -\frac{(x-y)\otimes(x-y)}{4t^2}e^{-|x-y|^2/4t} + \frac{1}{2t}e^{-|x-y|^2/4t} I, \tag{4.12}$$

we have

$$\nabla_1\nabla_2 K_t(x,x) = (4\pi t)^{-n/2}\Big(\frac{1}{2t}I + O(1)\Big). \tag{4.13}$$

Thus, for

$$\widetilde{\mathcal{C}}_{t,x} = (4\pi t)^{n/2}\mathcal{C}_{t,x}, \tag{4.14}$$

we have

$$\widetilde{\mathcal{C}}_{t,x} = \begin{pmatrix} I + O(t) & \nabla_2 A_0(x,x) + O(t) \\ \nabla_1 A_0(x,x) + O(t) & (2t)^{-1}I + O(1) \end{pmatrix}. \tag{4.15}$$

Consequently

$$\begin{pmatrix} 1 & 0 \\ 0 & 2t \end{pmatrix}\widetilde{\mathcal{C}}_{t,x} = \begin{pmatrix} I & \nabla_2 A_0(x,x) \\ 0 & I \end{pmatrix} + O(t). \tag{4.16}$$

It follows that, for $t>0$ sufficiently small, $\widetilde{\mathcal{C}}_{t,x}$ is invertible (hence positive definite) and

$$\widetilde{\mathcal{C}}_{t,x}^{-1}\begin{pmatrix} 1 & 0 \\ 0 & (2t)^{-1} \end{pmatrix} = \begin{pmatrix} I & \beta(x) \\ 0 & I \end{pmatrix} + O(t), \tag{4.17}$$

with $\beta(x)\in\mathcal{L}(E_x\oplus\mathcal{L}(T_x,E_x))$, depending smoothly on x. Then

$$\widetilde{\mathcal{C}}_{t,x}^{-1} = \begin{pmatrix} I & 2t\beta(x) \\ 0 & 2tI \end{pmatrix} + \begin{pmatrix} O(t) & O(t^2) \\ O(t) & O(t^2) \end{pmatrix}. \tag{4.18}$$

It follows that, when $\varphi(\lambda)=e^{-t\lambda^2/2}$, and $t>0$ is sufficiently small, then (3.2) holds for $\Gamma_x^\varphi = \Gamma_{x,t}$, rewritten as

$$d\Gamma_{x,t}(v,A) = c_t(x)e^{-\gamma_{t,x}(v,A)}\,dv\,dA, \tag{4.19}$$

where

$$\begin{aligned} \gamma_{t,x}(v,A) &= (v,A)C_{t,x}\begin{pmatrix} v \\ A \end{pmatrix} \\ &= \frac{1}{2}(v,A)\mathcal{C}_{t,x}^{-1}\begin{pmatrix} v \\ A \end{pmatrix} \\ &= \frac{1}{2}(4\pi t)^{n/2}(v,A)\widetilde{\mathcal{C}}_{t,x}^{-1}\begin{pmatrix} v \\ A \end{pmatrix}, \end{aligned} \tag{4.20}$$

hence

$$\gamma_{t,x}(0,A) = \frac{1}{2}(4\pi t)^{n/2}\Big(2t\|A\|^2 + O(t^2)\Big). \tag{4.21}$$

Also,

$$\begin{aligned} c_t(x) = \alpha(C_{t,x}) &= \pi^{-m/2}(\det C_{t,x})^{1/2} \\ &= \pi^{-m/2}\Big(\frac{1}{2}(4\pi t)^{n/2}\Big)^{m/2}(2t+O(t^2))^{\nu/2}, \end{aligned} \tag{4.22}$$

with

$$\begin{aligned} m &= \dim E_x\oplus\mathcal{L}(T_x,E_x) = k+nk, \\ \nu &= \dim\mathcal{L}(T_x,E_x) = nk. \end{aligned} \tag{4.23}$$

In this setting, (2.16) yields

$$\mathbb{E}\Big[\mathcal{H}^{n-k}\mathcal{Z}(F^{\varphi})\Big] = \int\limits_M \kappa(t,x)\,dV(x), \tag{4.24}$$

where

$$\begin{aligned}\kappa(t,x) &= c_t(x) \int\limits_{\mathcal{L}(\mathbb{R}^n,\mathbb{R}^k)} e^{-\gamma_{t,x}(0,A)} L(A)\,dA \\ &= (2\pi)^{-m/2}(4\pi t)^{mn/4}(2t+O(t^2)) \int\limits_{\mathcal{L}(\mathbb{R}^n,\mathbb{R}^k)} e^{-(4\pi t)^{n/2}(t\|A\|^2+O(t^2))} L(A)\,dA.\end{aligned} \tag{4.25}$$

If we set

$$B = (4\pi t)^{n/4} t^{1/2} A, \tag{4.26}$$

we get

$$\begin{aligned}\kappa(t,x) &= (2\pi)^{-m/2}(4\pi t)^{mn/4}(2t+O(t^2))^{\nu/2}(4\pi t)^{-n\nu/4}t^{-\nu/2} \\ &\quad\times \int\limits_{\mathcal{L}(\mathbb{R}^n,\mathbb{R}^k)} e^{-(\|B\|^2+O(t))}(4\pi t)^{-nk/4}t^{-k/2}L(B)\,dB \\ &= (2\pi)^{-m/2}2^{\nu/2}t^{-k/2}(1+O(t)) \int\limits_{\mathcal{L}(\mathbb{R}^n,\mathbb{R}^k)} e^{-\|B\|^2} L(B)\,dB,\end{aligned} \tag{4.27}$$

which, to leading order, is independent of x. Consequently, with

$$\gamma(n,k) = \int\limits_{\mathcal{L}(\mathbb{R}^n,\mathbb{R}^k)} e^{-\|B\|^2} (\det BB^t)^{1/2}\,dB, \tag{4.28}$$

we have the following conclusion.

THEOREM 4.1. *Let $E \to M$ be a rank k real vector bundle over M, with $0 \le k \le n$. Let L be a second order, strongly elliptic, self-adjoint operator on sections of E, with scalar principal symbol. Then*

$$\mathbb{E}\Big[\mathcal{H}^{n-k}\mathcal{Z}(F^{\varphi_t})\Big] = (2\pi)^{-m/2}2^{\nu/2}\gamma(n,k)t^{-k/2}(\mathit{Vol}\ M)(1+O(t)), \tag{4.29}$$

as $t \searrow 0$, when F^{φ_t} is given by (1.9) with $\varphi_t(\lambda) = e^{-t\lambda^2/2}$.

REMARK. In the formulas above, $\|A\|^2$ and $\|B\|^2$ denote the squared Hilbert-Schmidt norms of these elements of $\mathcal{L}(\mathbb{R}^n,\mathbb{R}^k)$.

5. Other directions

In the previous sections, we took L to be a second-order, strongly elliptic differential operator, with scalar principal symbol, acting on sections of a vector bundle $E \to M$ over a compact, n-dimensional Riemannian manifold without boundary, and we used spectral cut-offs of the form e^{-tL}. Here we describe various ways one might extend the scope of these investigations.

I. More general operators L

One natural extension involves treating a strongly elliptic, self-adjoint differential operator L of order 2 whose principal symbol is not scalar, but rather takes values in the set of positive definite linear transformations on the fibers E_x. Methods of pseudodifferential operator calculus are well suited to construct parametrices for the semigroup e^{-tL} in this situation, and one might investigate extensions of Theorem 4.1 to cover such cases.

In fact, such pseudodifferential operator techniques are effective in constructing parametrices for e^{-tL} whenever L is a strongly elliptic pseudodifferential operator, of order $m > 0$, acting on sections of $E \to M$. An example is the Dirichlet-to-Neumann map,

$$\Lambda f = \frac{\partial}{\partial \nu} \operatorname{PI} f, \tag{5.1}$$

arising when $M = \partial\Omega$ and Ω is a smoothly bounded Riemannian manifold of dimension $n+1$. Here, $\operatorname{PI} f = u$ solves

$$Pu = 0, \quad u|_{\partial\Omega} = f, \tag{5.2}$$

perhaps with $P = -\Delta$, or some more general second order strongly elliptic differential operator on Ω. Then Λ is an elliptic pseudodifferential operator on M of order 1, whose spectral theory is of interest.

Returning to cases where L is a second order strongly elliptic differential operator, one can allow M to be a compact Riemannian manifold with nonempty boundary ∂M, and impose the Dirichlet boundary condition (or perhaps some other boundary condition) on ∂M. Via the method of layer potentials, parametrices for e^{-tL} are available in this more general setting, and they provide valuable information on the spectral theory of L.

II. Other families of spectral cut-offs

We have concentrated on the family of spectral cut-offs

$$\varphi_t(\lambda) = \varphi(t^{1/2}\lambda), \quad \text{with } \varphi(\lambda) = e^{-\lambda^2/2}, \tag{5.3}$$

which tend to 1 as $t \searrow 0$. It is also of interest to take functions $\varphi(\lambda)$ that cut off low frequencies as well as high frequencies, so that $\varphi(\lambda) \to 0$ both as $\lambda \searrow 0$ and as $\lambda \nearrow \infty$. Some examples that the techniques of §4 can readily handle include

$$\varphi(\lambda) = e^{-\lambda^2/2} - e^{-\lambda^2}. \tag{5.4}$$

In such a case, $\psi_t(\lambda) = \varphi_t(\lambda)^2$ yields

$$\psi_t(\lambda) = (e^{-t\lambda^2/2} - e^{-t\lambda^2})^2, \tag{5.5}$$

giving rise to

$$\begin{aligned} \psi_t(\sqrt{L}) &= (e^{-tL/2} - e^{-tL})^2 \\ &= e^{-tL} + e^{-2tL} - 2e^{-(3/2)tL}. \end{aligned} \tag{5.6}$$

For a stronger cut-off of low frequencies, one could take

$$\varphi(\lambda) = (e^{-\lambda^2/2} - e^{-\lambda^2})^\ell, \quad \ell \in \{2, 3, \dots\}, \tag{5.7}$$

leading to $\psi_t(\lambda) = \varphi_t(\lambda)^2$ with

$$\begin{aligned} \psi_t(\sqrt{L}) &= (e^{-tL/2} - e^{-tL})^{2\ell} \\ &= \sum_{j=0}^{2\ell} \binom{2\ell}{j} (-1)^j e^{-(2\ell - j/2)tL}. \end{aligned} \tag{5.8}$$

When the hypotheses of §4 apply, and more generally when one can produce a sufficiently useful parametrix for e^{-tL}, one can expect to be able to extend Theorem 4.1 to a result on the asymptotic behavior of $\mathbb{E}[\mathcal{H}^{n-k}\,\mathcal{Z}(F^{\varphi_t})]$ for such a family of functions $\varphi_t(\lambda)$. Similar statements hold for functions of the form

$$\varphi(\lambda) = \lambda^\ell e^{-\lambda^2/2}, \quad \ell \in \mathbb{N}, \tag{5.9}$$

giving rise to

$$\begin{aligned} \psi_t(\sqrt{L}) &= t^\ell L^\ell e^{-tL} \\ &= t^\ell \Big(-\frac{d}{dt}\Big)^\ell e^{-tL}. \end{aligned} \tag{5.10}$$

Such spectral cut-offs arise in Littlewood-Paley theory (cf. [**13**]) and hence are of great interest.

III. Wave equation approaches

A still finer family of spectral cut-offs has the form

$$\varphi_R(\lambda) = \varphi(\lambda - R), \tag{5.11}$$

with

$$\varphi(0) = 1, \quad \varphi \geq 0, \quad \hat{\varphi} \in C_0^\infty(-T, T), \tag{5.12}$$

for some $T > 0$. When A is a positive self-adjoint operator, the spectral cut-offs $\varphi_R(A)$ can be synthesized from the unitary group e^{itA}, via

$$\varphi_R(A) = \frac{1}{\sqrt{2\pi}} \int_{-T}^{T} \hat{\varphi}(t) e^{-iRt} e^{itA}\, dt. \tag{5.13}$$

When A is a positive, elliptic, scalar pseudodifferential operator of order 1, on a compact manifold without boundary, [**6**] constructed a parametrix for e^{itA} and initiated a study of $\varphi_R(A)$. In case $A = \sqrt{-\Delta}$, work of [**3**] and [**4**] carried this further and applied it to the study of the zero set of $F_\omega^{\varphi_R}$, including a study of its $(n-1)$-dimensional Hausdorff measure. It would be of interest to make a parallel investigation of other scalar pseudodifferential operators, such as the Dirichlet-to-Neumann map. More generally, one might consider first order elliptic pseudodifferential operators acting on sections of a vector bundle, in case they have scalar principal symbol.

It would also be interesting to tackle cases where the principal symbol is not scalar, though wave equation techniques tend to be less robust than heat equation techniques, due to phenomena associated with multiple characteristics, such as conical refraction.

There are also substantial technical difficulties in applying wave equation techniques when M has nonempty boundary, arising from rays of geometrical optics tangent to the boundary. The most accessible case is the case of diffractive boundary, where parametrices described in [**10**] are available. The analysis in [**9**] might point to an approach relevant to an extension of Theorem 4.1.

Appendix A. Remarks on $\gamma(n,k)$

The coefficients $\gamma(n,k)$ arose in the asymptotic formula (4.29), and were given by (4.28), which we recall is

$$\gamma(n,k) = \int\limits_{\mathcal{L}(\mathbb{R}^n,\mathbb{R}^k)} e^{-\|B\|^2} (\det BB^t)^{1/2}\, dB. \tag{A.1}$$

Recall that $\|B\|$ denotes the Hilbert-Schmidt norm of B, and we are assuming $1 \le k \le n$. We have the following formulas for the two extreme cases.

First,

$$\begin{aligned} \gamma(n,1) &= \int\limits_{\mathbb{R}^n} e^{-|x|^2} |x|\, dx \\ &= A_{n-1} \int_0^\infty e^{-r^2} r^n\, dr \\ &= \frac{1}{2} A_{n-1} \int_0^\infty e^{-s} s^{(n-1)/2}\, ds \\ &= \frac{1}{2} A_{n-1} \Gamma\Big(\frac{n+1}{2}\Big) \\ &= \pi^{n/2} \frac{\Gamma(\frac{n+1}{2})}{\Gamma(\frac{n}{2})}, \end{aligned} \tag{A.2}$$

where A_{n-1} denotes the area of the unit sphere $S^{n-1} \subset \mathbb{R}^n$.

At the other extreme,

$$\gamma(n,n) = \int\limits_{\mathcal{L}(\mathbb{R}^n)} e^{-\|B\|^2} |\det B|\, dB, \tag{A.3}$$

and using (15.4.12) of [**8**], we obtain

$$\begin{aligned} \gamma(n,n) &= \pi^{n/2} \prod_{j=1}^n \frac{\Gamma(\frac{1+j}{2})}{\Gamma(\frac{j}{2})} \\ &= \pi^{n/2} \frac{\Gamma(\frac{n+1}{2})}{\Gamma(\frac{1}{2})}. \end{aligned} \tag{A.4}$$

I do not have a calculation of $\gamma(n,k)$ for $1 < k < n$, though one might guess a pattern from (A.2) and (A.4).

References

[1] R. J. Adler and J. E. Taylor, *Random fields and geometry*, Springer Monographs in Mathematics, Springer, New York, 2007. MR2319516

[2] M. Berger, P. Gauduchon, and E. Mazet, *Le spectre d'une variété riemannienne* (French), Lecture Notes in Mathematics, Vol. 194, Springer-Verlag, Berlin-New York, 1971. MR0282313

[3] Y. Canzani and B. Hanin, *Scaling limit for the kernel of the spectral projector and remainder estimates in the pointwise Weyl law*, Anal. PDE **8** (2015), no. 7, 1707–1731, DOI 10.2140/apde.2015.8.1707. MR3399136

[4] Y. Canzani and B. Hanin, Local universality for zeros and critical points of monochromatic random waves, Preprint, arXiv:1610.09438.

[5] Y. Canzani and P. Sarnak, *Topology and nesting of the zero set components of monochromatic random waves*, Comm. Pure Appl. Math. **72** (2019), no. 2, 343–374, DOI 10.1002/cpa.21795. MR3896023

[6] L. Hörmander, *The spectral function of an elliptic operator*, Acta Math. **121** (1968), 193–218, DOI 10.1007/BF02391913. MR0609014

[7] D. Marinucci and G. Peccati, *Random fields on the sphere*, London Mathematical Society Lecture Note Series, vol. 389, Cambridge University Press, Cambridge, 2011. MR2840154

[8] M. L. Mehta, *Random matrices*, 3rd ed., Pure and Applied Mathematics (Amsterdam), vol. 142, Elsevier/Academic Press, Amsterdam, 2004. MR2129906

[9] R. B. Melrose, *Weyl's conjecture for manifolds with concave boundary*, Geometry of the Laplace operator (Proc. Sympos. Pure Math., Univ. Hawaii, Honolulu, Hawaii, 1979), Proc. Sympos. Pure Math., XXXVI, Amer. Math. Soc., Providence, R.I., 1980, pp. 257–274. MR573438

[10] R. Melrose and M. Taylor, Boundary Problems for Wave Equations with Grazing and Gliding Rays, Manuscript, available at:
http://mtaylor.web.unc.edu/notes (item 3).

[11] L. Nicolaescu, On the Kac-Rice formula, Lecture Notes, 2014,
http://www3.nd.edu/~lnicolae_Rice.pdf

[12] L. I. Nicolaescu, *Complexity of random smooth functions on compact manifolds*, Indiana Univ. Math. J. **63** (2014), no. 4, 1037–1065, DOI 10.1512/iumj.2014.63.5321. MR3263921

[13] E. M. Stein, *Topics in harmonic analysis related to the Littlewood-Paley theory.*, Annals of Mathematics Studies, No. 63, Princeton University Press, Princeton, N.J., University of Tokyo Press, Tokyo, 1970. MR0252961

[14] M. E. Taylor, *Pseudodifferential operators*, Princeton Mathematical Series, vol. 34, Princeton University Press, Princeton, N.J., 1981. MR618463

[15] M. E. Taylor, *Partial differential equations: Basic theory*, Texts in Applied Mathematics, vol. 23, Springer-Verlag, New York, 1996. MR1395147

[16] M. Taylor, Random fields: stationarity, ergodicity, and spectral behavior,
Lecture Notes, http://mtaylor.web.unc.edu/notes (item 6).

[17] S. Zelditch, *Real and complex zeros of Riemannian random waves*, Spectral analysis in geometry and number theory, Contemp. Math., vol. 484, Amer. Math. Soc., Providence, RI, 2009, pp. 321–342, DOI 10.1090/conm/484/09482. MR1500155

DEPARTMENT OF MATHEMATICS, UNIVERSITY OF NORTH CAROLINA, CHAPEL HILL NC, 27599
Email address: `met@math.unc.edu`

Selected Published Titles in This Series

736 **Jane Hawkins, Rachel L. Rossetti, and Jim Wiseman, Editors,** Dynamical Systems and Random Processes, 2019

734 **Peter Kuchment and Evgeny Semenov, Editors,** Differential Equations, Mathematical Physics, and Applications, 2019

733 **Peter Kuchment and Evgeny Semenov, Editors,** Functional Analysis and Geometry, 2019

732 **Samuele Anni, Jay Jorgenson, Lejla Smajlović, and Lynne Walling, Editors,** Automorphic Forms and Related Topics, 2019

731 **Robert G. Niemeyer, Erin P. J. Pearse, John A. Rock, and Tony Samuel, Editors,** Horizons of Fractal Geometry and Complex Dimensions, 2019

730 **Alberto Facchini, Lorna Gregory, Sonia L'Innocente, and Marcus Tressl, Editors,** Model Theory of Modules, Algebras and Categories, 2019

729 **Daniel G. Davis, Hans-Werner Henn, J. F. Jardine, Mark W. Johnson, and Charles Rezk, Editors,** Homotopy Theory: Tools and Applications, 2019

728 **Nicolás Andruskiewitsch and Dmitri Nikshych, Editors,** Tensor Categories and Hopf Algebras, 2019

727 **André Leroy, Christian Lomp, Sergio López-Permouth, and Frédérique Oggier, Editors,** Rings, Modules and Codes, 2019

726 **Eugene Plotkin, Editor,** Groups, Algebras and Identities, 2019

725 **Shijun Zheng, Marius Beceanu, Jerry Bona, Geng Chen, Tuoc Van Phan, and Avy Soffer, Editors,** Nonlinear Dispersive Waves and Fluids, 2019

724 **Lubjana Beshaj and Tony Shaska, Editors,** Algebraic Curves and Their Applications, 2019

723 **Donatella Danielli, Arshak Petrosyan, and Camelia A. Pop, Editors,** New Developments in the Analysis of Nonlocal Operators, 2019

722 **Yves Aubry, Everett W. Howe, and Christophe Ritzenthaler, Editors,** Arithmetic Geometry: Computation and Applications, 2019

721 **Petr Vojtěchovský, Murray R. Bremner, J. Scott Carter, Anthony B. Evans, John Huerta, Michael K. Kinyon, G. Eric Moorhouse, and Jonathan D. H. Smith, Editors,** Nonassociative Mathematics and its Applications, 2019

720 **Alexandre Girouard, Editor,** Spectral Theory and Applications, 2018

719 **Florian Sobieczky, Editor,** Unimodularity in Randomly Generated Graphs, 2018

718 **David Ayala, Daniel S. Freed, and Ryan E. Grady, Editors,** Topology and Quantum Theory in Interaction, 2018

717 **Federico Bonetto, David Borthwick, Evans Harrell, and Michael Loss, Editors,** Mathematical Problems in Quantum Physics, 2018

716 **Alex Martsinkovsky, Kiyoshi Igusa, and Gordana Todorov, Editors,** Surveys in Representation Theory of Algebras, 2018

715 **Sergio R. López-Permouth, Jae Keol Park, S. Tariq Rizvi, and Cosmin S. Roman, Editors,** Advances in Rings and Modules, 2018

714 **Jens Gerlach Christensen, Susanna Dann, and Matthew Dawson, Editors,** Representation Theory and Harmonic Analysis on Symmetric Spaces, 2018

713 **Naihuan Jing and Kailash C. Misra, Editors,** Representations of Lie Algebras, Quantum Groups and Related Topics, 2018

712 **Nero Budur, Tommaso de Fernex, Roi Docampo, and Kevin Tucker, Editors,** Local and Global Methods in Algebraic Geometry, 2018

711 **Thomas Creutzig and Andrew R. Linshaw, Editors,** Vertex Algebras and Geometry, 2018

For a complete list of titles in this series, visit the
AMS Bookstore at **www.ams.org/bookstore/conmseries/**.